Universitext

Universitext

Universitext is a series of textbooks that presents material from a wide variety of mathematical disciplines at master's level and beyond. The books, often well class-tested by their author, may have an informal, personal even experimental approach to their subject matter. Some of the most successful and established books in the series have evolved through several editions, always following the evolution of teaching curricula, to very polished texts.

Thus as research topics trickle down into graduate-level teaching, first textbooks written for new, cutting-edge courses may make their way into *Universitext*.

More information about this series at http://www.springer.com/series/223

Benjamin Steinberg

Representation Theory
of Finite Monoids

Benjamin Steinberg
Department of Mathematics
The City College of New York (CCNY)
New York, NY, USA

ISSN 0172-5939   ISSN 2191-6675   (electronic)
Universitext
ISBN 978-3-319-43930-3   ISBN 978-3-319-43932-7   (eBook)
DOI 10.1007/978-3-319-43932-7

Library of Congress Control Number: 2016954642

Printed on acid-free paper

This Springer imprint is published by Springer Nature
The registered company is Springer International Publishing AG
The registered company address is: Gewerbestrasse 11, 6330 Cham, Switzerland

For Ariane and Nicholas

Preface

This text aims to provide an introduction to the representation theory of finite monoids for advanced graduate students and researchers in algebra, algebraic combinatorics, automata theory, and probability theory. As this is the first presentation of the theory in the form of a book, I have made no attempt to be encyclopedic. I also did not hesitate to make assumptions on the ground field that are stronger than strictly necessary in order to keep things simpler. Given that I am targeting a fairly broad audience, a lot of the technical jargon of semigroup theory is deliberately avoided to make the text accessible to as wide a readership as possible.

Much of the impetus behind recent research in the representation theory of finite monoids has come from applications to Markov chains, automata theory, and combinatorics. I have devoted three chapters to some of these applications, which I think of as the apex of the text. The first of these chapters proves a theorem of Berstel and Reutenauer [BR90] on the rationality of zeta functions of cyclic regular languages and symbolic dynamical systems. The second is concerned with synchronization problems in automata theory and the Černý conjecture [Vol08] and also touches on the subject of synchronizing permutation groups [ACS15]. The third of these chapters concerns applications of representation theory to probability theory, namely to Markov chains. It develops Brown's theory of left regular band random walks [Bro00, Bro04] as well as more recent results [Ste06, Ste08, ASST15b]. In particular, the eigenvalues of the transition matrices for the Tsetlin library, the riffle shuffle, and the Ehrenfest urn model are computed using monoid representation theory. The chapter also includes my previously unpublished, simple proof [Ste10a] of Brown's diagonalizability theorem for left regular band walks. Some nice applications that I have not touched upon here is the work of Malandro and Rockmore on fast Fourier transforms for inverse monoids with applications to the spectral analysis of partially ranked data [MR10, Mal10, Mal13].

My viewpoint on the representation theory of finite monoids differs from the classical one found in [CP61, Chapter 5] and [LP69, McA71, RZ91, Okn91].

These authors worked very much with semigroup algebras (not necessarily unital) and explicit matrix representations. In particular, they relied heavily on computations in the algebra of a 0-simple semigroup, expressed as a Rees matrix semigroup (so-called Munn algebras). In my opinion, this both obscured what was really happening and, at the same time, rendered the material essentially inaccessible to nonspecialists.

The approach here is highly module theoretic and follows the modern flavor of the theory of quasi-hereditary algebras [CPS88] and stratified algebras [CPS96]. Putcha should be credited with first making the connection between monoid representation theory and quasi-hereditary algebras [Put98]. The viewpoint taken here is most closely reflected in the literature in my joint work with Ganyushkin and Mazorchuk [GMS09]. See also the work of Kuhn [Kuh94b].

The background that is expected throughout the book is comfort with module theory (including categories, functors, tensor products, exact sequences, and projective and injective modules) and familiarity both with group representation theory in the non-modular setting and with the basics of Wedderburn theory (e.g., semisimple algebras and modules, radicals, and composition series). Occasionally, some facts from the representation theory of symmetric groups are needed. Some of the later chapters require more advanced techniques from homological algebra and the theory of finite dimensional algebras. I do not expect the reader to have previously studied semigroup theory. A certain amount of mathematical maturity is required for the book. Some of the more technical algebraic results can be accepted as black boxes for those who are primarily interested in the applications.

The text is divided into seven parts and three appendices. The first part is a streamlined introduction to finite semigroup theory. No prior knowledge beyond elementary group theory is assumed, and the approach is a departure from those of standard texts because I am primarily targeting a reader interested in monoids as a tool and not necessarily as an end in themselves. For instance, I have emphasized monoid actions in the proofs of a number of results in which they do not usually appear explicitly to highlight the similarity with ring theory. Chapter 1 develops the basic theory of monoids. This is followed by Chapter 2 studying $\mathscr{R}$-trivial monoids, which play a key role in applications to Markov chains. Chapter 3 develops the theory of inverse monoids, which form another important class of examples because they capture partial symmetries of mathematical objects.

The second part forms the core material of the book. It gives a modern, module theoretic treatment of the Clifford-Munn-Ponizovskiĭ theory connecting the irreducible representations of monoids with the irreducible representations of their maximal subgroups. The approach used here is based on the theory of idempotents, or what often goes under the name "recollement" [BBD82, CPS88, CPS96], connecting the representation theory of the algebras A, eAe, and A/AeA for an idempotent e. Chapter 4 develops this theory in the abstract in more detail than can be found in, for instance,

[Gre80, Chapter 6] or [Kuh94b]. This will be the most technically challenging of the essential chapters for readers who are not pure algebraists, and it may be worth accepting the results without being overly concerned with the proofs. Chapter 5 introduces monoid algebras and representations of monoids and then applies the general theory to monoid algebras to deduce the fundamental structure theorem of Clifford-Munn-Ponizovskiĭ for irreducible representations of finite monoids. I then provide Putcha's construction of the irreducible representations of the full transformation monoid as an example of the theory. Afterward, the classical result characterizing semisimplicity of monoid algebras (cf. [CP61, Chapter 5] or [Okn91]) is given a module theoretic proof that avoids entirely the dependence of earlier proofs on Munn algebras and non-unital (*a priori*) algebras. I then furnish the explicit description of the irreducible representations in terms of monomial representations (or Schützenberger representations) that can be found in [LP69, RZ91]. The chapter ends with a section proving the semisimplicity, in good characteristic, of the algebra of the monoid of $n \times n$ matrices over a finite field following Kovács [Kov92].

The third part of the book concerns character theory. Chapter 6 studies the Grothendieck ring of a monoid algebra over an arbitrary field. The case of the complex field was studied by McAlister [McA72]. It is shown that the Grothendieck ring of the monoid algebra is isomorphic to the direct product of the Grothendieck rings of its maximal subgroups (one per regular $\mathscr{J}$-class). I pay careful attention to the subring spanned by the one-dimensional simple modules. This subring encodes the triangularizable representations of a monoid and is important for applications to Markov chains. Chapter 7 is concerned with the character theory of monoids over the complex numbers as developed by McAlister [McA72] and, independently, Rhodes and Zalcstein [RZ91]. Although much of the theory could be developed over more general fields [MQS15], both the applications and the desire to keep things simple have led me to stick to this restricted setting. The key result here is that the character table is block upper triangular with group character tables as the diagonal blocks. This means that the character table is invertible. The inverse of the character table is a crucial tool for computing the composition factors of a module from its character. The final two sections of Chapter 7 are optional and are included primarily to show the applicability of monoid character theory despite the lack of the orthogonality relations that seem so crucial in most applications of group characters.

The fourth part of the book is devoted to the representation theory of inverse monoids. Just as groups abstract the notion of permutation groups, inverse monoids abstract the notion of partial permutation monoids and hence form a very natural and important generalization of groups. Like groups, they have semisimple algebras in good characteristic. The approach taken here follows my papers [Ste06, Ste08]. Namely, it is shown that the algebra of an inverse monoid is, in fact, the algebra of an associated groupoid and hence is isomorphic to a direct product of matrix algebras over group algebras.

The isomorphism is very explicit, using Möbius inversion, and so one can very efficiently convert group theoretic results into inverse monoid theoretic ones. In more detail, Chapter 8 develops the basics of the representation theory of finite categories. This theory is essentially equivalent to the representation theory of finite monoids, as can be seen from the results of Webb and his school [Web07]. I explicitly state and prove here a parametrization of the simple modules for a finite category via Clifford-Munn-Ponizovskiĭ theory. The theory is then restricted to groupoid algebras. The actual representation theory of inverse monoids is broached in Chapter 9. A highlight is a formula for decomposing a representation into its irreducible constituents from its character using only group character tables and the Möbius function of the lattice of idempotents.

In Part V, I provide an exposition of the theory of the Rhodes radical. This is the congruence on a monoid corresponding to the direct sum of all its irreducible representations. It was first studied over the complex numbers by Rhodes [Rho69b], whence the name, and in general by Almeida, Margolis, the author, and Volkov [AMSV09]. The Rhodes radical is crucial for determining which monoids have only one-dimensional irreducible representations or, equivalently, have basic algebras. These are precisely the monoids admitting a faithful upper triangular matrix representation. Such monoids are the ones whose representation theory is most useful for analyzing Markov chains. With a heavy heart, I have decided to restrict my attention primarily to the characteristic zero setting. The results in positive characteristic are stated without proof. The reason for this choice is that the applications mostly involve fields of characteristic zero and also the characteristic $p > 0$ case requires a bit more structural semigroup theory. The treatment here differs from [AMSV09] (and in some sense has quite a bit in common with [Rho69b]) in that I take advantage of the bialgebra structure on a monoid algebra and R. Steinberg's lovely theorem [Ste62] on the tensor powers of a faithful representation. Although the language of bialgebras is not used explicitly, Chapter 10 essentially develops the necessary bialgebraic machinery to prove Steinberg's theorem following the approach of Rieffel [Rie67] (see also [PQ95, Pas14]). Chapter 11 then develops the theory of the Rhodes radical in connection with nilpotent bi-ideals. The characterization of monoids triangularizable over an algebraically closed field of characteristic zero, or equivalently having basic algebras, is given. Here, many of the lines of investigation developed throughout the text are finally interwoven.

Part VI consists of the three chapters on applications that I have already discussed previously. They are Chapter 12 on the rationality of zeta functions of cyclic regular languages and symbolic dynamical systems, Chapter 13 on transformation monoids and applications to automata theory, and Chapter 14 on Markov chains.

The seventh, and final, part concerns advanced topics. These chapters expect the reader to be familiar with some homological algebra, in particular, with the Ext-functor. Chapter 15 characterizes von Neumann regular monoids with a self-injective algebra, which I believe to be a new result. Chapter 16 proves Nico's upper bound on the global dimension of the algebra of a regular monoid [Nic71, Nic72]. The approach here is inspired by Putcha's observation [Put98] that regular monoids have quasi-hereditary algebras and the machinery of quasi-hereditary algebras [CPS88, DR89]. Chapter 17 presents the notion of quivers and recalls their role in modern representation theory. I then provide methods for computing the Gabriel quiver of the algebra of a left regular band and of a $\mathscr{J}$-trivial monoid. These are results of Saliola [Sal07] and of Denton, Hivert, Schilling and Thiéry [DHST11], respectively. They are, in fact, special cases of more general results of Margolis and myself, obtained using Hochschild cohomology [MS12a], that I do not attempt to present in the current text. The quivers of the algebras of several natural families of monoids are computed as examples. Chapter 18 is a survey, without proofs, of some further developments not covered in the book.

Appendices A–C cover background material concerning finite dimensional algebras, group representation theory, and Möbius inversion for posets. The purpose of Appendices A–C is twofold: both to review some concepts and results with which I hope that the reader is already familiar, or is willing to accept, and to provide in one location specific statements that can be referenced throughout the book. No attempt is made here to be self-contained: proofs are mostly omitted, but references are given to standard texts. Appendix A reviews those elements of the theory of finite dimensional algebras that are used throughout. Most of the book uses nothing more than Wedderburn theory and the Jordan-Hölder theorem. Primitive idempotents are occasionally used. Morita theory and duality appear at a couple of points in the text, most of which can be skipped by readers primarily interested in applications. Projective indecomposable modules and projective covers are only used in the final chapters. Appendix B reviews group representation theory, again without many proofs. Modular representation theory does not appear at all and I have mostly restricted character theory to the field of complex numbers. The last section surveys aspects of the representation theory of the symmetric group in characteristic zero. Appendix C briefly introduces Rota's theory of Möbius inversion for partially ordered sets [Rot64]. This is a fundamental technique in algebraic combinatorics. The poset of idempotents of a monoid plays a key role in monoid representation theory, and Möbius inversion is an important tool for decoding the information hidden in the fixed subspaces of the idempotents under the representation.

One thing that is not discussed in this book is applications of representation theory to monoid theory, itself. My focus instead has been on how monoid representation theory is of use in other areas of mathematics, and this has been the direction of most research in the subject. Much of the investigation in finite monoid theory has been motivated by applications to automata theory, which naturally leads to studying certain "equational" classes of finite monoids rather than individual monoids; see [Eil76, Pin86, Alm94, Pin97, RS09]. In this setting, there have been some results proved using representation theory. For example, there is the result of Rhodes computing the Krohn-Rhodes complexity (defined in [KR68]) of a finite completely regular monoid using character theory [Rho69b]. Together with Almeida, Margolis, and Volkov, I provided representation theoretic proofs of some deep results in automata theory involving unambiguous marked products of regular languages and marked products with modular counters; see [AMSV09]. These results are beyond the scope of this text. Monoid representation theory has also provided deeper understanding of the structure of finite monoids of Lie type [OP91, Put94, Put95, Okn98, Put99, Ren05].

The chapter dependencies are roughly as follows. All the chapters rely on Chapters 1–3, to varying degrees. Chapter 5 depends on Chapter 4. Chapter 6 depends on Chapter 5, whereas Chapter 7 requires Chapters 5 and 6. Chapter 8 relies on Chapters 4 and 5. Chapter 9 depends on Chapters 5–8. Chapter 10 requires Chapters 5 and 6. Chapter 11 depends on Chapters 5, 9, and 10. Chapter 12 mostly uses results from Chapters 6 and 7; some of the examples make use of results from Chapters 5 and 9. Chapter 13 is relatively self-contained, using only at the end some results from Chapter 5. Chapter 14 mostly relies on Chapters 7 and 11, as well as on the description of the simple modules for $\mathscr{R}$-trivial monoids in Chapter 5. Chapters 15 and 16 depend on Chapters 4 and 5. Chapter 17 requires Chapters 5, 7, 9, and 11 as well as some of the examples from Chapter 14; it also expects familiarity with path algebras and algebras of posets, as per Chapter 8.

As always, I am indebted to John Rhodes who taught me virtually everything I know about finite semigroups and who first interested me in semigroup representation theory. This book is also very much influenced by my collaborators Stuart Margolis, Volodymyr Mazorchuk, Franco Saliola, Anne Schilling, and Nicolas Thiéry. Mohan Putcha has been a continuing inspiration to my viewpoint on semigroup representation theory. Discussions with Nicholas Kuhn and Lex Renner have also enlightened me. A wonderful conversation with Persi Diaconis at Banff encouraged my, then, nascent interest in applications of monoid representation theory to Markov chain theory. The following people have provided invaluable feedback concerning the text, itself: Darij Grinberg, Michael Kinyon, Stuart Margolis, Volodymyr Mazorchuk, Don McAlister, Mohan Putcha, Christophe Reutenauer, Anne Schilling, and Peter Webb. I would also like to thank the anonymous referees for their careful reading of the manuscript and for their many helpful suggestions.

The writing of this book was supported in part by a grant from the Simons Foundation (#245268 to Benjamin Steinberg) and the Binational Science Foundation of Israel and the USA (#2012080). Also, I was supported in part by CUNY Collaborative Research Incentive Grant #2123, a PSC-CUNY grant, and NSA MSP grant #H98230-16-1-0047.

New York, NY, USA   *Benjamin Steinberg*
June 28, 2016

Contents

Part VII Advanced Topics

Introduction

In the preface to the first edition of his book, *Theory of Groups of Finite Order* [Bur55], W. Burnside explains that he does not treat the representation theory of finite groups because "it would be difficult to find a result that could be most directly obtained by the consideration of groups of linear transformations." In the second edition, produced a few years later, group representation theory plays a central role, in part because of Burnside's proof that a group whose order involves at most two primes is solvable; it took many years before this result was proved without representation theory [Gol70].

D. B. McAlister wrote a survey [McA71] in 1971 on semigroup representation theory. In the introduction he says "although the representation theory of semigroups has given rise to many important concepts in semigroup theory, it has not yet proved nearly as useful as has group representation theory." He goes on to write, referring to results of J. Rhodes [KR68, Rho69b]: "The only real application of representation theory so far has arisen in the complexity theory of finite semigroups and machines."

Forty-five years later, we see many more applications of the representation theory of finite monoids, including applications to combinatorics and symbolic dynamics [BR90, Sol02, AM06], to automata theory and the Černý conjecture [Kar03, AMSV09, AS09, Ste10b, Ste10c, Ste11, Per13], to probability theory and the analysis of Markov chains [BD98, BHR99, BBD99, Bro00, Bro04, Bjö08, Bjö09, AD10, CG12, Sal12, AKS14b, ASST15a, ASST15b], data analysis and fast Fourier transforms [MR10, Mal10, Mal13], homotopy theory [HK88], and, most recently, to generic representation theory [Kuh15]. Nonetheless, no book dedicated to the representation theory of finite monoids and its applications has previously been written, thereby rendering it difficult for nonspecialists to access the literature and learn the techniques. For instance, Brown was forced to redevelop from scratch a very special case of monoid representation theory in order to analyze Markov chains [Bro00, Bro04].

The direct catalyst for writing this treatise was the following question, asked on MathOverflow on August 30, 2010 [Mik10]: *Why aren't representations of monoids studied so much?* This book may very well be the longest answer ever given to a MathOverflow question! The short answer is that the representation theory of monoids has been intensively studied by semigroup theorists since the 1940s, but has never been made available to the general mathematical public in an accessible form; the only books containing even a snippet of the subject consider the representation theory of finite monoids as a special case of the representation theory of more general monoids [CP61, Okn91, Okn98]. The long answer is more complicated.

There are several notable reasons that the representation theory of finite monoids took so long to develop and become applicable. As pointed out by McAlister [McA71], the algebra of a finite monoid is seldom semisimple. With the exception of the modular representation theory of finite groups, which has its own particular nature, there was not a well-developed theory of non-semisimple finite dimensional algebras until Gabriel's foundational work on the representation theory of quivers [Gab72]. Nowadays there is a fairly well-developed theory [DK94, GR97, ARS97, Ben98, ASS06].

In the context of the pre-Gabriel era of finite dimensional algebras, semigroup theorists, in particular Clifford [Cli42], Munn [Mun55, Mun57b], and Ponizovskiĭ [Pon58], quickly proceeded to describe the irreducible representations of a finite monoid in terms of group representation theory and to determine which finite monoids have semisimple algebras. The definitive summary of their approach is the chapter of Clifford and Preston [CP61, Chapter 5]. The character theory was further developed independently by McAlister [McA72] and by Rhodes and Zalcstein (whose work was only published many years after it was written [RZ91]). But then semigroup theorists were left somewhat at a loss as of what to do next in this non-semisimple setting; and by the late 1970s, progress in the representation theory of finite monoids had grounded to a halt. However, if one looks back at the pioneering work of Munn and Ponizovskiĭ [Pon58, Mun57b], Hewitt and Zuckerman [HZ57], and Nico [Nic71, Nic72], one already sees many of the ideas and constructions that appeared later on in the theory of quasi-hereditary algebras [CPS88, DR89] and cellular algebras [GL96].

There has been a flurry of renewed interest in monoid representation theory, beginning in the 1990s. The theory of linear algebraic monoids, developed principally by Putcha and Renner (see the monographs [Put88, Ren05]), led to a number of questions in the representation theory of finite monoids, in particular concerning the semisimplicity of the complex algebra of the monoid of $n \times n$ matrices over a finite field and of the algebras of more general finite monoids of Lie type [OP91, Kov92, Put99]. Also during this period, Putcha [Put96] completed the work begun by Hewitt and Zuckerman [HZ57] in determining explicitly the character table and irreducible representations of the full transformation monoid (the monoid of all self-maps of a finite set). In the late 1990s, Putcha [Put98] recognized that the algebra of a von Neumann regular monoid is quasi-hereditary, making the first connection with the modern representation theory of finite dimensional algebras. He also made use of the representation theory of quivers to show that the full transformation monoid of degree at least 5 has infinite representation type; the representation type in degrees 1, 2, and 3 had been shown to be finite earlier on by Ponizovskiĭ [Pon87, Pon93], whereas degree 4 was tackled afterward by Ringel [Rin00], who again established finite representation type.

Further interest in monoid representation theory was sparked by the work of Bidigare, Hanlon and Rockmore [BHR99] and Diaconis and Brown [BD98, Bro00, Bro04], which found a major application of monoid representation theory in the analysis of certain finite state Markov chains including the Tsetlin library, the top-to-random shuffle, and the riffle shuffle. This development was highlighted by Diaconis in his ICM lecture [Dia98]; group representation theory had already been applied to Markov chain theory for many years at that point, cf. [DS81, Dia88, CSST08]; however, it should be noted that group representation theory can only be applied to Markov chains with a uniform stationary distribution, whereas monoid representation theory is generally applicable. The applications of monoid representation theory to Markov chains and Solomon's descent algebra [Bro04, Sch06, AM06, Sal08, Sal10] brought interest in the subject to probabilists and combinatorialists. It also motivated the author to work out a more explicit and combinatorial version of the representation theory of finite inverse monoids [Ste06, Ste08] in order to describe the multiplicities of the eigenvalues of random walks on triangularizable monoids.

Since then, there has been rapid progress in the representation theory of finite monoids with many new papers appearing in the intervening years, e.g., [Sal07, Sch08, Sal09, GMS09, DHST11, Den11, MS11, BBBS11, MS12a, MS12c, HST13, MSS15a, MSS15b, Ste16a, Ste16b]. In particular, progress has been made in computing quivers of monoid algebras and global dimension. It has been observed recently by various authors (e.g., [MS12a, LS12]) that the Clifford-Munn-Ponizovskiĭ theory applies equally well to the representation theory of finite categories and this has been exploited in a number of ways [BD12, Lin14, DE15, BD15].

Much of the representation theory of finite monoids has been guided by the exploration of naturally arising examples coming both from within semigroup theory and also from outside of it via combinatorics, automata theory, or probability theory. Putcha's detailed study of the representation theory of the full transformation monoid played an important role in interfacing with the representation theory of quivers and quasi-hereditary algebras [Put96, Put98]. The quest by Kovács, Okniński, and Putcha [OP91, Kov92, Put99] to establish the semisimplicity of the algebra of the monoid of $n \times n$ matrices over a finite field has led to interesting connections with generic representation theory [Kuh15].

Solomon's combinatorial approach [Sol02] to the representation theory of the symmetric inverse monoid (also known as the rook monoid), refining to some extent Munn's original approach [Mun57a], led to the author's combinatorial approach to the representation theory of finite inverse monoids [Ste06, Ste08], as well as a host of work by researchers in algebraic combinatorics on related algebras, e.g., [HR01, Gro02, DHP03, Hal04, Gro06, Pag06, AO08]. Partition monoids, Brauer monoids, and their ilk are further examples of important monoids from other areas of mathematics whose representation theory will certainly influence future developments.

Saliola's study of the representation theory of hyperplane face monoids [Sal09], motivated by the applications in probability theory [BD98, BHR99, Bro00, Bro04] and algebraic combinatorics [Sch06, AM06, Sal08, Sal10], led to a very interesting and satisfactory approach to the representation theory of left regular bands, establishing connections with algebraic topology and discrete geometry [Sal07, MSS15a, MSS15b].

Developments in the representation theory of $\mathscr{J}$-trivial monoids and $\mathscr{R}$-trivial monoids [Sch08, DHST11, BBBS11, MS12a] were very much motivated by the study of the 0-Hecke algebra of a finite Coxeter group, which is in fact a monoid algebra, cf. [Nor79, Car86, Fay05, Den11]. The techniques developed then proved useful in the analysis of a number of Markov chains [AKS14b, ASST15a, ASST15b].

A deeper study of the homological properties of regular monoids, with further applications to computing quivers, was performed by Margolis and the author [MS11] in order to compute the quiver of a monoid algebra introduced by Hsiao [Hsi09] in connection with the Mantaci-Reutenauer descent algebra of a wreath product [MR95].

As examples have motivated much of the theory, we have made sure to include a fair share in this text. We have included the most explicit construction to date of the irreducible representations of the full transformation monoid, clarifying the work of Putcha, and we have provided its character table. A detailed proof of the semisimplicity of the algebra of the monoid of $n \times n$ matrices over a finite field in good characteristic, following the approach of Kovács, is also in the text. Two descriptions of the character table of the symmetric inverse monoid (or rook monoid) are given. The irreducible representations of the free left regular band and the hyperplane face monoids of the braid and boolean arrangements are described. In addition we compute the quivers of the algebras of the free left regular band, the hyperplane face monoid of the boolean arrangement, and the Catalan monoid; we also provide a new proof that the algebra of the latter monoid is isomorphic to an incidence algebra.

Applications play a crucial role in this book, as they form the primary motivation for writing it. Therefore, there are three chapters dedicated to applications outside of monoid theory. The first of these chapters provides a beautiful application of monoid representation theory to symbolic dynamics, due to Berstel and Reutenauer [BR90]. A classical result of Bowen and Lanford shows that the zeta function of a shift of finite type is rational. For instance, the Ihara zeta function of a graph [Ter99] is the zeta function of a shift of finite type. Manning then extended the result to show that zeta functions of sofic shifts are rational [Man71]; this extension is much more difficult to prove than the original result for shifts of finite type. See [LM95] for background in symbolic dynamics and the classical approach to these results.

Berstel and Reutenauer found an elegant proof of a generalization of Manning's result using the representation theory of finite monoids. The original proof of Berstel and Reutenauer is inductive; here we deduce it as an immediate consequence of McAlister's characterization of the virtual characters of a finite monoid [McA72].

The second chapter of applications is primarily devoted to automata theory [Eil74], a branch of theoretical computer science that has also interacted fruitfully with geometric group theory via Thurston's theory of automatic groups [ECH+92] and via automaton groups and self-similar groups [Nek05], as studied by Grigorchuk and others [GNS00]. An automaton is a machine with a finite number of internal states that can change state each time it processes an input symbol according to a specified transition rule. Often automata are given initial and terminal states so that they can accept a language; languages accepted by finite state automata are called regular languages. One can view each input letter as a mapping on the set of states and so the automaton can be viewed fruitfully as a transformation monoid with a distinguished set of generators. The algebraic theory of automata, cf. [Eil76, Pin86, Str94], tries to use monoid theory to solve problems about regular languages that arise in computer science. Just as a permutation group gives rise to a permutation module, a transformation monoid gives rise to a transformation module. Therefore, a finite state automaton has a module associated with it, something first exploited by Schützenberger and which led to his theory of rational power series in noncommuting variables [BPR10, BR11]. This module can often be used to obtain information relevant to automata theory. Here we focus on applications to the Černý conjecture (see [Vol08] for a survey), a popular problem in automata theory concerning synchronization that has been open for nearly fifty years and is the subject of many articles.

The final chapter of applications presents monoid representation theory as a tool in the analysis of finite state Markov chains, a branch of probability theory. The spirit of this application is similar to that of the application to automata theory. A finite state Markov chain can be viewed as consisting of a finite state space and a finite set of transitions of the states together with a probability distribution on the transitions. The system evolves by the application of transitions at random according to this distribution. (This is called a random mapping representation of the Markov chain [LPW09].) The transition matrix of the Markov chain is a convex combination of the operators on the transformation module coming from the transitions. Thus submodules give invariant subspaces for the transition matrix. This can be exploited to compute the eigenvalues of the transition matrix if the transformation module is sufficiently nice. Here we provide the general theory for random walks on triangularizable monoids. Our principal examples are left regular band walks of the sort studied by Brown [Bro00, Bro04], such as the Tsetlin library, the riffle shuffle, and the Ehrenfest urn model. A highlight of the chapter is the author's previously unpublished, simple proof [Ste10a] of Brown's diagonalizability theorem for transition matrices of left regular band walks. The interested reader should consult [ASST15b] for further developments outside of the left regular band realm and for applications of monoid theory to other aspects of Markov chain theory besides computing eigenvalues.

There are also applications of monoid representation theory that we do not consider in this text. For example, Malandro has applied the representation theory of the symmetric inverse monoid to the analysis of partially ranked data [Mal13]. In order to do the analysis efficiently, a fast Fourier transform was needed for the symmetric inverse monoid [MR10, Mal10].

Hopefully, the reader will find the representation theory of finite monoids a rich and intriguing mélange of group representation theory, finite dimensional algebras, and algebraic combinatorics that is both applicable to various areas of mathematics and also a subject of intrinsic beauty.

Part I

Elements of Monoid Theory

1

The Structure Theory of Finite Monoids

This chapter contains those elements of the structure theory of finite monoids that we shall need for the remaining chapters. It also establishes some notation that will be used throughout. More detailed sources for finite semigroup theory include [KRT68, Eil76, Lal79, Alm94, RS09]. Introductory books on the algebraic theory of semigroups, in general, are [CP61, CP67, Hig92, How95]. A detailed study of some of the most important transformation monoids can be found in [GM09]. In this book, all semigroups and monoids will be finite except for endomorphism monoids of vector spaces and free monoids. On a first reading, it may be advisable to skip the proofs in this chapter.

1.1 Basic notions

A *semigroup* is a set S, possibly empty, with an associative binary operation, usually called *multiplication* and denoted multiplicatively by juxtaposition or by ".". A *monoid* is a semigroup M with an *identity* element, often written as 1. The identity is necessarily unique. In particular, a monoid is nonempty. We shall primarily be interested here in monoids, but sometimes we shall also need to work with semigroups. A *subsemigroup* of a semigroup is a subset, possibly empty, which is closed under multiplication. A *submonoid* of a monoid is a subsemigroup containing the identity. It is traditional in semigroup theory to call a subsemigroup G of a semigroup S, which is a group with respect to the induced binary operation, a *subgroup*. We do *not* require the identity of G to coincide with the identity of S in the case that S is a monoid. This is because most of the subgroups of a monoid that arise in practice will not share its identity. In fact, a recurrent theme in this text is that a finite monoid is a collection of finite groups loosely tied together by a partially ordered set.

© Springer International Publishing Switzerland 2016
B. Steinberg, *Representation Theory of Finite Monoids*,
Universitext, DOI 10.1007/978-3-319-43932-7_1

An element $m \in M$ is called a *unit* if there exists $m^{-1} \in M$ such that $mm^{-1} = 1 = m^{-1}m$. The units form a group called the *group of units* of M. The group of units of M is the unique maximal subgroup of M which is also a submonoid.

An element z of a semigroup S is called a *zero element* (or simply a *zero*) if $sz = z = zs$ for all $s \in S$. It is easy to check that zero elements, if they exist, are unique. Often, a zero element is denoted in the literature by 0. However, we shall typically not follow this practice in this text to avoid confusion between the zero of a monoid and that of its monoid algebra.

As usual, if S is a semigroup and $A, B \subseteq S$, then

$$AB = \{ab \mid a \in A, b \in B\}.$$

If S is a semigroup, then S^{op} denotes the *opposite semigroup*. It has the same underlying set as S, but the binary operation $*$ is given by $s * t = ts$. (Similar notation will also be used for opposite rings, posets, and categories.)

A *homomorphism* $\varphi \colon M \longrightarrow N$ of monoids is a mapping such that

(i) $\varphi(1) = 1$
(ii) $\varphi(m_1 m_2) = \varphi(m_1)\varphi(m_2)$

for all $m_1, m_2 \in M$. A bijective homomorphism is called an *isomorphism*. The inverse of an isomorphism is again an isomorphism. Two monoids are said to be *isomorphic* if there is an isomorphism between them. Semigroup homomorphisms and isomorphisms are defined analogously.

A *congruence* on a monoid M is an equivalence relation $\equiv$ with the property that $m_1 \equiv m_2$ implies $um_1v \equiv um_2v$ for all $u, v \in M$. It is straightforward to verify that $M/{\equiv}$ then becomes a monoid with respect to the product

$$[m_1]_\equiv \cdot [m_2]_\equiv = [m_1 m_2]_\equiv,$$

where $[m]_\equiv$ denotes the equivalence (or *congruence*) class of m, and that the projection $\pi \colon M \longrightarrow M/{\equiv}$ is a surjective homomorphism. If $\varphi \colon M \longrightarrow N$ is a homomorphism, then $\ker \varphi$ is the congruence defined by $m_1 \ker \varphi \, m_2$ if $\varphi(m_1) = \varphi(m_2)$. One easily verifies that $\varphi(M) \cong M/\ker\varphi$.

Given a subset A of a monoid M, the submonoid $\langle A \rangle$ *generated* by A is the smallest submonoid of M containing A. The subsemigroup generated by a subset of a semigroup is defined (and denoted) analogously.

If S, T are semigroups, then their *direct product* $S \times T$ is a semigroup with coordinate-wise binary operation $(s, t)(s', t') = (ss', tt')$. The direct product of two monoids is again a monoid.

An element e of a semigroup S is an *idempotent* if $e = e^2$. The set of idempotents of a subset $X \subseteq S$ is denoted $E(X)$. Of course, $1 \in E(M)$ for any monoid M. If $e \in E(S)$, then eSe is a monoid with respect to the binary operation inherited from S with identity e (because $eese = ese = esee$). The group of units of eSe is denoted by G_e and is called the *maximal subgroup* of

S at e. It is the unique subgroup with identity e that is maximal with respect to containment, whence the name. We put $I_e = eSe \setminus G_e$. Note that if M is a monoid, then G_1 is the group of units of M.

There is a *natural partial order* on the set $E(S)$ of idempotents of a semigroup S defined by putting $e \leq f$ if $ef = e = fe$. Equivalently, $e \leq f$ if and only if $eSe \subseteq fSf$.

The reader should verify that the image of an idempotent under a homomorphism is always an idempotent.

1.2 Cyclic semigroups

This section describes the structure of a finite cyclic semigroup. These results will be used later when we study the character theory of a finite monoid.

Fix a finite semigroup S and let $s \in S$. Then there exists a smallest positive integer c, called the *index* of s, such that $s^c = s^{c+d}$ for some $d > 0$. The smallest possible choice of d is called the *period* of s. Notice that $s^c = s^{c+qd}$ for any $q \geq 0$.

Proposition 1.1. *Let $s \in S$ have index c and period d. Then $s^i = s^j$ if and only if $i = j$ or $i, j \geq c$ and $i \equiv j \bmod d$.*

Proof. Without loss of generality, assume that $i < j$. If $i, j \geq c$ and $j = i + qd$ with $q > 0$, then $s^j = s^{c+qd+(i-c)} = s^{c+i-c} = s^i$. Conversely, if $s^i = s^j$ with $i < j$, then by definition of the index, we have $i \geq c$. Also if $j - i = qd + r$ with $0 \leq r < d$ and $q \geq 0$, then since $s^i = s^j$ we have that

$$s^{c+r} = s^{c+qd+r} = s^{c+j-i} = s^{c+id+j-i} = s^j s^{c+id-i} = s^i s^{c+id-i} = s^{c+id} = s^c.$$

By definition of the period, we conclude that $r = 0$ and so $i \equiv j \bmod d$. □

Let $\langle s \rangle = \{s^n \mid n \geq 1\}$ be the *cyclic subsemigroup* generated by s. It follows from the proposition that $\langle s \rangle = \{s, s^2, \ldots, s^{c+d-1}\}$, where c is the index of s and d is the period, and that these elements are distinct.

Corollary 1.2. *Let $s \in S$ have index c and period d. Then the subsemigroup $C = \{s^n \mid n \geq c\}$ of $\langle s \rangle$ is a cyclic group of order d. The identity of C, denoted s^ω, is the unique idempotent in $\langle s \rangle$ and is given by s^m where $m \geq c$ and $m \equiv 0 \bmod d$. If $s^{\omega+1} = s^\omega s$, then $C = \langle s^{\omega+1} \rangle$.*

Proof. Clearly C is a subsemigroup. The mapping $\varphi \colon C \longrightarrow \mathbb{Z}/d\mathbb{Z}$ given by $\varphi(s^n) = n + d\mathbb{Z}$ for $n \geq c$ is well defined and injective by Proposition 1.1. It is surjective because $n + d\mathbb{Z} = \varphi(s^{cd+n})$ for $n \geq 0$. Trivially, φ is a homomorphism. Thus φ is an isomorphism of semigroups and hence C is a cyclic group of order d with identity s^m where $m \geq c$ and $m \equiv 0 \bmod d$.

If $s^k \in E(\langle s \rangle)$, then $s^k = s^{2k}$ and hence $k \geq c$ by definition of the index. Because C is a group, it has a unique idempotent and so $s^k = s^\omega$. The final statement follows because φ is an isomorphism and hence s^k is a generator of C for any $k \geq c$ with $k \equiv 1 \bmod d$. But if $m \geq c$ with $m \equiv 0 \bmod d$, then $s^{\omega+1} = s^m s = s^{m+1}$ and, moreover, $m+1 \geq c$ and $m+1 \equiv 1 \bmod d$. Thus $C = \langle s^{\omega+1} \rangle$. $\qquad \square$

Notice that $\langle s \rangle$ is a group if and only if $s^{\omega+1} = s$.

Remark 1.3. If S is a finite semigroup of order n, then $s^{n!} = s^\omega$ for all $s \in S$. Indeed, if s is of index c and period d, then obviously $c \leq n$ and $d \leq n$. Thus $n! \geq c$ and $n! \equiv 0 \bmod d$ and so $s^{n!} = s^\omega$ by Corollary 1.2. Consequently, $s^{\omega+1} = s^{n!+1}$ for all $s \in S$.

A fundamental consequence of Corollary 1.2 is that nonempty finite semigroups have idempotents.

Corollary 1.4. *A nonempty finite semigroup contains an idempotent.*

Proof. If S is a finite semigroup and $s \in S$, then $\langle s \rangle \subseteq S$ contains the idempotent s^ω. $\qquad \square$

Corollary 1.4 admits a refinement, sometimes known as the *pumping lemma*.

Lemma 1.5. *Let S be a semigroup of order $n > 0$. Then $S^n = SE(S)S$. Equivalently, $s \in S^n$ if and only if $s = uev$ with $e \in S$ an idempotent and $u, v \in S$.*

Proof. The equivalence of the two statements is clear. Trivially, $SE(S)S \subseteq S^n$. Suppose $s \in S^n$. Write $s = s_1 \cdots s_n$ with $s_i \in S$ for $i = 1, \ldots, n$. If the elements $s_1, s_1 s_2, \ldots, s_1 s_2 \cdots s_n$ of S are all distinct, then they constitute all the elements of S. Because S contains an idempotent by Corollary 1.4, we have that $s_1 \cdots s_i \in E(S)$ for some i and hence $s = (s_1 \cdots s_i e)(e s_{i+1} \cdots s_n)$ with $e = s_1 \cdots s_i$ an idempotent. So assume that they are not all distinct and hence there exist $i < j$ such that $s_1 \cdots s_i = s_1 \cdots s_i s_{i+1} \cdots s_j$. Then by induction $s_1 \cdots s_i = s_1 \cdots s_i (s_{i+1} \cdots s_j)^k$ for all $k \geq 0$. In particular, if $e = (s_{i+1} \cdots s_j)^\omega$, then $s_1 \cdots s_i e = s_1 \cdots s_i$. Therefore, we have that $s = s_1 \cdots s_i s_{i+1} \cdots s_n = (s_1 \cdots s_i e)(e s_{i+1} \cdots s_n)$, as required. $\qquad \square$

We end this section with another useful consequence of the existence of idempotents in finite semigroups.

Lemma 1.6. *Let $\varphi \colon S \longrightarrow T$ be a surjective homomorphism of finite semigroups. Then $\varphi(E(S)) = E(T)$.*

Proof. The inclusion $\varphi(E(S)) \subseteq E(T)$ is clear. To establish the reverse inclusion, let $e \in E(T)$. Then $\varphi^{-1}(e)$ is a nonempty finite semigroup and hence contains an idempotent f. $\qquad \square$

1.3 The ideal structure and Green's relations

An important role in monoid theory is played by ideals and the closely related notion of Green's relations [Gre51]. Let M be a finite monoid throughout this section.

A *left ideal* (respectively, *right ideal*) of M is a nonempty subset I such that $MI \subseteq I$ (respectively, $IM \subseteq I$). A (two-sided) *ideal* is a nonempty subset $I \subseteq M$ such that $MIM \subseteq I$. Any left, right, or two-sided ideal of M is a subsemigroup and hence contains an idempotent by Corollary 1.4. If $I_1, \ldots, I_n$ are all the ideals of M, then $I_1 I_2 \cdots I_n$ is an ideal of M and is contained in all other ideals. Consequently, each finite monoid M has a unique *minimal ideal*. Notice that the union of two ideals is again an ideal.

If $m \in M$, then Mm, mM, and MmM are the *principal*, respectively, left, right, and two-sided ideals generated by m. It will be convenient to put

$$I(m) = \{s \in M \mid m \notin MsM\}.$$

If $I(m) \neq \emptyset$, then it is an ideal. Also, $I(m) = \emptyset$ if and only if m belongs to the minimal ideal of M.

We shall need the following three equivalence relations, called *Green's relations*, associated with the ideal structure of M. Put, for $m_1, m_2 \in M$,

(i) $m_1 \mathscr{J} m_2$ if and only if $Mm_1M = Mm_2M$;
(ii) $m_1 \mathscr{L} m_2$ if and only if $Mm_1 = Mm_2$;
(iii) $m_1 \mathscr{R} m_2$ if and only if $m_1M = m_2M$.

The $\mathscr{J}$-class of an element m is denoted by J_m, and similarly the $\mathscr{L}$-class and $\mathscr{R}$-class of m are denoted L_m and R_m, respectively. Green's relations (of which there are two others that appear implicitly, but not explicitly, throughout this text) were introduced by Green in his celebrated Annals paper [Gre51] and have played a crucial role in all of semigroup theory ever since.

For example, if $\Bbbk$ is a field, then two $n \times n$ matrices $A, B \in M_n(\Bbbk)$ are $\mathscr{J}$-equivalent if and only if they have the same rank. They are $\mathscr{L}$-equivalent if and only if they are row equivalent and they are $\mathscr{R}$-equivalent if and only if they are column equivalent. If T_n denotes the monoid of all self-maps of an n-element set (called the *full transformation monoid of degree n*), then two mappings f, g are $\mathscr{J}$-equivalent if and only if they have the same *rank* (where the rank of a mapping is the cardinality of its image). They are $\mathscr{L}$-equivalent if and only if they induce the same partition into fibers over their images and they are $\mathscr{R}$-equivalent if and only if they have the same image (or range).

A monoid M is called $\mathscr{R}$-*trivial* if $mM = nM$ implies $m = n$, that is, the $\mathscr{R}$-relation is equality. The notion of $\mathscr{L}$-triviality is defined dually. A monoid is $\mathscr{J}$-*trivial* if $MmM = MnM$ implies that $m = n$. Note that $\mathscr{J}$-trivial monoids are both $\mathscr{R}$-trivial and $\mathscr{L}$-trivial. The converse is also true by Corollary 1.14, below.

A useful property of Green's relations is that they are compatible with passing from M to eMe with $e \in E(M)$.

Lemma 1.7. *Let $e \in E(M)$ and let $m_1, m_2 \in eMe$. If $\mathscr{K}$ is one of Green's relations $\mathscr{J}$, $\mathscr{L}$, or $\mathscr{R}$, then $m_1 \mathscr{K} m_2$ in M if and only if $m_1 \mathscr{K} m_2$ in eMe.*

Proof. It is clear that if $(eMe)m_1(eMe) = (eMe)m_2(eMe)$, then $Mm_1M = Mm_2M$. Conversely, if $Mm_1M = Mm_2M$, then

$$(eMe)m_1(eMe) = eMm_1Me = eMm_2Me = (eMe)m_2(eMe).$$

This handles the case of $\mathscr{J}$. Similarly, $eMem_1 = eMem_2$ implies $Mm_1 = Mm_2$, and $Mm_1 = Mm_2$ implies $eMem_1 = eMm_1 = eMm_2 = eMem_2$, establishing the result for $\mathscr{L}$. The result for $\mathscr{R}$ is dual. □

To continue our investigation of the ideal structure and Green's relations, it will be helpful to introduce the notion of an M-set. A (left) M-*set* consists of a set X together with a mapping $M \times X \longrightarrow X$, written $(m, x) \mapsto mx$ and called a left *action*, such that

(i) $1x = x$
(ii) $m_1(m_2 x) = (m_1 m_2)x$

for all $x \in X$ and $m_1, m_2 \in M$. Right M-sets are defined dually or, alternatively, can be identified with left M^{op}-sets. For example, each left ideal of M is naturally an M-set. An M-set is said to be *faithful* if $mx = m'x$, for all $x \in X$, implies $m = m'$ for $m, m' \in M$.

A mapping $\varphi\colon X \longrightarrow Y$ of M-sets is said to be M-*equivariant* if $\varphi(mx) = m\varphi(x)$ for all $m \in M$ and $x \in X$. If φ is bijective, we say that φ is an *isomorphism of M-sets*. The inverse of an isomorphism is again an isomorphism. As usual, X is said to be *isomorphic* to Y, denoted $X \cong Y$, if there is an isomorphism between them. We write $\mathrm{Hom}_M(X, Y)$ for the set of all M-equivariant mappings from X to Y. There is, of course, a category of M-sets and M-equivariant mappings. We use the obvious notation for endomorphism monoids and automorphism groups of M-sets.

Proposition 1.8. *Let $e \in E(M)$ and let X be an M-set. Then $\mathrm{Hom}_M(Me, X)$ is in bijection with eX via the mapping $\varphi \mapsto \varphi(e)$. Moreover, one has that $\mathrm{End}_M(Me) \cong (eMe)^{op}$ and $\mathrm{Aut}_M(Me) \cong G_e^{op}$.*

Proof. Let $\varphi\colon Me \longrightarrow X$ be M-equivariant and let $m \in Me$. Then $\varphi(m) = \varphi(me) = m\varphi(e)$. In particular, $\varphi(e) = e\varphi(e) \in eX$ and φ is uniquely determined by $\varphi(e)$. It remains to show that if $x \in eX$, then there exists $\varphi\colon Me \longrightarrow X$ with $\varphi(e) = x$. Write $x = ey$ with $y \in X$. Define $\varphi(m) = mx$ for $m \in Me$. Then $\varphi(m'm) = (m'm)x = m'(mx) = m'\varphi(m)$ for $m \in Me$ and $m' \in M$. This establishes that φ is M-equivariant. Also we have that $\varphi(e) = ex = eey = ey = x$. This proves the first statement.

The second statement follows from the first and the observation that if $\varphi, \psi \in \mathrm{End}_M(Me)$, then $\varphi(\psi(e)) = \varphi(\psi(e)e) = \psi(e)\varphi(e)$ and hence $\alpha \mapsto \alpha(e)$ provides an isomorphism of $\mathrm{End}_M(Me)$ and $(eMe)^{op}$. □

Remark 1.9. The second part of Proposition 1.8 can be reformulated as saying that eMe acts on the right of Me via endomorphisms of the left action of M and each endomorphism of Me arises from a unique element of eMe.

Since L_e is the set of generators of Me and $\mathrm{Aut}_M(Me) \cong G_e^{op}$, it follows that G_e acts on the right of L_e. More precisely, we have the following proposition, where we recall that a group G acts *freely* on a set X if the stabilizer of each element of X is trivial.

Proposition 1.10. *If $e \in E(M)$, then G_e acts freely on both the right of L_e and the left of R_e via multiplication.*

Proof. Let $m \in L_e$ and $g \in G_e$. Then $mge = mg$ and so $Mmg \subseteq Me$. If $ym = e$, then $g^{-1}ymg = g^{-1}eg = e$ and hence $Me = Mmg$. Therefore, $mg \in L_e$ and so G_e acts on the right of L_e via right multiplication. To see that the action is free, suppose that $mg = m$ and $e = ym$. Then $g = eg = ymg = ym = e$. Thus the action is free. The result for R_e is dual. $\qquad\square$

We now connect the ideal structure near an idempotent with M-sets.

Theorem 1.11. *Let $e, f \in E(M)$. Then the following are equivalent.*

(i) $Me \cong Mf$ as left M-sets.
(ii) $eM \cong fM$ as right M-sets.
(iii) There exist $a, b \in M$ with $ab = e$ and $ba = f$.
(iv) There exist $x, x' \in M$ with $xx'x = x$, $x'xx' = x'$, $x'x = e$ and $xx' = f$.
(v) $MeM = MfM$.

Proof. We prove the equivalence of (i), (iii), and (iv) by establishing that (i) implies (iv) implies (iii) implies (i). The equivalence of (ii) with (iii) and (iv) is then dual.

We begin by proving that (i) implies (iv). Assume that $Me \cong Mf$ and let $\varphi \colon Me \longrightarrow Mf$ and $\psi \colon Mf \longrightarrow Me$ be inverse isomorphisms. Put $x' = \varphi(e) \in eMf$ and $x = \psi(f) \in fMe$ (by Proposition 1.8). Then $x'x = x'\psi(f) = \psi(x'f) = \psi(x') = \psi(\varphi(e)) = e$ and $xx' = x\varphi(e) = \varphi(xe) = \varphi(x) = \varphi(\psi(f)) = f$. Note that $x'xx' = ex' = x'$ and $xx'x = xe = x$ because $x' \in eMf$ and $x \in fMe$. This yields (iv). Clearly, (iv) implies (iii) by taking $a = x'$ and $b = x$.

Next suppose that (iii) holds. If $m \in Me$, then $ma = mea = maba = maf$ and so $ma \in Mf$. Thus we can define $\varphi \colon Me \longrightarrow Mf$ by $\varphi(m) = ma$ and the mapping is clearly M-equivariant. Similarly, we can define an M-equivariant mapping $\psi \colon Mf \longrightarrow Me$ by $\psi(m) = mb$. Then $\psi(\varphi(m)) = mab = me = m$ for $m \in Me$ and $\varphi(\psi(m)) = mba = mf = m$ for $m \in Mf$. Thus φ and ψ are inverse isomorphisms yielding (i).

Trivially, (iii) implies $MeM = MababM \subseteq MbaM = MfM$ and dually $MfM \subseteq MeM$. Therefore, (iii) implies (v). Assume now that $MeM = MfM$. Write $f = xey$ with $x, y \in M$. Then observe that $f = xeyf$ and hence

$Mf = Meyf$. By Proposition 1.8, there is a unique M-equivariant mapping $\varphi\colon Me \longrightarrow Mf$ with $\varphi(e) = eyf \in eMf$. As $\varphi(Me) = M\varphi(e) = Meyf = Mf$, we conclude that φ is surjective and hence $|Mf| \leq |Me|$. A symmetric argument shows that $|Me| \leq |Mf|$. Therefore, φ is in fact a bijection and hence an isomorphism. Thus we have that (v) implies (i), thereby completing the proof of the theorem. $\qquad\square$

Isomorphic M-sets obviously have isomorphic endomorphism monoids and automorphism groups. Proposition 1.8 and Theorem 1.11 therefore have the following important consequence.

Corollary 1.12. *Let $e, f \in E(M)$ with $MeM = MfM$. Then $eMe \cong fMf$ and $G_e \cong G_f$.*

The next theorem exhibits another crucial property of finite monoids, called *stability*. It provides a certain degree of cancellativity in finite monoids.

Theorem 1.13. *Let $m, x \in M$. Then we have that*

$$MmM = MxmM \iff Mm = Mxm; \tag{1.1}$$
$$MmM = MmxM \iff mM = mxM. \tag{1.2}$$

In other words, for $m \in M$, one has that $J_m \cap Mm = L_m$ and $J_m \cap mM = R_m$.

Proof. We just prove (1.1), as (1.2) is dual. Trivially, $Mm = Mxm$ implies $MmM = MxmM$. For the converse, assume that $m = uxmv$. Clearly, we have $Mxm \subseteq Mm \subseteq Mxmv$. On the other hand, $|Mxmv| \leq |Mxm|$ because $z \mapsto zv$ is a surjective map from Mxm to $Mxmv$. It follows that both of the above containments are equalities and hence $Mxm = Mm$. The final statement is clearly equivalent to (1.1) and (1.2). $\qquad\square$

In other words, Theorem 1.13 says that if $xm \in J_m$, then $xm \in L_m$ and if $mx \in J_m$, then $mx \in R_m$.

Our first consequence is a reformulation of $\mathscr{J}$-equivalence.

Corollary 1.14. *Let $m_1, m_2 \in M$. Then the following are equivalent.*

(i) $Mm_1M = Mm_2M$.
(ii) There exists $r \in M$ such that $Mm_1 = Mr$ and $rM = m_2M$.
(iii) There exists $s \in M$ such that $m_1M = sM$ and $Ms = Mm_2$.

Proof. It is clear that both (ii) and (iii) imply (i). Let us show that (i) implies (ii), as the argument for (i) implies (iii) is similar. Write $m_1 = um_2v$ and $m_2 = xm_1y$. Let $r = xm_1$. Then $uryv = m_1$ and so $Mxm_1M = MrM = Mm_1M$. Stability (Theorem 1.13) then yields $Mr = Mxm_1 = Mm_1$. Also $m_2 = ry$ and $r = xm_1 = xum_2v$ and so $MrM = Mm_2M = MryM$. Theorem 1.13 then implies that $rM = ryM = m_2M$. This completes the proof. $\qquad\square$

Another important consequence of stability is that the set of non-units of a finite monoid is an ideal (if nonempty).

Corollary 1.15. *Let M be a finite monoid with group of units G. Then $G = J_1$ and $M \setminus G$, if nonempty, is an ideal of M.*

Proof. Trivially, $G \subseteq J_1$. Suppose that $m \in J_1$, that is, $MmM = M1M = M$. Then since $1m = m = m1$, it follows from stability (Theorem 1.13) that $mM = M = Mm$. Thus we can write $xm = 1 = my$ for some $x, y \in M$ and so m has both a left and a right inverse. Therefore, $m \in G$ and so $J_1 = G$. Note that since $MmM \subseteq M = M1M$ for all $m \in M$, it follows that

$$I(1) = \{m \in M \mid 1 \notin MmM\} = M \setminus J_1 = M \setminus G,$$

completing the proof.   □

Recall that if $e \in E(M)$, then $I_e = eMe \setminus G_e$.

Corollary 1.16. *Let $e \in E(M)$. Then $J_e \cap eMe = G_e$. In particular, $I_e = eI(e)e$ and is an ideal of eMe (if nonempty).*

Proof. By Lemma 1.7, we have that $J_e \cap eMe$ is the $\mathscr{J}$-class of e in eMe. But this is the group of units G_e of eMe by Corollary 1.15. This establishes that $J_e \cap eMe = G_e$. Clearly, $eI(e)e$ is an ideal of eMe (if nonempty) and $eI(e)e \subseteq I_e$. Conversely, if $m \in I_e = eMe \setminus G_e$, then $m \notin J_e$ by the first part of the corollary. Since $m \in MeM$, we conclude that $e \notin MmM$. Therefore, $m \in I(e)$ and so $m = eme \in eI(e)e$.   □

As a consequence, we can now describe the orbits of G_e on L_e and R_e.

Corollary 1.17. *Let $e \in E(M)$. Then two elements $m, n \in L_e$ belong to the same right G_e-orbit if and only if $mM = nM$. Similarly, $m, n \in R_e$ belong to the same left G_e-orbit if and only if $Mm = Mn$.*

Proof. We prove only the first statement. Trivially, if $mG_e = nG_e$, then $mM = nM$. Suppose that $mM = nM$ and write $m = nu$. Then, as $me = m$ and $ne = n$, we conclude that $m = n(eue)$. Therefore, $MeM = MmM = MneueM \subseteq MeueM \subseteq MeM$ and hence $eue \in J_e$. But then $eue \in eMe \cap J_e = G_e$ and so $mG_e = nG_e$.   □

Corollary 1.16 also illuminates the structure of the minimal ideal of a finite monoid.

Remark 1.18. The minimal ideal of a monoid is always a $\mathscr{J}$-class. Indeed, if I is the minimal ideal of M, then, for all $m \in I$, we have $MmM \subseteq I$ and hence $MmM = I$ by minimality. Thus all elements of I generate the same principal ideal.

Corollary 1.19. *Let I be the minimal ideal of M and let $e \in E(I)$. Then $eIe = eMe = G_e$.*

Proof. Since I is an ideal, $eMe \subseteq I$ and so $eMe = e(eMe)e \subseteq eIe \subseteq eMe$. Therefore, $eMe = eIe \subseteq I$. Since $I = J_e$ by Remark 1.18 we have $G_e = eMe \cap J_e = eIe \cap I = eIe$ by Corollary 1.16. This concludes the proof. $\square$

Another important concept that we shall need from ideal theory is that of a principal series. A *principal series* for M is an unrefinable chain of ideals

$$\emptyset = I_0 \subsetneq I_1 \subsetneq \cdots \subsetneq I_s = M \tag{1.3}$$

where are including $\emptyset$ for convenience, even though it is not an ideal. Note that I_1 must be the minimal ideal of M. Of course, principal series exist for any finite monoid. It turns out that the differences $I_k \setminus I_{k-1}$, with $1 \le k \le s$, are precisely the $\mathscr{J}$-classes of M.

Proposition 1.20. *Let (1.3) be a principal series for M. Then each difference $I_k \setminus I_{k-1}$ with $1 \le k \le s$ is a $\mathscr{J}$-class and each $\mathscr{J}$-class arises for exactly one choice of k.*

Proof. First note that if $m, m' \in I_k \setminus I_{k-1}$, then we have by definition of a principal series that $MmM \cup I_{k-1} = I_k = Mm'M \cup I_{k-1}$ and hence $m \in Mm'M$ and $m' \in MmM$. Therefore, $MmM = Mm'M$. If $MmM = MnM$, then clearly $n \in I_k \setminus I_{k-1}$ and thus $I_k \setminus I_{k-1} = J_m$.

If $m \in M$ and j is minimal with $m \in I_j$, then $m \in I_j \setminus I_{j-1}$ and so $J_m = I_j \setminus I_{j-1}$ by the previous paragraph. Since the sets $I_k \setminus I_{k-1}$ with $1 \le k \le s$ are obviously disjoint, we conclude that each $\mathscr{J}$-class occurs for exactly one choice of k. $\square$

For example, if $M = M_n(\mathbb{F})$ is the monoid of $n \times n$ matrices over a finite field $\mathbb{F}$ and I_r denotes the ideal of matrices of rank at most r, then

$$\emptyset \subsetneq I_0 \subsetneq I_1 \subsetneq \cdots \subsetneq I_n = M$$

is a principal series for M.

1.4 von Neumann regularity

A fundamental role in semigroup theory is played by the notion of von Neumann regularity. We will formulate things for a monoid M, although much of the same holds for semigroups. An element $m \in M$ is said to be (von Neumann) *regular* if $m = mam$ for some $a \in M$, i.e., $m \in mMm$. We say that the monoid M is (von Neumann) *regular* if all of its elements are regular. Of course, every group G is regular because $gg^{-1}g = g$ for $g \in G$. Many naturally occurring monoids are regular, such as the monoid of all self-maps on a finite set and the monoid of all endomorphisms of a finite dimensional vector space. We continue to assume that M is a finite monoid.

Proposition 1.21. *Let $e \in E(M)$ and $m \in eMe$. Then m is regular in M if and only if it is regular in eMe. Consequently, M regular implies eMe regular.*

Proof. This follows from the observation that $mMm = meMem$. □

Next we describe some equivalent conditions to regularity.

Proposition 1.22. *The following are equivalent for $m \in M$.*

(i) m is regular.
(ii) $Mm = Me$ for some idempotent $e \in E(M)$.
(iii) $mM = eM$ for some idempotent $e \in E(M)$.
(iv) $MmM = MeM$ for some idempotent $e \in E(M)$.

Proof. Assume that m is regular, say that $m = mam$. Then $ma = mama$ and $am = amam$ are idempotents and $Mm = Mam$, $mM = maM$. Thus (i) implies (ii) and (iii). If $Mm = Me$ with $e \in E(M)$, then write $m = ae$ and $e = bm$. Then $mbm = me = aee = ae = m$ and so m is regular. Thus (ii) implies (i) and the proof that (iii) implies (i) is similar. Trivially, (ii) implies (iv). Suppose that (iv) holds. Then we have by Corollary 1.14 that there exists $r \in M$ such that $Mm = Mr$ and $rM = eM$. But then r is regular by (iii) implies (i) and hence $Mm = Mr = Mf$ for some $f \in E(M)$ by (i) implies (ii). Therefore, m is regular by (ii) implies (i). This completes the proof. □

A $\mathscr{J}$-class containing an idempotent is called a *regular $\mathscr{J}$-class*. Proposition 1.22 says that a $\mathscr{J}$-class is regular if and only if all its elements are regular.

Proposition 1.23. *Let I be an ideal of M. Then $I^2 = I$ if and only if I is generated (as an ideal) by idempotents, that is, $I = ME(I)M = IE(I)I$.*

Proof. The equality $IE(I)I = ME(I)M$ is straightforward. If $I = ME(I)M$, then trivially $I = I^2$ because $uev = (ue)(ev) \in I^2$ for $e \in E(I)$. Conversely, suppose that $I^2 = I$ and let $n = |I|$. Then $I = I^n = IE(I)I$ by Lemma 1.5. This completes the proof. □

Corollary 1.24. *If $m \in M$, then $(MmM)^2 = MmM$ if and only if $MmM = MeM$ for some $e \in E(M)$, that is, if and only if m is regular.*

Proof. Assume that $(MmM)^2 = MmM$. Proposition 1.23 then implies that $m \in MeM$ for some $e \in E(MmM)$ and so $MeM = MmM$. Thus m is regular by Proposition 1.22. The converse is clear. □

A consequence of the previous three results is that M is regular if and only if each of its ideals is idempotent.

Corollary 1.25. *A finite monoid M is regular if and only if $I = I^2$ for every ideal I of M.*

1.5 Exercises

1.1. Prove that a nonempty semigroup S is a group if and only if $aS = S = Sa$ for all $a \in S$.

1.2. Let S be a nonempty finite semigroup satisfying both the left and right cancellation laws stating that $xy = xz$ implies $y = z$ and $yx = zx$ implies $y = z$. Prove that S is a group.

1.3. Prove that a finite monoid is a group if and only if it has a unique idempotent.

1.4. Let $\varphi \colon M \longrightarrow N$ be a monoid homomorphism. Prove that $M/\ker\varphi \cong \varphi(M)$.

1.5. Let $\varphi \colon M \longrightarrow N$ be a monoid homomorphism and let $e \in E(M)$. Putting $f = \varphi(e) \in E(N)$, prove that:

(a) $\varphi(eMe) \subseteq fNf$;
(b) $\varphi|_{eMe} \colon eMe \longrightarrow fNf$ is a homomorphism of monoids;
(c) $\varphi(G_e) \subseteq G_f$.

1.6. Let I be an ideal of a monoid M. Define an equivalence relation $\equiv_I$ on M by $m \equiv_I n$ if and only if $m = n$ or $m, n \in I$. Prove that $\equiv_I$ is a congruence and that the class of I is a zero element of the quotient. The quotient of M by $\equiv_I$ is denoted M/I and called the *Rees quotient* of M by I.

1.7. Prove that there are five semigroups of order 2 up to isomorphism.

1.8. A monoid M is said to be *congruence-free* if the only congruences on M are the universal equivalence relation and equality. Prove that a finite monoid M is congruence-free if and only if it is either a finite simple group or isomorphic to $\{0, 1\}$ under multiplication.

1.9. Suppose that $a, b, c \in M$. Prove that $a \mathscr{L} b$ implies $ac \mathscr{L} bc$ and $a \mathscr{R} b$ implies $ca \mathscr{R} cb$.

1.10. Let N be a submonoid of M and let a, b be regular elements of N. Prove that $Na = Nb$ if and only if $Ma = Mb$. Give an example showing that the hypothesis that a, b are regular in N is necessary.

1.11. Prove that if M is a finite monoid and $a, b \in M$, then $(ab)^\omega \mathscr{J} (ba)^\omega$.

1.12. Let M be a finite monoid. Prove that $\mathscr{J}$ is the join of $\mathscr{R}$ and $\mathscr{L}$ in the lattice of equivalence relations on M.

1.13. Let M be a monoid and suppose that $x, x' \in M$ with $xx'x = x$ and $x'xx' = x'$.

(a) Prove that $e = xx'$ and $f = x'x$ are idempotents.

(b) Prove that $\varphi\colon eMe \longrightarrow fMf$ given by $\varphi(m) = x'mx$ is an isomorphism of monoids with inverse given by $\psi(m) = xmx'$.

1.14. Let T_n be the monoid of all self-maps on an n-element set. Prove that if $f, g \in T_n$, then $f \mathrel{\mathscr{J}} g$ if and only if they have the same rank, $f \mathrel{\mathscr{R}} g$ if and only if they have the same image and $f \mathrel{\mathscr{L}} g$ if and only if they induce the same partition into fibers over their images.

1.15. Let $A, B \in M_n(\Bbbk)$ for a field $\Bbbk$. Show that $A \mathrel{\mathscr{J}} B$ if and only if they have the same rank.

1.16. Prove that T_n and $M_n(\Bbbk)$ (where $\Bbbk$ is a finite field) are regular.

1.17. Prove that if M is a monoid and $a \in M$, then a is regular if and only if there exists $b \in M$ with $aba = a$ and $bab = b$.

1.18. Let J be a $\mathscr{J}$-class of a finite monoid M and suppose that $R \subseteq J$ is an $\mathscr{R}$-class and $L \subseteq J$ is an $\mathscr{L}$-class. Prove that $R \cap L \neq \emptyset$.

1.19. Show that if $e \in E(M)$, then $G_e = R_e \cap L_e$.

1.20. Let M be a finite monoid and $e \in E(M)$. Prove that $J_e \subseteq L_e R_e$.

1.21. Let M be a finite monoid and suppose that $a, b \in M$ with $a \mathrel{\mathscr{L}} b$. Suppose that $xa = b$ and $yb = a$ with $x, y \in M$. Define $\varphi_x\colon R_a \longrightarrow R_b$ by $\varphi_x(m) = xm$ and $\varphi_y\colon R_b \longrightarrow R_a$ by $\varphi_y(m) = ym$. This is well defined by Exercise 1.9. Prove that φ_x is a bijection with inverse φ_y such that $\varphi_x(m) \mathrel{\mathscr{L}} m$ for all $m \in R_a$.

1.22. Let M be a finite monoid and $a, b \in M$ with $MaM = MbM$. Prove that $|L_a \cap R_a| = |L_b \cap R_b|$ using Exercise 1.21 and its dual.

1.23. Let M be a finite monoid and let $e \in E(M)$. Prove that $|J_e| = mn|G_e|$ where m is the number of $\mathscr{R}$-classes contained in J_e and n is the number of $\mathscr{L}$-classes contained in J_e. (Hint: you might find Exercises 1.18, 1.19, and 1.22 useful.)

1.24 (Clifford-Miller). Let M be a finite monoid and $a \mathrel{\mathscr{J}} b$. Prove that $ab \mathrel{\mathscr{J}} a \mathrel{\mathscr{J}} b$ if and only if $L_a \cap R_b$ contains an idempotent.

1.25 (Rhodes). Let $\varphi\colon M \longrightarrow N$ be a surjective homomorphism of finite monoids and let J be a $\mathscr{J}$-class of N. Let X be the set of all $\mathscr{J}$-classes J' of M such that $\varphi(J') \subseteq J$ and partially order X by $J' \leq J''$ if $MJ'M \subseteq MJ''M$.

(a) Prove that if $J' \in X$ is minimal, then $\varphi(J') = J$.
(b) Prove that if J is regular, then X has a unique minimal element J' and that J' is regular.

1.26. Let M be a finite monoid such that each regular $\mathscr{J}$-class of M is a subsemigroup. Prove that each regular $\mathscr{J}$-class of each quotient of M is also a subsemigroup. (Hint: use Exercise 1.25.)

1.27. This exercise proves a special case of Rees's theorem. Let M be a finite monoid with minimal ideal I. Fix $e \in E(M)$ and let $G = eIe$, which is a group by Corollary 1.19. Using Exercise 1.18, we can choose $\lambda_1, \ldots, \lambda_s \in L_e$ such that $R_{\lambda_1}, \ldots, R_{\lambda_s}$ are the $\mathscr{R}$-classes of I and $\rho_1, \ldots, \rho_r \in R_e$ such that $L_{\rho_1}, \ldots, L_{\rho_r}$ are the $\mathscr{L}$-classes of I. Define a matrix $P \in M_{rs}(\mathbb{C}G)$ by $P_{ij} = \rho_i \lambda_j \in eIe = G$. Let $S = \{gE_{ij} \in M_{sr}(\mathbb{C}G) \mid g \in G\}$ where E_{ij} is the elementary matrix unit with 1 in the ij-entry and 0 elsewhere.

(a) Verify that S is a semigroup with respect to the product $A \odot B = APB$ for $A, B \in S$.
(b) Prove that $\varphi \colon S \longrightarrow I$ defined by

$$\varphi(gE_{ij}) = \lambda_i g \rho_j$$

   is an isomorphism. (Hint: Exercise 1.20 may be helpful for surjectivity.)

1.28. Let M be a finite monoid and $a \in M$. Define a binary operation $\circ_a$ on $aM \cap Ma$ by $xa \circ_a ay = xay$ for $x, y \in M$.

(a) Show that $\circ_a$ is a well-defined associative operation.
(b) Show that $aM \cap Ma$ is a monoid with respect to $\circ_a$ with a as the identity.
(c) Show that $R_a \cap L_a$ is the group of units of $(aM \cap Ma, \circ_a)$.
(d) Show that if $e \in E(M)$, then $eM \cap Me$ equipped with $\circ_e$ is isomorphic to eMe.
(e) Prove that if $MaM = MbM$, then $(aM \cap Ma, \circ_a) \cong (bM \cap Mb, \circ_b)$.
(f) Deduce that if $MaM = MbM$, then $|R_a \cap L_a| = |R_b \cap L_b|$.

1.29. Let B be the monoid generated by the mappings $f, g \colon \mathbb{N} \longrightarrow \mathbb{N}$ given by $f(n) = n + 1$ and

$$g(n) = \begin{cases} n - 1, & \text{if } n > 0 \\ 0, & \text{if } n = 0. \end{cases}$$

Show that Theorem 1.13 fails for B.

2

$\mathscr{R}$-trivial Monoids

This chapter studies $\mathscr{R}$-trivial monoids, that is, monoids where Green's relation $\mathscr{R}$ is the equality relation. They form an important class of finite monoids, which have quite recently found applications in the analysis of Markov chains; see Chapter 14. The first section of this chapter discusses lattices and prime ideals of arbitrary finite monoids, as they shall play an important role in both the general theory and in the structure theory of $\mathscr{R}$-trivial monoids. The second section concerns $\mathscr{R}$-trivial monoids and, in particular, left regular bands, which are precisely the regular $\mathscr{R}$-trivial monoids.

2.1 Lattices and prime ideals

The results of this section can be found for arbitrary semigroups in the work of Petrich [Pet63,Pet64]. These results have their roots in Clifford's paper [Cli41]. Our approach roughly follows that of [MS12a].

Recall that a non-empty partially ordered set (*poset*) P is a *lattice* if any two elements $p, q \in P$ have both a *meet* (greatest lower bound) $p \wedge q$ and a *join* (least upper bound) $p \vee q$. A finite poset is a lattice if and only if it has a greatest element and each pair of elements has a meet, which occurs if and only if it has a least element and each pair of elements has a join. In this section all lattices will be finite.

A lattice can be viewed as a monoid with respect to either its join or its meet. It is customary in semigroup theory to use the meet operation as the default monoid structure and so often we write it as juxtaposition, unless we want to emphasize its order theoretic aspect. Notice that if P is a lattice, then P is a commutative monoid with $P = E(P)$ and that the lattice order is the natural partial order because $e \leq f$ if and only if $e \wedge f = e = f \wedge e$. Conversely, if M is a finite commutative monoid such that $M = E(M)$, then M is a lattice with respect to its natural partial order and the product is the meet operation. Indeed, we have $(ef)e = e(ef) = ef$ and so $ef \leq e$. Similarly,

B. Steinberg, *Representation Theory of Finite Monoids*,
Universitext, DOI 10.1007/978-3-319-43932-7_2

we have $ef \le f$. Moreover, if $m \le e, f$, then $efm = mef = mf = m$ and so $m \le ef$. Thus $ef = e \wedge f$. As the identity is the maximum element in the natural partial order, we conclude that M is a lattice. In summary, we have proved the following classical proposition.

Proposition 2.1. *Finite lattices are precisely finite commutative monoids in which each element is idempotent. The meet is the product and the order is the natural partial order. The maximum element is the identity.*

Semigroups in which each element is idempotent are called *bands*. If M is a finite monoid, then there is a universal homomorphism from M to a lattice. Momentarily, it will be shown that this lattice can be identified with the lattice of prime ideals of M.

An ideal P of a monoid M is said to be *prime* if its complement $M \setminus P$ is a submonoid or, equivalently, P is a proper ideal and $ab \in P$ implies $a \in P$ or $b \in P$. The union $P \cup Q$ of two prime ideals P, Q is again a prime ideal because $M \setminus (P \cup Q) = (M \setminus P) \cap (M \setminus Q)$ is a submonoid. By convention, we shall also declare the empty set to be a prime ideal. Of course, $M \setminus \emptyset = M$. With this convention, the prime ideals form a lattice $\mathrm{Spec}(M)$ with respect to the inclusion ordering with union as the join. Note that $I_1 = M \setminus G_1$ is the unique maximal prime ideal; the fact that it is an ideal, if non-empty, is part of Corollary 1.15.

Lemma 2.2. *Let M be a finite monoid. Let P be a prime ideal and put $M' = M \setminus P$.*

(i) If $m, n \in M'$, then $MmM \subseteq MnM$ if and only if $M'mM' \subseteq M'nM'$.
(ii) Each $\mathscr{J}$-class of M' is also a $\mathscr{J}$-class of M.
(iii) Let $J(P)$ be the minimal ideal of M'. If $e \in J(P)$ is an idempotent, then $J(P)$ is the $\mathscr{J}$-class of e in M and $P = I(e)$.
(iv) If I is an ideal of M, then $J(P) \subseteq I$ if and only if $I \nsubseteq P$.

Proof. Clearly if $M'mM' \subseteq M'nM'$, then $MmM \subseteq MnM$. For the converse, write $m = xny$ with $x, y \in M$. Then since $m \notin P$ and P is an ideal, we deduce that $x, y \notin P$ and so $x, y \in M'$. Thus $M'mM' \subseteq M'nM'$ establishing (i).

To prove (ii), first note that if $m \in P$, then $J_m \subseteq MmM \subseteq P$ and hence any $\mathscr{J}$-class of M that intersects P is contained in it. So it suffices to check that if $m, n \in M'$, then $MmM = MnM$ if and only if $M'mM' = M'nM'$, which is immediate from (i).

To establish (iii) observe that $J(P)$ is a $\mathscr{J}$-class of M' by Remark 1.18 and hence, by (ii), it is also a $\mathscr{J}$-class of M. Being a subsemigroup of M', it follows that $J(P)$ contains an idempotent e. Since $e \in J(P) \subseteq M'mM' \subseteq MmM$ for all $m \in M'$, we conclude that $M' \subseteq M \setminus I(e)$. Conversely, if $m \in M \setminus I(e)$, then $e \in MmM$. As $e \notin P$, we conclude that $m \notin P$. Thus $M' = M \setminus I(e)$ and so $I(e) = P$.

To prove (iv), let I be an ideal of M. If $J(P) \subseteq I$, then clearly $I \nsubseteq P$. If $m \in I \setminus P$, then $J(P) \subseteq M'mM' \subseteq MmM \subseteq I$ because $J(P)$ is the minimal ideal of M'. This completes the proof. $\square$

Let us define an idempotent $e \in E(M)$ to be *coprime* if $I(e)$ is a prime ideal or, equivalently, $e \in MmM \cap Mm'M$ implies $e \in Mmm'M$. Note that it is immediate from the definition that if $e, f \in E(M)$ are $\mathscr{J}$-equivalent, that is, $MeM = MfM$, then e is coprime if and only if f is coprime. Thus it is natural to call a $\mathscr{J}$-class J *coprime* if it contains a coprime idempotent, in which case all its idempotents are coprime. Notice that every prime ideal is of the form $I(e)$ for some coprime idempotent e by Lemma 2.2.

We put
$$\Lambda(M) = \{MeM \mid e \text{ is coprime}\}.$$

It is a poset with respect to inclusion. Note that we can identify $\Lambda(M)$ with the set of coprime $\mathscr{J}$-classes ordered by $J \leq J'$ if $MJM \subseteq MJ'M$. Let us characterize the coprime $\mathscr{J}$-classes.

Proposition 2.3. *Let M be a finite monoid. Then the following are equivalent for a $\mathscr{J}$-class J.*

(i) J is a coprime $\mathscr{J}$-class.
(ii) J is a subsemigroup.
(iii) J is the minimal ideal of $M \setminus P$ where P is a prime ideal.

Proof. If J is a coprime $\mathscr{J}$-class, then J contains an idempotent f with $P = I(f)$ a prime ideal by definition. We claim that J is the minimal ideal $J(P)$ of $M' = M \setminus P = M \setminus I(f)$. Indeed, if $e \in J(P)$ is an idempotent, then $e \in M'fM' \subseteq MfM$ because $J(P)$ is the minimal ideal of M' and $f \in M'$. But $e \notin I(f)$ implies that $f \in MeM$. Thus $e \in J$ and so $J = J(P)$ because $J(P)$ is a $\mathscr{J}$-class of M by Lemma 2.2. This establishes that (i) implies (iii).

Since the minimal ideal of any monoid is a subsemigroup, (iii) implies (ii). Finally, if J is a subsemigroup, then J contains an idempotent e by Corollary 1.4. If $e \in MmM \cap Mm'M$, then $umv = e = xm'y$ with $u, v, x, y \in M$. Then $e = eumv = xm'ye$. Therefore, $eum, m'ye \in J$ and hence $eumm'ye \in J$ because J is a subsemigroup. Thus $e \in Mmm'M$ and so $I(e)$ is a prime ideal, that is, J is a coprime $\mathscr{J}$-class. This completes the proof. $\square$

As a corollary, it follows that prime ideals are in bijection with coprime $\mathscr{J}$-classes. More precisely, we have the following.

Theorem 2.4. *There is an isomorphism of posets $\Lambda(M) \cong \mathrm{Spec}(M)$. Consequently, $\Lambda(M)$ is a lattice.*

Proof. For a prime ideal P of M, we continue to denote by $J(P)$ the minimal ideal of $M \setminus P$. By Proposition 2.3, $J(P)$ is a coprime $\mathscr{J}$-class. Define $\psi \colon \mathrm{Spec}(M) \longrightarrow \Lambda(M)$ by $\psi(P) = MJ(P)M$. We claim that ψ is a bijection with inverse $\tau \colon \Lambda(M) \longrightarrow \mathrm{Spec}(M)$ given by $\tau(MeM) = I(e)$ for e a coprime idempotent. Indeed, $\tau(\psi(P)) = I(e)$ where $e \in J(P)$ is an idempotent. But then $P = I(e)$ by Lemma 2.2(iii) and hence $\tau\psi$ is the identity on $\mathrm{Spec}(M)$.

On the other hand, if e is a coprime idempotent and $P = I(e)$, then $\psi(\tau(MeM)) = MJ(P)M$. But $J(P) \subseteq MeM$ by Lemma 2.2(iv) because $e \notin I(e)$. On the other hand, $J(P) \subseteq M \setminus I(e)$ implies by definition that $e \in MJ(P)M$. Thus $MeM = MJ(P)M$ and so ψ and τ are inverse bijections.

We next observe that ψ and τ are order-preserving. Indeed, if $MeM \subseteq MfM$, then $e \notin MmM$ implies $f \notin MmM$ and hence $I(e) \subseteq I(f)$. Thus τ is order-preserving. Conversely, if P, Q are prime ideals with $P \subseteq Q$, then $M \setminus Q \subseteq M \setminus P$ and hence $J(Q) \subseteq M \setminus P$. Therefore, $J(P) \subseteq MJ(Q)M$ by Lemma 2.2(iv) and so $\psi(P) \subseteq \psi(Q)$. Thus ψ is order-preserving. This completes the proof that $\Lambda(M)$ is isomorphic to $\mathrm{Spec}(M)$. $\square$

Let us define a mapping $\sigma \colon M \longrightarrow \Lambda(M)$ by

$$\sigma(m) = \bigvee_{\substack{MeM \in \Lambda(M), \\ MeM \subseteq MmM}} MeM, \tag{2.1}$$

that is, $\sigma(m)$ is the largest element of $\Lambda(M)$ contained MmM. Note that if $e \in E(M)$ belongs to the minimal ideal of M, then $I(e) = \emptyset$ is a prime ideal and so e is coprime, whence $MeM \in \Lambda(M)$. As $MeM \subseteq MmM$, the join in (2.1) is over a non-empty index set.

Theorem 2.5. *The mapping $\sigma \colon M \longrightarrow \Lambda(M)$ is a surjective homomorphism. Moreover, given any homomorphism $\varphi \colon M \longrightarrow P$ with P a lattice, there is a unique homomorphism $\varphi' \colon \Lambda(M) \longrightarrow P$ such that*

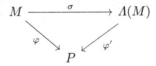

commutes.

Proof. First note that σ is surjective because if $e \in E(M)$ is coprime, then $\sigma(e) = MeM$ by definition. As 1 is coprime (because $J_1 = G_1$ is a subsemigroup by Corollary 1.15), $\sigma(1) = M$ is the maximum of $\Lambda(M)$. If $e \in E(M)$ is coprime, then

$$
\begin{aligned}
MeM \subseteq \sigma(m) \wedge \sigma(m') &\iff MeM \subseteq \sigma(m) \text{ and } MeM \subseteq \sigma(m') \\
&\iff MeM \subseteq MmM \text{ and } MeM \subseteq Mm'M \\
&\iff MeM \subseteq MmM \cap Mm'M \\
&\iff e \in MmM \cap Mm'M \\
&\iff e \in Mmm'M \\
&\iff MeM \subseteq Mmm'M \\
&\iff MeM \subseteq \sigma(mm').
\end{aligned}
$$

It follows that $\sigma(mm') = \sigma(m) \wedge \sigma(m')$.

To prove the second statement it suffices to show that if $\varphi(m) \neq \varphi(m')$, then $\sigma(m) \neq \sigma(m')$. Without loss of generality, assume that $\varphi(m) \not\geq \varphi(m')$. Let $I = \{p \in P \mid p \not\geq \varphi(m')\}$. It is easy to check that I is a prime ideal of P and hence $Q = \varphi^{-1}(I)$ is a prime ideal of M with $m \in Q$ and $m' \notin Q$. Therefore, $J(Q) \not\subseteq MmM$ (as $MmM \subseteq Q$) and $J(Q) \subseteq Mm'M$ by Lemma 2.2(iv). Then $MJ(Q)M = \psi(Q)$ (using the notation of the proof of Theorem 2.4) belongs to $\Lambda(M)$ and satisfies $MJ(Q)M \not\subseteq MmM$ and $MJ(Q)M \subseteq Mm'M$. Therefore, we have that $MJ(Q)M \not\subseteq \sigma(m)$ and $MJ(Q)M \subseteq \sigma(m')$, and hence $\sigma(m) \neq \sigma(m')$. This completes the proof. $\qquad\square$

It turns out that if a regular $\mathscr{J}$-class consists of either a single $\mathscr{L}$-class or a single $\mathscr{R}$-class, then it is coprime.

Proposition 2.6. *Let J be a regular $\mathscr{J}$-class consisting of either a single $\mathscr{L}$-class or a single $\mathscr{R}$-class. Then J is a subsemigroup and hence coprime.*

Proof. We just handle the case of a single $\mathscr{L}$-class, as the other is dual. Let $a, b \in J$. Since b is regular, we have $bM = eM$ for some idempotent $e \in J$ by Proposition 1.22. Write $e = by$ with $y \in M$. Because J is an $\mathscr{L}$-class, it follows that $a = xe$ for some $x \in M$. Thus $aby = ae = xee = xe = a$ and so $abM = aM$, whence $ab \in J$. We conclude that J is a subsemigroup. $\qquad\square$

2.2 $\mathscr{R}$-trivial monoids and left regular bands

Recall that a monoid M is $\mathscr{R}$-trivial if $mM = nM$ implies $m = n$. These are the monoids, which up to now, have played a role in applications to Markov chains. Schocker studied them under the name "weakly ordered semigroups" [Sch08, BBBS11]. Important examples of $\mathscr{R}$-trivial monoids include 0-Hecke monoids [Nor79, Car86, Fay05, DHST11, Den11], Catalan monoids [HT09, DHST11, MS12b, GM14], hyperplane face monoids [BHR99, Bro04, Sal09] (see also Exercise 14.12) and oriented matroids [BLVS+99]. We shall now investigate $\Lambda(M)$ in this special case.

Corollary 2.7. *Let M be an $\mathscr{R}$-trivial monoid.*

(i) *Each regular $\mathscr{J}$-class of M is coprime (i.e., each idempotent of M is coprime) and hence*

$$\Lambda(M) = \{MeM \mid e \in E(M)\}.$$

(ii) *Each regular element of M is an idempotent.*
(iii) *If $e \in E(M)$, then $em = e$ if and only if $m \notin I(e)$, that is, $e \in MmM$.*
(iv) *The universal mapping $\sigma\colon M \longrightarrow \Lambda(M)$ is given by $\sigma(m) = Mm^\omega M$.*

Proof. For (i), it suffices, by Proposition 2.6, to observe that each $\mathscr{J}$-class is in fact an $\mathscr{L}$-class. Indeed, if $MmM = MnM$, then by Corollary 1.14 there exists $r \in M$ with $mM = rM$ and $Mr = Mn$. By $\mathscr{R}$-triviality, $r = m$ and so $Mm = Mn$. To prove (ii), if $m \in M$ is regular, then $mM = eM$ for some idempotent e by Proposition 1.22. But then $m = e$ by $\mathscr{R}$-triviality. For (iii), trivially $em = e$ implies $m \notin I(e)$. For the converse, assume that $m \notin I(e)$, that is, $e = xmy$ with $x, y \in M$. Then $e = e^2 = exmy$ and so $eM = exM$. Therefore, we have that $e = ex$ by $\mathscr{R}$-triviality and so $e = exmy = emy$. Thus $eM = emM$ and we conclude that $e = em$ by $\mathscr{R}$-triviality. To prove (iv), since $\Lambda(M)$ is a lattice, we have that $\sigma(m) = \sigma(m^\omega)$. But $Mm^\omega M$ is obviously the largest element of $\Lambda(M)$ contained in $Mm^\omega M$. This completes the proof. $\square$

An $\mathscr{R}$-trivial monoid M is called a *left regular band* if it is a regular monoid. In this case $M = E(M)$ by Corollary 2.7. Left regular bands are characterized by satisfying the identity $xyx = xy$ for all $x, y \in M$.

Lemma 2.8. *A monoid M is a left regular band if and only if it satisfies the identity $xyx = xy$.*

Proof. If M is a left regular band, then $xyxy = xy$ and so $xyxM = xyM$. We conclude that $xyx = xy$ by $\mathscr{R}$-triviality. Conversely, suppose that M satisfies the identity $xyx = xy$. Taking $y = 1$, we see that each element of M is idempotent and so, in particular, M is regular. If $mM = nM$, then $mu = n$ and $nv = m$ for some $u, v \in M$. As $uvu = uv$, we then have $m = muv = muvu = n$ and so M is $\mathscr{R}$-trivial. This completes the proof. $\square$

Since $M = E(M)$ in a left regular band, we have that

$$\Lambda(M) = \{MmM \mid m \in M\}$$

in this case. Actually, we have the following.

Proposition 2.9. *Let M be a left regular band.*

(i) $MmM = Mm$ for all $m \in M$.
(ii) $MmM \cap MnM = MmnM$.
(iii) The lattice operation on $\Lambda(M)$ is intersection.
(iv) $m \leq n$ in the natural partial order if and only if $mM \subseteq nM$.

Proof. We shall use Lemma 2.8 without comment throughout the proof. For (i), clearly $Mm \subseteq MmM$. For the converse, if $u, v \in M$, then $umv = umvm \in Mm$ and so $MmM \subseteq Mm$. By (i), to prove (ii) we need that $Mm \cap Mn = Mmn$. Clearly $Mmn \subseteq Mn$ and $Mmn = Mmnm \subseteq Mm$. Thus we have $Mmn \subseteq Mm \cap Mn$. Suppose that $x \in Mm \cap Mn$, say $x = um = vn$ with $u, v \in M$. Then we have $xmn = ummn = umn = xn = vnn = vn = x$ and so $x \in Mmn$. Item (iii) is immediate from (ii). To prove (iv), if $m \leq n$, then $nm = m$ and so $mM \subseteq nM$. Conversely, if $mM \subseteq nM$, then $m = nx$ for some $x \in M$ and so $nm = nnx = nx = m$ and $mn = mnm = mm = m$. Thus $m \leq n$. This completes the proof. $\square$

2.3 Exercises

2.1. Prove that the following are equivalent for a finite poset P.

(i) P is a lattice.
(ii) P has a maximum element and binary meets.
(iii) P has a minimum element and binary joins.

2.2. Let M be a finite monoid. Let $\tau\colon \Lambda(M) \longrightarrow \mathrm{Spec}(M)$ be the mapping from the proof of Theorem 2.4 and $\sigma\colon M \longrightarrow \Lambda(M)$ be the mapping from Theorem 2.5. Prove that

$$\tau(\sigma(m)) = \bigcup_{\substack{P \in \mathrm{Spec}(M), \\ m \notin P}} P$$

for $m \in M$.

2.3. Prove that a finite monoid M is $\mathscr{J}$-trivial if and only if it is both $\mathscr{L}$-trivial and $\mathscr{R}$-trivial.

2.4. Let M be a finite monoid. Prove that M is $\mathscr{R}$-trivial if and only if $(mn)^\omega m = (mn)^\omega$ for all $m, n \in M$.

2.5. Prove that a finite monoid M is $\mathscr{J}$-trivial if and only if $m^{\omega+1} = m^\omega$ and $(mn)^\omega = (nm)^\omega$ for all $m, n \in M$.

2.6. Let M be a finite monoid. Prove that M is $\mathscr{R}$-trivial if and only if each of its regular $\mathscr{R}$-classes is a singleton.

2.7. Prove that the class of finite $\mathscr{R}$-trivial monoids is closed under direct product, submonoids and homomorphic images.

2.8. Let M be a finite $\mathscr{R}$-trivial monoid, $m \in M$ and $e \in E(M)$. Prove that $MeM \subseteq MmM$ if and only if $Me \subseteq Mm$.

2.9. Let M be a finite $\mathscr{R}$-trivial monoid and let $e, f \in E(M)$. Prove that $Me \cap Mf = M(ef)^\omega = M(fe)^\omega$.

2.10. Use Exercises 2.8 and 2.9 to show that if M is a finite $\mathscr{R}$-trivial monoid, then $\Upsilon(M) = \{Me \mid e \in E(M)\}$ is a lattice with intersection as the meet and that $\Upsilon(M) \cong \Lambda(M)$ via $Me \mapsto MeM$.

2.11. Let M be a finite $\mathscr{L}$-trivial monoid and Ω a faithful left M-set. Let $e, f \in E(M)$.

(a) Prove that $eM \subseteq fM$ if and only if $e\Omega \subseteq f\Omega$.
(b) Prove that $e\Omega \cap f\Omega = (ef)^\omega\Omega = (fe)^\omega\Omega$ and hence $\{e\Omega \mid e \in E(M)\}$ is a submonoid of the power set of Ω equipped with intersection.

(c) Prove that $\Lambda(M) \cong \{e\Omega \mid e \in E(M)\}$ where the latter is ordered by inclusion. (Hint: use the dual of Exercise 2.10.)

2.12. Let M be a finite monoid. Prove that M is $\mathscr{L}$-trivial if and only if there is a faithful finite left M-set X and a linear ordering on X such that $mx \leq x$ for all $m \in M$ and $x \in X$.

3

Inverse Monoids

An important class of regular monoids is the class of inverse monoids. In fact, many semigroup theorists would assert that inverse monoids form the most important class of monoids outside of groups. They abstract the notion of partial symmetry in much the same way that groups abstract the notion of symmetry. For a detailed discussion of this viewpoint, see Lawson [Law98]. From the perspective of this book they provide a natural class of monoids whose representation theory we can understand as well as that of groups. Namely, we shall see in Chapter 9 that the algebra of an inverse monoid can be explicitly decomposed as a direct product of matrix algebras over the group algebras of its maximal subgroups (one per $\mathscr{J}$-class).

A good reference for inverse semigroup theory is the book of Lawson [Law98]. A more encyclopedic reference is Petrich [Pet84]. Paterson's book [Pat99] describes important connections between inverse semigroup theory and the theory of C^*-algebras. Here we develop just the basic theory.

3.1 Definitions, examples, and structure

A monoid M is called an *inverse monoid* if, for all $m \in M$, there exists a unique element $m^* \in M$, called the *inverse* of m, such that $mm^*m = m$ and $m^*mm^* = m^*$. Every group is an inverse monoid where $g^* = g^{-1}$. A lattice E is an inverse monoid with respect to its meet operation where $e^* = e$ for all $e \in E$. It is straightforward to verify that if $\varphi \colon M \longrightarrow N$ is a homomorphism of inverse monoids, then $\varphi(m^*) = \varphi(m)^*$ for all $m \in M$.

If X is a set, then a *partial mapping* or *partial transformation* $f \colon X \longrightarrow X$ is a mapping from a subset $\mathrm{dom}(f)$ of X, called the *domain* of f, to X. If $f \colon X \longrightarrow X$ and $g \colon X \longrightarrow X$ are partial mappings, then their composition $f \circ g \colon X \longrightarrow X$ has domain $g^{-1}(\mathrm{dom}(f))$ and $(f \circ g)(x) = f(g(x))$ for all $x \in g^{-1}(\mathrm{dom}(f))$. In other words, $f \circ g$ is defined where it makes sense to perform g followed by f. The composition of partial mappings is associative

© Springer International Publishing Switzerland 2016
B. Steinberg, *Representation Theory of Finite Monoids*,
Universitext, DOI 10.1007/978-3-319-43932-7_3

and hence the set PT_X of all partial transformations of X is a monoid; if $X = \{1, \ldots, n\}$, then we write PT_n for PT_X. The empty partial mapping, that is, the unique partial mapping with empty domain, is the zero element of PT_X. If $Y \subseteq X$, then we write 1_Y for the partial mapping with domain Y that fixes Y pointwise.

Example 3.1 (Symmetric inverse monoid). If X is a set, then the *symmetric inverse monoid* on X is the inverse monoid I_X of all partial injective mappings $f \colon X \longrightarrow X$ with respect to composition of partial mappings. We write I_n for the symmetric inverse monoid on $\{1, \ldots, n\}$. If $f \in I_X$ has domain A and range B, then $f \colon A \longrightarrow B$ is a bijection. The inverse bijection $f^{-1} \colon B \longrightarrow A$ can be viewed as a partial injective mapping on X and $f^* = f^{-1}$ in I_X. Note that $E(I_X) = \{1_Y \mid Y \subseteq X\}$. The *rank* $\operatorname{rk} f$ of a partial injective mapping f is the cardinality of its image (or, equivalently, of its domain).

The monoid I_n is isomorphic to the *rook monoid* R_n consisting of all $n \times n$ *partial permutation matrices*, that is, matrices with $0/1$-entries and at most one nonzero entry in each row and column [Sol02]. The name "rook monoid" is used because $|R_n|$ is the number of legal placements of rooks on an $n \times n$ chessboard. The reader should verify that if B denotes the Borel subgroup of all upper triangular matrices of the general linear group $GL_n(\Bbbk)$, then

$$M_n(\Bbbk) = \bigcup_{r \in R_n} BrB$$

with the union being disjoint. This is the analogue of the Bruhat-Chevalley decomposition for $M_n(\Bbbk)$ [Ren05].

A fundamental, but not entirely obvious, property of inverse monoids (due to Munn and Penrose [MP55]) is that their idempotents commute.

Theorem 3.2. *A monoid M is an inverse monoid if and only if it is regular and its idempotents commute.*

Proof. Suppose first that M is regular with commuting idempotents. Let $m \in M$. By regularity, $m = mxm$ for some $x \in M$. But if $u = xmx$, then $mum = mxmxm = mxm = m$ and $umu = xmxmxmx = xmxmx = xmx = u$. Suppose that $mvm = m$ and $vmv = v$, as well. Then using that um, vm, mu, mv are commuting idempotents, we have that

$$u = umu = umvmu = vmumu = vmu = vmvmu = vmumv = vmv = v.$$

It follows that M is an inverse monoid.

Conversely, suppose that M is an inverse monoid. Trivially, M is regular. Note that $e^* = e$ for any idempotent $e \in E(M)$ because $eee = e$. Let $e, f \in E(M)$ and put $x = f(ef)^*e$. Then $x^2 = f(ef)^*ef(ef)^*e = f(ef)^*e = x$ is an idempotent and hence $x^* = x$. But also we have

$$x(ef)x = f(ef)^*eeff(ef)^*e = f(ef)^*(ef)(ef)^*e = f(ef)^*e = x$$
$$(ef)x(ef) = eff(ef)^*eef = ef(ef)^*(ef) = ef.$$

Therefore, $ef = x^* = x$ and consequently ef is an idempotent. Exchanging the roles of e and f, it follows that fe is also an idempotent. Then we compute

$$(ef)(fe)(ef) = efef = ef$$
$$fe(ef)(fe) = fefe = fe$$

and so $fe = (ef)^* = ef$. This completes the proof.   □

Our first corollary is that $E(M)$ is a submonoid of M.

Corollary 3.3. *Let M be an inverse monoid. Then $E(M)$ is a commutative submonoid. Consequently, if M is finite, then $E(M)$ is a lattice with respect to the natural partial order and ef is the meet of $e, f \in E(M)$.*

Proof. If $e, f \in E(M)$, then $efef = eeff = ef$ and so $E(M)$ is a commutative submonoid. The rest of the corollary is the content of Proposition 2.1.   □

Another consequence of the fact that idempotents commute in an inverse monoid is that $m \mapsto m^*$ is an involution.

Corollary 3.4. *Let M be an inverse monoid and $m, n \in M$.*

(i) $(m^*)^* = m$.
(ii) $mm^*nn^* = nn^*mm^*$.
(iii) $(mn)^* = n^*m^*$.
(iv) mem^* is idempotent for $e \in E(M)$.

Proof. The first item is clear. Since mm^*, nn^* are idempotents, (ii) follows from Theorem 3.2. For (iii), we compute $mn(n^*m^*)mn = mm^*mnn^*n = mn$ and $n^*m^*(mn)n^*m^* = n^*nn^*m^*mm^* = n^*m^*$. Thus $(mn)^* = n^*m^*$. The final item follows because $mem^*mem^* = mm^*mem^* = mem^*$.   □

Quotients of inverse monoids are again inverse monoids.

Corollary 3.5. *Let M be an inverse monoid and let $\varphi\colon M \longrightarrow N$ be a surjective homomorphism of monoids. Then N is an inverse monoid.*

Proof. Regularity of M easily implies that $N = \varphi(M)$ is regular and so it remains to show that the idempotents of N commute. Let $e, f \in E(N)$ and suppose that $e = \varphi(x)$ and $f = \varphi(y)$. Then $e = ee^* = \varphi(x)\varphi(x^*) = \varphi(xx^*)$ and $f = ff^* = \varphi(y)\varphi(y^*) = \varphi(yy^*)$. It follows that $ef = \varphi(xx^*yy^*) = \varphi(yy^*xx^*) = fe$. This completes the proof.   □

A final corollary involves the operation of passing from M to eMe for an idempotent $e \in M$.

Corollary 3.6. *If M is an inverse monoid, then so is eMe for each idempotent $e \in E(M)$. Moreover, the equality*

$$G_e = \{m \in M \mid m^*m = e = mm^*\}$$

holds.

Proof. The monoid eMe is regular by Proposition 1.21 and clearly has commuting idempotents, whence eMe is an inverse monoid. For the second statement, $m^*m = e = mm^*$ implies that $eme = mm^*mm^*m = m$ and hence by the first statement, it suffices to consider the case $e = 1$. Clearly, if $g \in G_1$ is a unit, then $g^{-1} = g^*$ and hence $gg^* = 1 = g^*g$. Conversely, if $m^*m = 1 = mm^*$, then m is a unit. $\qquad\square$

Inverse monoids can also be characterized in terms of Green's relations. The hypothesis of finiteness in the following proposition is unnecessary.

Proposition 3.7. *Let M be a finite monoid. Then M is an inverse monoid if and only if each $\mathscr{L}$-class and $\mathscr{R}$-class of M contains a unique idempotent.*

Proof. Assume first that M is an inverse monoid and $m \in M$. Then $mM = mm^*M$ and $Mm = Mm^*m$ and so each $\mathscr{L}$-class and $\mathscr{R}$-class contains an idempotent. If e, f are $\mathscr{L}$-equivalent idempotents, then $e = xf$ implies $fe = ef = xff = xf = e$ and so $e \leq f$. A dual argument shows that $f \leq e$. Thus $e = f$. A similar argument shows that each $\mathscr{R}$-class contains at most one idempotent.

Next assume that M is a monoid all of whose $\mathscr{L}$-classes and $\mathscr{R}$-classes contain a unique idempotent. Regularity of M is immediate from Proposition 1.22. As in the proof of Theorem 3.2, if $m = mxm$ and $u = xmx$, then $mum = m$ and $umu = u$. Suppose that we have $mvm = m$, $vmv = v$, too. Then $um \mathscr{L} m \mathscr{L} vm$ and $mu \mathscr{R} m \mathscr{R} mv$, from which it follows by the hypothesis that $um = vm$ and $mu = mv$, as these elements are idempotents. Thus $u = umu = vmu = vmv = v$. It follow that M is an inverse monoid. $\qquad\square$

The symmetric inverse monoid I_X on a set X is naturally partial ordered by putting $f \leq g$ if the partial injective mapping f is a restriction of g. Observe that $f^*f = 1_{\mathrm{dom}(f)}$ and so $f \leq g$ if and only if $gf^*f = f$. This motivates defining an order on any inverse monoid via $m \leq n$ if $m = nm^*m$. The following lemma will be needed to show that this yields a well-behaved partial order.

Lemma 3.8. *Let M be an inverse monoid and $m, n \in M$. Then the following are equivalent.*

   (i) $m = ne$ with $e \in E(M)$.
  (ii) $m = fn$ with $f \in E(M)$.
 (iii) $m = nm^*m$.
 (iv) $m = mm^*n$.

Proof. Trivially, we have that (iii) implies (i) and (iv) implies (ii). If (i) holds, then $nm^*m = nen^*ne = nn^*ne = ne = m$ and so (iii) holds. Similarly, (ii) implies (iv). It remains to establish the equivalence of (i) and (ii). Suppose that (i) holds. Then $(nen^*)n = nn^*ne = ne = m$ and $nen^* \in E(M)$ by Corollary 3.4, yielding (ii). The proof that (ii) implies (i) is dual. $\qquad\square$

The *natural partial order* on an inverse monoid is defined by putting $m \leq n$ if the equivalent conditions of Lemma 3.8 hold. Notice that if $m \leq e$ with $e \in E(M)$, then $m = ef$ with $f \in E(M)$ and hence m is an idempotent because $E(M)$ is a submonoid. Moreover, the restriction of the natural partial order to $E(M)$ is the usual partial order.

Proposition 3.9. *Let M be an inverse monoid. Then $\leq$ is a partial order, enjoying the following properties.*

(i) $m \leq n$ if and only if $m^* \leq n^*$.
(ii) $m \leq n$ and $m' \leq n'$ implies $mm' \leq nn'$.

Proof. From $m = mm^*m$, we have $m \leq m$. If $m \leq n$ and $n \leq m$, then $m = nm^*m$ and $n = mn^*n$ and so $m = (mn^*n)m^*m = mm^*mn^*n = mn^*n = n$. Finally, if $m \leq m' \leq m''$, then there exist $e, f \in E(M)$ such that $m = m'e$ and $m' = m''f$. Then $m = m''fe$ and $fe \in E(M)$. Thus $m \leq m''$. This completes the proof that $\leq$ is a partial order.

If $m = ne$ with $e \in E(M)$, then $m^* = (ne)^* = e^*n^* = en^*$. Thus $m \leq n$ implies $m^* \leq n^*$. The reverse implication is proved dually. Suppose now that $m \leq n$ and $m' \leq n'$. Then $m = fn$ and $m' = n'e$ with $e, f \in E(M)$. Therefore, $mm' = mn'e \leq mn'$ and $mn' = fnn' \leq nn'$. Thus $mm' \leq nn'$. $\qquad\square$

Green's relations $\mathscr{L}$ and $\mathscr{R}$ take on a particularly pleasant form for inverse monoids. This will be exploited later in the text to associate a groupoid to each inverse monoid.

Proposition 3.10. *Let M be an inverse monoid and $m_1, m_2 \in M$.*

(i) $m_1 \mathscr{L} m_2$ if and only if $m_1^*m_1 = m_2^*m_2$.
(ii) $m_1 \mathscr{R} m_2$ if and only if $m_1m_1^* = m_2m_2^*$.

Proof. We just prove (i). Clearly, $m_1^*m_1 = m_2^*m_2$ implies $Mm_1 = Mm_1^*m_1 = Mm_2^*m_2 = Mm_2$. The converse follows because if $m_1 = xm_2$, then $m_1^*m_1 = m_2^*x^*xm_2 \leq m_2^*m_2$ and dually $m_2^*m_2 \leq m_1^*m_1$. Thus $m_1^*m_1 = m_2^*m_2$. $\qquad\square$

As a corollary, we can shed some light on the nature of commutative inverse monoids.

Proposition 3.11. *Let M be a commutative inverse monoid. Then the $\mathscr{J}$-classes of M are its maximal subgroups.*

Proof. If $m \in M$, then by commutativity we have $mm^* = m^*m$ and hence if we call this common idempotent e, then $m \in G_e$ by Corollary 3.6. Since M is commutative, we have that $\mathscr{R} = \mathscr{L} = \mathscr{J}$. Hence Proposition 3.10 and Corollary 3.6 immediately imply that G_e is the $\mathscr{J}$-class of m. $\qquad\square$

A direct consequence is that any submonoid of a finite commutative inverse monoid is again an inverse monoid.

Corollary 3.12. *Let M be a finite commutative inverse monoid. Then any submonoid N of M is also a commutative inverse monoid.*

Proof. Trivially, N has commuting idempotents and so it remains to show that N is regular. Let $m \in N$. By Proposition 3.11 the $\mathscr{J}$-class of m in M is a maximal subgroup G of M. Let $n = |G|$. Then $m^{n-1} \in N$ and $mm^{n-1}m = m$. We deduce that N is regular. This concludes the proof. $\qquad\square$

In a finite inverse monoid $\mathscr{J}$-equivalent elements are incomparable.

Proposition 3.13. *Let M be a finite inverse monoid and $m, n \in M$ with $MmM = MnM$ and $m \leq n$. Then $m = n$.*

Proof. Since $m = nm^*m$, we have that $mM = nM$ by stability (Theorem 1.13). Thus $mm^* = nn^*$ by Proposition 3.10. But then we have $m = mm^*n = nn^*n = n$. This completes the proof. $\qquad\square$

The analogue of Cayley's theorem for inverse monoids is the *Preston-Wagner theorem*, which is markedly less trivial because it is not so obvious how to make M act on itself by partial injections. Although we do not use this result in the sequel, we provide a proof to give the reader a flavor of the theory of inverse monoids.

Theorem 3.14. *Let M be an inverse monoid. Then M embeds in I_M.*

Proof. For each $m \in M$, define a mapping $\varphi_m \colon m^*mM \longrightarrow mm^*M$ by $\varphi_m(x) = mx$. Note that $\varphi_{m^*} \colon mm^*M \longrightarrow m^*mM$ satisfies $\varphi_{m^*}(\varphi_m(x)) = m^*mx = x$ for $x \in m^*mM$ and similarly $\varphi_m \circ \varphi_{m^*}$ is the identity on mm^*M. Therefore, φ_m, φ_{m^*} are mutual inverses and so φ_m can be viewed as a partial injective map on M with domain m^*mM and range mm^*M, that is, $\varphi_m \in I_M$.

Define $\varphi \colon M \longrightarrow I_M$ by $\varphi(m) = \varphi_m$. We must verify that $\varphi_{mn} = \varphi_m \circ \varphi_n$ as partial maps to show that φ is a homomorphism. Note that the domain of φ_{mn} is $(mn)^*(mn)M = n^*m^*mnM$. Moreover, for $x \in n^*m^*mnM$, one has that $\varphi_{mn}(x) = mnx$. Clearly, $n^*m^*mnM = n^*n(n^*m^*mnM) \subseteq n^*nM$ and if $x \in n^*m^*mnM$, then we have

$$\varphi_n(x) = nx \in nn^*m^*mnM = m^*mnn^*nM \subseteq m^*mM.$$

Therefore, $\varphi_m(\varphi_n(x))$ is defined and $\varphi_m(\varphi_n(x)) = mnx = \varphi_{mn}(x)$. Conversely, if $\varphi_m(\varphi_n(x))$ is defined, then $x \in n^*nM$ and $nx = \varphi_n(x) \in m^*mM$.

Therefore, $m^*mnx = nx$ and so $n^*m^*mnx = n^*nx = x$. Thus $x \in n^*m^*mnM$ and $\varphi_{mn}(x) = mnx = \varphi_m(\varphi_n(x))$. This proves the $\varphi_{mn} = \varphi_m \circ \varphi_n$ as partial mappings, and so φ is a homomorphism.

Suppose that $\varphi(m) = \varphi(n)$, that is, $\varphi_m = \varphi_n$. Then the latter two maps have the same domain, whence $m^*mM = n^*nM$. Moreover, $nm^*m = \varphi_n(m^*m) = \varphi_m(m^*m) = mm^*m = m$ and so we have that $m \leq n$. A symmetric argument yields $n \leq m$ and hence $m = n$. Thus the mapping φ is injective. This establishes the theorem. $\qquad\square$

The following property will be reformulated in the language of groupoids in Chapter 9.

Proposition 3.15. *Let M be an inverse monoid and suppose that $m^*m = nn^*$. Then $(mn)^*mn = n^*n$ and $mn(mn)^* = mm^*$.*

Proof. We compute $(mn)^*mn = n^*m^*mn = n^*nn^*n = n^*n$. The second equality is proved in a similar fashion. $\qquad\square$

The next lemma will be crucial in describing the structure of the algebra of an inverse monoid.

Lemma 3.16. *Let M be an inverse monoid and let $m, n \in M$. Then there exist unique $m_0 \leq m$ and $n_0 \leq n$ such that $m_0n_0 = mn$ and $m_0^*m_0 = n_0n_0^*$.*

Proof. Let $m_0 = mnn^*$ and $n_0 = m^*mn$. Then $m_0n_0 = mnn^*m^*mn = mm^*mnn^*n = mn$ and $m_0 \leq m$, $n_0 \leq n$. Also, we compute that $m_0^*m_0 = nn^*m^*mnn^* = nn^*m^*m$ and $n_0n_0^* = m^*mnn^*m^*m = nn^*m^*m = m_0^*m_0$, establishing the existence of m_0, n_0. Suppose that $m_1 \leq m$, $n_1 \leq n$ and $m_1n_1 = mn$ with $m_1^*m_1 = n_1n_1^*$. By Proposition 3.15, we have that

$$m_1m_1^* = (m_1n_1)(m_1n_1)^* = (mn)(mn)^* = mnn^*m^*. \qquad (3.1)$$
$$n_1^*n_1 = (m_1n_1)^*(m_1n_1) = (mn)^*(mn) = n^*m^*mn \qquad (3.2)$$

By Lemma 3.8 and (3.1), we have that

$$m_1 = m_1m_1^*m = mnn^*m^*m = mm^*mnn^* = mnn^* = m_0$$

and, similarly,

$$n_1 = nn_1^*n_1 = nn^*m^*mn = m^*mnn^*n = m^*mn = n_0$$

using (3.2). This establishes the uniqueness. $\qquad\square$

3.2 Conjugacy in the symmetric inverse monoid

Let X be a finite set. Two elements $f, g \in I_X$ are said to be *conjugate* if there exists h in the symmetric group S_X with $f = hgh^{-1}$. Conjugacy is an equivalence relation on I_X. The description of conjugacy for the symmetric

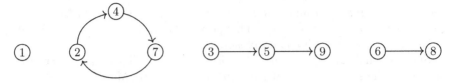

Fig. 3.1. The orbits of f

group can be generalized to the symmetric inverse monoid in a natural way; a detailed exposition can be found in the monograph [Lip96]. The results of this section will be used when we prove that the algebra of the monoid of all $n \times n$ matrices over a finite field is semisimple over any field of characteristic zero.

Fix $f \in I_X$. If $n > 0$, let us put $f^{-n} = (f^*)^n$. Then the *orbit* of $x \in X$ under f is

$$\mathcal{O}(x) = \{f^n(x) \mid n \in \mathbb{Z}\}.$$

The orbits of f clearly form a partition of X because f is injective.

Let us say that $x \in X$ is a *recurrent* point of f if $f^n(x) = x$ for some $n > 0$. If $x \in X$ is not recurrent, then we call x *transient*. Note that each orbit consists either entirely of recurrent points or entirely of transient points; the former are called *recurrent orbits* and the latter *transient orbits*. Let $\mathrm{rec}(f)$ denote the set of recurrent points of f. If x is recurrent, then

$$\mathcal{O}(x) = \{f^n(x) \mid n > 0\}$$

and f acts on $\mathcal{O}(x)$ as a cyclic permutation. In other words, $f|_{\mathrm{rec}(f)}$ is a permutation and the orbits of recurrent points yield the cycle decomposition of $f|_{\mathrm{rec}(f)}$. A transient element $x \in X$ is called a *source* if f^{-1} is undefined at x. By finiteness of X and injectivity of f, each transient orbit contains a unique source and if $x \in X$ is a source, then

$$\mathcal{O}(x) = \{f^n(x) \mid n \geq 0\}.$$

The action of f on a transient orbit of size m is nilpotent with $f^m = 1_\emptyset$. Note that each transient orbit of f contains a unique element that does not belong to the domain of f.

Example 3.17. Consider $f \in I_9$ given by

$$f = \begin{pmatrix} 1\,2\,3\,4\,5\,6\,7\,8\,9 \\ 4\,5\,7\,9\,8\,2 \end{pmatrix}.$$

The orbits of f are $\{1\}$, $\{2,4,7\}$, $\{3,5,9\}$, and $\{6,8\}$. The orbit $\{2,4,7\}$ is recurrent and f acts on it as the cyclic permutation $(2\ 4\ 7)$. The remaining orbits are transient with sources 1, 3 and 6. Notice that 3 is not of the form $f^n(5)$ with $n \geq 0$ and so the negative powers of f are really needed in the definition of an orbit. See Figure 3.1.

If $f \in I_X$, then

$$X \supseteq f(X) \supseteq f^2(X) \supseteq \cdots$$

and the sequence stabilizes by finiteness of X. Moreover, if $f^m(X) = f^{m+1}(X)$, then it is straightforward to verify that $\mathrm{rec}(f) = f^m(X)$; see Exercise 3.12. Define the *rank sequence* of $f \in I_X$ to be

$$\vec{r}(f) = (\mathrm{rk}\, f^0, \mathrm{rk}\, f^1, \mathrm{rk}\, f^2, \ldots).$$

Notice that $\vec{r}(f)$ is a weakly decreasing sequence of nonnegative integers, with first entry $|X|$ and which becomes constant as soon as two consecutive entries are equal. Moreover, this eventually constant value is $|\mathrm{rec}(f)|$ by the preceding discussion. Note that if $f, g \in I_X$ and $\vec{r}(f) = \vec{r}(g)$, then $|\mathrm{rec}(f)| = |\mathrm{rec}(g)|$.

Recall that the *cycle type* of a permutation g of an n-element set is the partition $(\lambda_1, \ldots, \lambda_r)$ of n where $\lambda_1 \geq \lambda_2 \geq \cdots \geq \lambda_r$ are the sizes of the orbits of g. The cycle type of the identity map on the empty set is the empty partition ().

Example 3.18. The partial bijection $f \in I_9$ from Example 3.17 has rank sequence $\vec{r}(f) = (9, 6, 4, 3, 3, \ldots)$ and $f|_{\mathrm{rec}(f)}$ has cycle type (3).

Theorem 3.19. *Let X be a finite set and $f, g \in I_X$. Then f, g are conjugate if and only if*

(i) $\vec{r}(f) = \vec{r}(g)$;
(ii) $f|_{\mathrm{rec}(f)}$ has the same cycle type as $g|_{\mathrm{rec}(g)}$.

Proof. It is clear that f, g conjugate implies that $\vec{r}(f) = \vec{r}(g)$. Also, if $f = hgh^{-1}$ with $h \in S_X$, then $\mathrm{rec}(f) = h(\mathrm{rec}(g))$ and h provides a size-preserving bijection between the orbits of g on $\mathrm{rec}(g)$ and f on $\mathrm{rec}(f)$. Thus $f|_{\mathrm{rec}(f)}$ has the same cycle type as $g|_{\mathrm{rec}(g)}$.

For the converse, we have by (ii) that there is a bijection

$$k \colon \mathrm{rec}(g) \longrightarrow \mathrm{rec}(f)$$

taking m-cycles of g to m-cycles of f preserving the cyclic ordering. Hence $f|_{\mathrm{rec}(f)} = kg|_{\mathrm{rec}(g)}k^{-1}$. We must extend k to transient points.

Note that if $s \in I_X$, then, for $i \geq 0$, $\mathrm{rk}(s^i) - \mathrm{rk}(s^{i+1})$ is the number of transient orbits of s on X of size greater than i. Hence f, g have the same number of transient orbits of each size. Thus we can find a bijection

$$k' \colon X \setminus \mathrm{rec}(g) \longrightarrow X \setminus \mathrm{rec}(f)$$

sending the source of each transient orbit of g to the source of a transient orbit of f of the same size and such that if $x \in X \setminus \mathrm{rec}(g)$ has source x' for its orbit and $g^m(x') = x$ with $m \geq 0$, then $k'(x) = f^m(k'(x'))$. It follows that if we define $h \colon X \longrightarrow X$ by

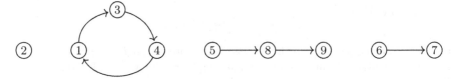

Fig. 3.2. The orbits of g

$$h(x) = \begin{cases} k(x), & \text{if } x \in \text{rec}(g) \\ k'(x), & \text{else,} \end{cases}$$

then $f = hgh^{-1}$. This completes the proof. □

Example 3.20. Let

$$g = \begin{pmatrix} 1\ 2\ 3\ 4\ 5\ 6\ 7\ 8\ 9 \\ 3\ \ \ 4\ 1\ 8\ 7\ \ \ 9 \end{pmatrix}$$

be in I_9. Then $\text{rec}(g) = \{1,3,4\}$, $\vec{r}(g) - (9,6,4,3,3,\ldots)$, and $g|_{\text{rec}(g)}$ has cycle type (3). See Figure 3.2. Hence g is conjugate to f from Example 3.17. In fact, $f = hgh^{-1}$ where

$$h = \begin{pmatrix} 1\ 2\ 3\ 4\ 5\ 6\ 7\ 8\ 9 \\ 2\ 1\ 4\ 7\ 3\ 6\ 8\ 5\ 9 \end{pmatrix}.$$

Let us say that $f \in I_X$ is *semi-idempotent* if f fixes $\text{rec}(f)$. An immediate consequence of Theorem 3.19 is its following special case.

Corollary 3.21. *Let X be a finite set and let $f, g \in I_X$ be semi-idempotent. Then f, g are conjugate if and only if $\vec{r}(f) = \vec{r}(g)$.*

Proof. Since each orbit of $f|_{\text{rec}(f)}$ and $g|_{\text{rec}(g)}$ is a singleton, the corollary is immediate from Theorem 3.19. □

3.3 Exercises

3.1. Prove that an inverse monoid is a group if and only if 1 is its unique idempotent.

3.2. Prove that if $\varphi \colon M \longrightarrow N$ is a homomorphism between inverse monoids, then $\varphi(m^*) = \varphi(m)^*$.

3.3. Prove that a direct product of inverse monoids is an inverse monoid.

3.4. Let M be a monoid with commuting idempotents. Prove that the regular elements of M form a submonoid of M, which is an inverse monoid.

3.5. Let M be an inverse monoid and $a, b \in M$. Show that $a \mathcal{L} b$ if and only if $a^* \mathcal{R} b^*$. Deduce that in each $\mathcal{J}$-class of a finite inverse monoid the number of $\mathcal{R}$-classes is equal to the number of $\mathcal{L}$-classes.

3.6. Let $f, g \in I_n$. Prove that $f \mathcal{J} g$ if and only if they have the same rank, $f \mathcal{L} g$ if and only if they have the same domain and $f \mathcal{R} g$ if and only if they have the same image.

3.7. Let $n \geq 0$. Prove that

$$|I_n| = \sum_{r=0}^{n} \binom{n}{r}^2 \cdot r!.$$

3.8. Prove that the symmetric inverse monoid I_n is isomorphic to the rook monoid R_n (cf. Example 3.1).

3.9. Let X be a finite set. Prove that PT_X is a regular monoid.

3.10. Let G be a group. Define the *wreath product* $G \wr I_n$ to consist of all pairs (f, σ) such that $\sigma \in I_n$ and $f \colon \mathrm{dom}(\sigma) \longrightarrow G$ is a mapping. The product is given by $(f, \sigma)(g, \tau) = (h, \sigma\tau)$ where $h(x) = f(\tau(x))g(x)$ for $x \in \mathrm{dom}(\sigma\tau)$. Prove that $G \wr I_n$ is an inverse monoid.

3.11. Let X be a finite set and $f \in I_X$. Prove that the orbits of f form a partition of X.

3.12. Let X be a finite set and $f \in I_X$. Prove that if $f^m(X) = f^{m+1}(X)$ with $m > 0$, then $\mathrm{rec}(f) = f^m(X)$.

3.13. Let X be a finite set and $f \in I_X$. Prove that $\mathrm{rec}(f) = f^\omega(X)$.

3.14. Let M be a finite inverse monoid. Prove that M has central idempotents if and only if each $\mathcal{J}$-class of M is a maximal subgroup.

3.15. Let E be a finite lattice and let $\{G_e \mid e \in E\}$ be an E-indexed family of disjoint finite abelian groups. Suppose that one has homomorphisms $\rho_e^f \colon G_f \longrightarrow G_e$ whenever $e \leq f$ such that:

(i) $\rho_e^e = 1_{G_e}$;
(ii) $\rho_e^f \circ \rho_f^h = \rho_e^h$ whenever $e \leq f \leq h$.

(a) Prove that $M = \bigcup_{e \in E} G_e$ is a finite commutative inverse monoid with respect to the product

$$mn = \rho_{e \wedge f}^e(m) \cdot \rho_{e \wedge f}^f(n)$$

for $m \in G_e$ and $n \in G_f$.
(b) Prove that $E(M) \cong E$.

3.16. Prove that every finite commutative inverse monoid is isomorphic to one of the form constructed in Exercise 3.15.

3.17. Let M be an inverse monoid. Define a relation $\sim$ on M by $m \sim m'$ if there exists $n \in M$ with $n \leq m$ and $n \leq m'$.

(a) Prove that $\sim$ is a congruence.
(b) Prove that $M/\!\sim$ is a group.
(c) Let $\rho: M \longrightarrow M/\!\sim$ be the quotient map. Prove that if $\varphi: M \longrightarrow G$ is a homomorphism with G a group, then there is a unique homomorphism $\psi: M/\!\sim \longrightarrow G$ such that the diagram

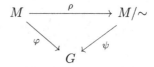

commutes.

3.18. Let M be an inverse monoid and define $\rho: M \longrightarrow I_{E(M)}$ by

$$\rho(m)(e) = \begin{cases} mem^*, & \text{if } e \leq m^*m \\ \text{undefined}, & \text{else} \end{cases}$$

for $m \in M$ and $e \in E(M)$.

(a) Prove that ρ is a homomorphism. (It is called *Munn representation of M* [Law98].)
(b) Prove that $\ker \rho$ is the largest congruence $\equiv$ on M (with respect to containment) such that $e \equiv f$ implies $e = f$ for all $e, f \in E(M)$.
(c) Prove that $\rho(m)$ is an idempotent if and only if m commutes with all idempotents of M.

3.19. A matrix $A \in M_n(\mathbb{C})$ is a *partial isometry* if $A = AA^*A$ where A^* is the Hermitian adjoint.

(a) Prove that a submonoid of $M_n(\mathbb{C})$ consisting entirely of partial isometries and closed under the Hermitian adjoint is an inverse monoid.
(b) Prove that every finite inverse monoid is isomorphic to one of the form in (a). (Hint: turn the Preston-Wagner representation of Theorem 3.14 into a linear representation by partial isometries.)

3.20. Let G be a group. Let $K(G)$ be the set of all cosets gH with $H \leq G$ a subgroup.

(a) Define a binary operation $\odot$ on $K(G)$ by

$$g_1 H_1 \odot g_2 H_2 = g_1 g_2 \langle g_2^{-1} H_1 g_2, H_2 \rangle.$$

Prove that $K(G)$ is an inverse monoid.

(b) Prove that if $\mathbb{K}$ is a finite Galois extension of $\mathbb{F}$ and G is the Galois group of $\mathbb{K}$ over $\mathbb{F}$, then $K(G)$ is isomorphic to the inverse monoid of all partial field automorphisms of $\mathbb{K}$ over $\mathbb{F}$, i.e., the inverse monoid of all field homomorphisms $f \colon \mathbb{L} \longrightarrow \mathbb{K}$ with $\mathbb{F} \subseteq \mathbb{L} \subseteq \mathbb{K}$, fixing $\mathbb{F}$ pointwise, under composition of partial mappings. (Familiarity with Galois theory is required for this problem.)

Part II

Irreducible Representations

4

Recollement: The Theory of an Idempotent

In this chapter we provide an account of the theory connecting the category of modules of a finite dimensional algebra A with the module categories of the algebras eAe and A/AeA, for an idempotent $e \in A$, known as *recollement* [BBD82, CPS88, CPS96]. We first learned of this subject from the monograph of Green [Gre80, Chapter 6]. A presentation much closer in spirit to ours is that of Kuhn [Kuh94b]. In the next chapter, we shall apply this theory to construct the irreducible representations of a finite monoid and in a later chapter we shall extend the results to finite categories.

In this text all rings are unital. Throughout the book, we shall be considering a finite dimensional algebra A over a field $\Bbbk$. Modules will be assumed to be left modules unless otherwise mentioned. We shall denote by A-mod the category of finite dimensional A-modules or, equivalently, finitely generated A-modules. If A^{op} denotes the *opposite algebra* of A (i.e., it has the same underlying vector space but with product $a * b = ba$), then A^{op}-mod can be identified with the category of finite dimensional right A-modules. We shall use $M_n(R)$ to denote the ring of $n \times n$ matrices over a ring R.

For this chapter we fix a finite dimensional algebra A over a field $\Bbbk$ and an idempotent $e \in A$. As is usual, if $X, Y \subseteq A$, then

$$XY = \left\{ \sum_{i=1}^n x_i y_i \mid x_i \in X, y_i \in Y \right\}$$

With this notation, AeA is an ideal of A and eAe is an algebra with identity e. In fact, $eAe \cong \mathrm{End}_A(Ae)^{op}$ (cf. Proposition A.20).

© Springer International Publishing Switzerland 2016
B. Steinberg, *Representation Theory of Finite Monoids*,
Universitext, DOI 10.1007/978-3-319-43932-7_4

4.1 A miscellany of functors

In this section we consider several functors associated with the idempotent e. Note that eA is an eAe-A-bimodule and Ae is an A-eAe-bimodule via left and right multiplication. This allows us to define functors

$$\mathrm{Ind}_e: eAe\text{-mod} \longrightarrow A\text{-mod}$$
$$\mathrm{Coind}_e: eAe\text{-mod} \longrightarrow A\text{-mod}$$
$$\mathrm{Res}_e: A\text{-mod} \longrightarrow eAe\text{-mod}$$
$$T_e: A\text{-mod} \longrightarrow A\text{-mod}$$
$$N_e: A\text{-mod} \longrightarrow A\text{-mod}$$

by putting

$$\mathrm{Ind}_e(V) = Ae \otimes_{eAe} V$$
$$\mathrm{Coind}_e(V) = \mathrm{Hom}_{eAe}(eA, V)$$
$$\mathrm{Res}_e(V) = eV \cong \mathrm{Hom}_A(Ae, V) \cong eA \otimes_A V$$
$$T_e(V) = AeV$$
$$N_e(V) = \{v \in V \mid eAv = 0\}.$$

We remark that the isomorphism $eV \cong \mathrm{Hom}_A(Ae, V)$ is from Proposition A.20, whereas the isomorphism $eV \cong eA \otimes_A V$ is given by the mapping $eA \otimes_A V \longrightarrow eV$ defined on basic tensors by $a \otimes v \mapsto av$ with inverse mapping given by $v \mapsto e \otimes v$.

We call Ind_e the *induction functor*, Coind_e the *coinduction functor* and Res_e the *restriction functor*. The submodule $T_e(V)$ is called the *trace* of the projective module Ae in V (cf. Exercise 4.5). Readers familiar with category theory will observe that Ind_e is left adjoint to Res_e and Coind_e is right adjoint to Res_e. More precisely, the usual hom-tensor adjunction yields the following proposition (see Exercise 4.1).

Proposition 4.1. *Let V be an A-module and W an eAe-module. Then there are natural isomorphisms* $\mathrm{Hom}_A(\mathrm{Ind}_e(W), V) \cong \mathrm{Hom}_{eAe}(W, \mathrm{Res}_e(V))$ *and* $\mathrm{Hom}_A(V, \mathrm{Coind}_e(W)) \cong \mathrm{Hom}_{eAe}(\mathrm{Res}_e(V), W)$.

The category A/AeA-mod can be identified with the full subcategory of A-mod consisting of those modules V with $eV = 0$. Observe that $N_e(V)$ is the largest submodule of V annihilated by e, i.e., the largest submodule that is an A/AeA-module. Thus $N_e: A\text{-mod} \longrightarrow A/AeA\text{-mod}$ can be viewed as the right adjoint of the inclusion functor A/AeA-mod $\longrightarrow A$-mod. On the other hand, one can show that the assignment $V \mapsto V/T_e(V)$ gives a functor $A\text{-mod} \longrightarrow A/AeA\text{-mod}$ which is left adjoint to the inclusion functor. We shall not prove these assertions, as they shall not be used in the sequel. The diagram of adjoint functors

$$\xleftarrow{\;(-)/T_e(-)\;} \qquad \xleftarrow{\;\mathrm{Ind}_e\;}$$

$$A/AeA\text{-mod} \xrightarrow{\quad} A\text{-mod} \xrightarrow{\;\mathrm{Res}_e\;} eAe\text{-mod}$$

$$\xleftarrow{\;N_e\;} \qquad \xleftarrow{\;\mathrm{Coind}_e\;}$$

is an example of what is called a *recollement* of abelian categories (see, for example, [Kuh94b]).

We turn next to exactness properties of these functors.

Proposition 4.2. *The functor* Res_e *is exact. If* Ae *(respectively,* eA*) is a flat right (respectively, projective left)* eAe*-module, then* Ind_e *(respectively,* Coind_e*) is exact.*

Proof. Exactness of Res_e follows because Ae is a projective A-module and $\mathrm{Res}_e(-) \cong \mathrm{Hom}_A(Ae, -)$ as functors. The second statement follows from the definitions.   □

A consequence of the preceding two propositions is that Ind_e preserves projectivity and Coind_e preserves injectivity of modules.

Proposition 4.3. *The functor* Ind_e *takes projective modules to projective modules and the functor* Coind_e *takes injective modules to injective modules.*

Proof. Let P be a projective eAe-module. Then we have that

$$\mathrm{Hom}_A(\mathrm{Ind}_e(P), -) \cong \mathrm{Hom}_{eAe}(P, \mathrm{Res}_e(-)).$$

Because P is projective and Res_e is exact, we conclude that $\mathrm{Hom}_A(\mathrm{Ind}_e(P), -)$ is exact. Therefore, $\mathrm{Ind}_e(P)$ is projective. An entirely similar argument shows that if I is an injective eAe-module, then $\mathrm{Coind}_e(I)$ is injective.   □

The restriction functor interacts nicely with tensor products and hom functors.

Proposition 4.4. *Let* A, B *be a finite dimensional algebras and* e *an idempotent of* A*. Suppose that* U *is a* B-A*-bimodule and* V *is an* A-B*-bimodule. Then* Ue *is a* B-eAe*-bimodule and* eV *is an* eAe-B*-bimodule by restricting the actions and there are natural isomorphisms of* eAe*-modules*

$$e(V \otimes_B W) \cong eV \otimes_B W$$
$$e\,\mathrm{Hom}_B(U, W) \cong \mathrm{Hom}_B(Ue, W)$$

for every B*-module* W*.*

Proof. We handle tensor products first. Note that there is an eAe-B-bimodule isomorphism $eA \otimes_A V \longrightarrow eV$ given on basic tensors by $a \otimes v \mapsto av$ for $a \in eA$ and $v \in V$ with inverse $v \mapsto e \otimes v$ for $v \in eV$. Hence we have eAe-module isomorphisms

$$e(V \otimes_B W) \cong eA \otimes_A (V \otimes_B W) \cong (eA \otimes_A V) \otimes_B W \cong eV \otimes_B W$$

establishing the result for tensor products.

Similarly, there is a B-eAe-bimodule isomorphism $U \otimes_A Ae \longrightarrow Ue$ defined on basic tensors by $u \otimes a \mapsto ua$ for $u \in U$ and $a \in Ae$ with inverse $u \mapsto u \otimes e$ for $u \in Ue$. We then have isomorphisms of eAe-modules

$$e \operatorname{Hom}_B(U, W) \cong \operatorname{Hom}_A(Ae, \operatorname{Hom}_B(U, W))$$
$$\cong \operatorname{Hom}_B(U \otimes_A Ae, W)$$
$$\cong \operatorname{Hom}_B(Ue, W)$$

as required. This completes the proof.

Because of Proposition 4.4, we shall sometimes identify $e(V \otimes_B W)$ with $eV \otimes_B W$ when no confusion will arise. An important consequence of Proposition 4.4 is that Ind_e and Coind_e are right quasi-inverse to Res_e.

Proposition 4.5. *There are natural isomorphisms*

$$\operatorname{Res}_e \circ \operatorname{Ind}_e(V) \cong V \cong \operatorname{Res}_e \circ \operatorname{Coind}_e(V)$$

for any eAe-module V.

Proof. We have by Proposition 4.4 natural isomorphisms

$$\operatorname{Res}_e(\operatorname{Ind}_e(V)) = e(Ae \otimes_{eAe} V) \cong eAe \otimes_{eAe} V \cong V$$
$$\operatorname{Res}_e(\operatorname{Coind}_e(V)) = e \operatorname{Hom}_{eAe}(eA, V) \cong \operatorname{Hom}_{eAe}(eAe, V) \cong V$$

as required.

Explicitly, the isomorphism $\operatorname{Res}_e(\operatorname{Ind}_e(V)) \longrightarrow V$ sends $e\left(\sum a_i \otimes v_i\right)$ to $\sum ea_i v_i$ for $a_i \in Ae$ and $v_i \in V$ and the isomorphism $\operatorname{Res}_e(\operatorname{Coind}_e(V)) \longrightarrow V$ takes φ to $\varphi(e)$ for $\varphi \in e \operatorname{Coind}_e(V)$. $\square$

We now deduce that Ind_e and Coind_e are fully faithful (i.e., induce isomorphisms of hom-sets).

Proposition 4.6. *The functors Ind_e and Coind_e are fully faithful.*

Proof. We just handle the case of Ind_e, as the other case is similar. Let V, W be eAe-modules. Then the $\Bbbk$-linear mapping

$$\psi \colon \operatorname{Hom}_{eAe}(V, W) \longrightarrow \operatorname{Hom}_A(\operatorname{Ind}_e(V), \operatorname{Ind}_e(W))$$

given by $f \mapsto \operatorname{Ind}_e(f) = 1_{Ae} \otimes f$ is injective because $\operatorname{Res}_e \circ \operatorname{Ind}_e$ is isomorphic to the identity functor on eAe-mod by Proposition 4.5. But by Proposition 4.1 and Proposition 4.5, we have that

$$\operatorname{Hom}_{eAe}(V, W) \cong \operatorname{Hom}_{eAe}(V, \operatorname{Res}_e(\operatorname{Ind}_e(W))) \cong \operatorname{Hom}_A(\operatorname{Ind}_e(V), \operatorname{Ind}_e(W))$$

and so ψ is an isomorphism by dimension considerations. Thus Ind_e is fully faithful. $\square$

Let us also determine how restriction interacts with T_e and N_e.

Corollary 4.7. *Let V be an A-module. Then we have $\mathrm{Res}_e(N_e(V)) = 0$, $\mathrm{Res}_e(T_e(V)) = \mathrm{Res}_e(V)$, $\mathrm{Res}_e(V/N_e(V)) \cong \mathrm{Res}_e(V)$ and $\mathrm{Res}_e(V/T_e(V)) \cong 0$.*

Proof. First we have $\mathrm{Res}_e(N_e(V)) = eN_e(V) = 0$ by definition. Next we compute that $e(AeV) \subseteq eV \subseteq eAeV = e(AeV)$ and so $\mathrm{Res}_e(T_e(V)) = \mathrm{Res}_e(V)$. Finally, as the restriction functor is exact (Proposition 4.2), we have that $\mathrm{Res}_e(V/N_e(V)) \cong \mathrm{Res}_e(V)/\mathrm{Res}_e(N_e(V)) \cong \mathrm{Res}_e(V)$ because $\mathrm{Res}_e(N_e(V)) = 0$ and $\mathrm{Res}_e(V/T_e(V)) \cong \mathrm{Res}_e(V)/\mathrm{Res}_e(T_e(V)) \cong 0$ because $\mathrm{Res}_e(T_e(V)) = \mathrm{Res}_e(V)$.   □

It will be important to link the functors T_e and N_e with Ind_e and Coind_e.

Proposition 4.8. *Let V be an eAe-module. Then $T_e(\mathrm{Ind}_e(V)) = \mathrm{Ind}_e(V)$ and $N_e(\mathrm{Coind}_e(V)) = 0$.*

Proof. The first equality follows because

$$Ae(Ae \otimes_{eAe} V) = Ae(eAe) \otimes_{eAe} V = Ae \otimes_{eAe} V$$

as submodules of $Ae \otimes_{eAe} V$. For the second equality, let us suppose that $\varphi \colon eA \longrightarrow V$ is an eAe-module homomorphism with $eA\varphi = 0$. Then, for $a \in A$, we have $\varphi(ea) = \varphi(eea) = (ea\varphi)(e) = 0$ because $ea\varphi = 0$. Therefore, $\varphi = 0$ and so $N_e(\mathrm{Coind}_e(V)) = 0$. This completes the proof.   □

The induction and coinduction functors also preserve indecomposability. We prove this in two steps.

Proposition 4.9. *Let V be an A-module with eV an indecomposable eAe-module. If either $T_e(V) = V$ or $N_e(V) = 0$, then V is indecomposable.*

Proof. Suppose that $V = V_1 \oplus V_2$ is a direct sum decomposition of V as an A-module. Then $eV = eV_1 \oplus eV_2$. By indecomposability of eV, either $eV_1 = 0$ or $eV_2 = 0$. Let us assume without loss of generality that the latter occurs. If $N_e(V) = 0$, then we have $V_2 \subseteq N_e(V) = 0$ and so $V = V_1$. If $T_e(V) = V$, then $V = AeV = AeV_1 \oplus AeV_2 \subseteq V_1$ because $eV_2 = 0$. Therefore, we again have $V = V_1$. We conclude that V is indecomposable.   □

Putting Proposition 4.9 together with Proposition 4.5 and Proposition 4.8, we are able to deduce the following corollary.

Corollary 4.10. *Let V be an indecomposable eAe-module. Then $\mathrm{Ind}_e(V)$ and $\mathrm{Coind}_e(V)$ are indecomposable A-modules.*

If V is an A-module, then, by Proposition 4.1, the identity mapping $1_{eV} \in \mathrm{Hom}_{eAe}(\mathrm{Res}_e(V), \mathrm{Res}_e(V))$ corresponds to a pair of homomorphisms

$$\alpha \colon \mathrm{Ind}_e(eV) \longrightarrow V \quad \text{and} \quad \beta \colon V \longrightarrow \mathrm{Coind}_e(eV).$$

We connect $T_e(V)$ and $N_e(V)$ to these maps.

Proposition 4.11. *Let V be an A-module. Let $\alpha\colon \mathrm{Ind}_e(eV) \longrightarrow V$ and $\beta\colon V \longrightarrow \mathrm{Coind}_e(eV)$ be given by*

$$\alpha(a \otimes v) = av, \ \text{for } a \in Ae, \ v \in eV$$
$$\beta(v)(a) = av, \ \text{for } a \in eA, \ v \in V.$$

Then $T_e(V) = \alpha(\mathrm{Ind}_e(eV))$ and $N_e(V) = \ker \beta$. Furthermore, $\ker \alpha \subseteq N_e(\mathrm{Ind}_e(eV))$ and $T_e(\mathrm{Coind}_e(eV)) \subseteq \beta(V)$.

Proof. By Proposition 4.8, we have that

$$\alpha(\mathrm{Ind}_e(eV)) = \alpha(T_e(\mathrm{Ind}_e(eV))) \subseteq T_e(V).$$

On the other hand, if $a \in A$ and $v \in V$, then $aev = \alpha(ae \otimes ev)$. It follows that the image of α is $AeV = T_e(V)$. From $e(a \otimes v) = ea \otimes v = e \otimes eav = e \otimes e\alpha(a \otimes v)$ for $a \in Ae$ and $v \in eV$, we deduce that $ex = e \otimes e\alpha(x)$ for all $x \in \mathrm{Ind}_e(eV)$. It follows that if $x \in \ker \alpha$ and $a \in A$, then $eax = e \otimes e\alpha(ax) = e \otimes ea\alpha(x) = 0$ and so $\ker \alpha \subseteq N_e(\mathrm{Ind}_e(eV))$.

Similarly, by Proposition 4.8, we have that

$$\beta(N_e(V)) \subseteq N_e(\mathrm{Coind}_e(eV)) = 0$$

and so $N_e(V) \subseteq \ker \beta$. But if $v \in \ker \beta$, then $eAv = \beta(v)(eA) = 0$. Thus $v \in N_e(V)$ and so $\ker \beta = N_e(V)$. Let $\varphi \in e\,\mathrm{Coind}_e(eV)$ (so that $e\varphi = \varphi$) and put $v = \varphi(e) \in eV$. Then, for $a \in eA$, we have $\varphi(a) = (e\varphi)(a) = \varphi(ae) = ae\varphi(e) = a\varphi(e) = av = \beta(v)(a)$ and so $\varphi = \beta(v)$. Therefore, $e\,\mathrm{Coind}_e(eV) \subseteq \beta(V)$ and hence $T_e(\mathrm{Coind}_e(eV)) = Ae\,\mathrm{Coind}_e(eV) \subseteq A\beta(V) = \beta(V)$. This completes the proof. $\qquad\square$

If W is an eAe-module, then $e\,\mathrm{Ind}_e(W) \cong W \cong e\,\mathrm{Coind}_e(W)$ by Proposition 4.5 and so the identity map 1_W corresponds to a homomorphism $\varphi\colon \mathrm{Ind}_e(W) \longrightarrow \mathrm{Coind}_e(W)$, which in fact is the same map with respect to either isomorphism in Proposition 4.1.

Corollary 4.12. *Let W be an eAe-module and let*

$$\varphi\colon \mathrm{Ind}_e(W) \longrightarrow \mathrm{Coind}_e(W)$$

be given by

$$\varphi(a \otimes w)(b) = baw$$

for $a \in Ae$, $b \in eA$, and $w \in W$. Then $\varphi(\mathrm{Ind}_e(W)) = T_e(\mathrm{Coind}_e(W))$ and $\ker \varphi = N_e(\mathrm{Ind}_e(W))$. Thus there is an isomorphism

$$\mathrm{Ind}_e(W)/N_e(\mathrm{Ind}_e(W)) \cong T_e(\mathrm{Coind}_e(W))$$

of A-modules.

Proof. This follows by observing that $e \operatorname{Ind}_e(W) \cong W \cong e \operatorname{Coind}_e(W)$ and by applying Proposition 4.11 twice: once with $V = \operatorname{Coind}_e(W)$ and $\varphi = \alpha$ and once with $V = \operatorname{Ind}_e(W)$ and $\varphi = \beta$.

Indeed, the isomorphism $W \longrightarrow e \operatorname{Coind}_e(W)$ sends $w \in W$ to the mapping $\varphi_w \colon eA \longrightarrow W$ given by $\varphi_w(b) = bew$ for $b \in eA$. Then $\alpha(a \otimes \varphi_w) = a\varphi_w$ for $a \in Ae$. But $(a\varphi_w)(b) = \varphi_w(ba) = baew = baw$ and so $\alpha(a \otimes \varphi_w) = \varphi(a \otimes w)$.

Similarly, W is isomorphic to $e \operatorname{Ind}_e(W)$ via $w \mapsto e \otimes w$. Then, for $a \in Ae$ and $b \in eA$, we have that $\beta(a \otimes w)(b) = ba \otimes w = e \otimes baw = e \otimes \varphi(a \otimes w)(b)$. This completes the proof. $\qquad \square$

We end this section with a criterion for when Ind_e is an equivalence of categories (and hence A is Morita equivalent to eAe). The reader is referred to Section A.4 for the definitions of an equivalence of categories and Morita equivalence.

Theorem 4.13. *Let A be a finite dimensional algebra and $e \in A$ an idempotent. Then the functor*

$$\operatorname{Ind}_e \colon eAe\text{-mod} \longrightarrow A\text{-mod}$$

is an equivalence of categories if and only if $A = AeA$. Moreover, if $A = AeA$, then Res_e is a quasi-inverse of Ind_e.

Proof. Assume first that Ind_e is an equivalence of categories. Then $A \cong \operatorname{Ind}_e(V)$ for some eAe-module V. Proposition 4.8 yields $T_e(\operatorname{Ind}_e(V)) = \operatorname{Ind}_e(V)$ and so $AeA = T_e(A) = A$.

Conversely, if $AeA = A$, then $T_e(V) = AeV = AeAV = AV = V$ for all A-modules V. Also, if $v \in V$ with $eAv = 0$, then $Av = AeAv = 0$ and so $v = 0$. Therefore, $N_e(V) = 0$ for all A-modules V. Thus the natural map $\alpha \colon \operatorname{Ind}_e(eV) \longrightarrow V$ from Proposition 4.11 is an isomorphism by the selfsame proposition. It follows that $\operatorname{Ind}_e \circ \operatorname{Res}_e$ is naturally isomorphic to the identity functor on A-mod. Since we already know from Proposition 4.5 that $\operatorname{Res}_e \circ \operatorname{Ind}_e$ is naturally isomorphic to the identity functor on eAe-mod, this completes the proof. $\qquad \square$

The following corollary will be used later to describe the algebra of $M_n(\mathbb{F}_q)$ over a field $\Bbbk$ whose characteristic is different from that of $\mathbb{F}_q$.

Corollary 4.14. *Let A be a finite dimensional algebra and $e \in A$ an idempotent. If $A = AeA$, then $A^{op} \cong \operatorname{End}_{eAe}(eA)$.*

Proof. By Theorem 4.13, we have that Res_e is an equivalence of categories (with quasi-inverse Ind_e) and so

$$A^{op} \cong \operatorname{End}_A(A) \cong \operatorname{End}_{eAe}(\operatorname{Res}_e(A)) = \operatorname{End}_{eAe}(eA)$$

as required. $\qquad \square$

4.2 Idempotents and simple modules

We continue to assume that A is a finite dimensional $\Bbbk$-algebra and that $e \in A$ is an idempotent. A crucial lemma is that Res_e takes simple modules to simple modules or to zero.

Lemma 4.15. *Let S be a simple A-module. Then either $eS = 0$ or eS is a simple eAe-module.*

Proof. If $0 \neq v \in eS$, then $eAev = eAv = eS$ because $Av = S$ by simplicity of S. Thus eS is simple. $\qquad\square$

Corollary 4.16. *If V is a semisimple A-module, then $\mathrm{Res}_e(V)$ is a semisimple eAe-module.*

As another corollary, we obtain the following.

Corollary 4.17. *Let V be a finite dimensional A-module with composition factors $S_1, \ldots, S_m$ (with multiplicities). Then the composition factors of eV as an eAe-module are the nonzero entries of the list $eS_1, \ldots, eS_m$ (with multiplicities).*

Proof. Let $0 = V_0 \subseteq V_1 \subseteq \cdots \subseteq V_m = V$ be a composition series with $V_i/V_{i-1} \cong S_i$ for $i = 1, \ldots, m$. Then since restriction is exact, we have that $eV_i/eV_{i-1} \cong e(V_i/V_{i-1}) = eS_i$. By Lemma 4.15, each nonzero eS_i is simple. It follows that after removing the repetitions from the series

$$eV_0 \subseteq eV_1 \subseteq \cdots \subseteq eV_m = eV,$$

we obtain a composition series for eV whose composition factors are the nonzero entries of the list $eS_1, \ldots, eS_m$. $\qquad\square$

The radical of a module V over a finite dimensional algebra is denoted by $\mathrm{rad}(V)$. Let us relate the radicals of A and eAe.

Proposition 4.18. *One has that $\mathrm{rad}(eAe) = e\,\mathrm{rad}(A)e$. In particular, if A is semisimple, then eAe is also semisimple.*

Proof. Put $R = \mathrm{rad}(A)$ and let S be a simple A-module. Then $eS = 0$ or eS is a simple eAe-module by Lemma 4.15. In either case, $\mathrm{rad}(eAe)S = \mathrm{rad}(eAe)eS = 0$. Therefore, we have $\mathrm{rad}(eAe) \subseteq R$ by Theorem A.5 and hence $\mathrm{rad}(eAe) \subseteq eRe$. For the converse, it is evident that eRe is an ideal of eAe. We show that it is nilpotent. Indeed, if $R^n = 0$, then $(eRe)^n \subseteq R^n = 0$. It follows that $eRe \subseteq \mathrm{rad}(eAe)$ by Theorem A.5, completing the proof. $\qquad\square$

Our next goal is to investigate the radical and socle of induced and coinduced modules.

Proposition 4.19. *Let V be a finite dimensional A-module.*

(i) *If $V = T_e(V)$, then $N_e(V) \subseteq \mathrm{rad}(V)$. Equality holds if and only if eV is semisimple.*

(ii) *If $N_e(V) = 0$, then $\mathrm{soc}(V) \subseteq T_e(V)$. Equality holds if and only if eV is semisimple.*

Proof. Let $R = \mathrm{rad}(A)$. To prove (i), let M be a maximal submodule of V. If $N_e(V) \not\subseteq M$, then $N_e(V) + M = V$ and hence $eV = e(N_e(V) + M) = eM$. From this we obtain $M \supseteq AeM = AeV = V$, a contradiction. Thus $N_e(V) \subseteq M$ and hence, since M was arbitrary, we have $N_e(V) \subseteq \mathrm{rad}(V)$. From $\mathrm{rad}(V) = RV$ (cf. Theorem A.5) and $eRe = \mathrm{rad}(eAe)$, we obtain that $e\,\mathrm{rad}(V) = eRV = eRAeV = eReeV = \mathrm{rad}(eAe)eV = \mathrm{rad}(eV)$ by Theorem A.5 and hence eV is semisimple if and only if $e\,\mathrm{rad}(V) = 0$. Therefore, $\mathrm{rad}(V) \subseteq N_e(V)$ if and only if eV is semisimple, completing the proof of the first item.

Turning to (ii), let $S \subseteq V$ be a simple module. Because $N_e(V) = 0$, we deduce $0 \neq eS \subseteq eV$ and hence $S = AeS \subseteq AeV = T_e(V)$. It follows that $\mathrm{soc}(V) \subseteq T_e(V)$. Observe that $e\,\mathrm{rad}(AeV) = eRAeV = eReeV = \mathrm{rad}(eAe)eV = \mathrm{rad}(eV)$, by Theorem A.5, and so eV is semisimple if and only if $e\,\mathrm{rad}(T_e(V)) = 0$, i.e., $\mathrm{rad}(T_e(V)) \subseteq N_e(V) = 0$. But $\mathrm{rad}(T_e(V)) = 0$ if and only if $T_e(V) \subseteq \mathrm{soc}(V)$ and so eV is semisimple if and only if $T_e(V) = \mathrm{soc}(V)$, as required. $\qquad\square$

Next we analyze, to some extent, A-modules V with $eV \neq 0$ simple.

Proposition 4.20. *Let V be an A-module.*

(i) *If $V = T_e(V)$ and eV is simple, then $V/\mathrm{rad}(V) = V/N_e(V)$ is simple.*

(ii) *If $N_e(V) = 0$ and eV is simple, then $\mathrm{soc}(V) = T_e(V)$ is simple.*

Proof. For (i), note that $N_e(V) = \mathrm{rad}(V)$ by Proposition 4.19. If $v \notin N_e(V)$, then $0 \neq eAv \subseteq eV$ is an eAe-submodule and so $eAv = eV$ by simplicity of eV. Moreover, $Av \supseteq AeAv = AeV = V$. Thus $A(v + N_e(V)) = V/N_e(V)$ and we conclude that $V/N_e(V)$ is simple.

To prove (ii), we have that $\mathrm{soc}(V) = T_e(V)$ by Proposition 4.19. If $0 \neq v \in T_e(V) = AeV$, then $eAv \neq 0$ because $N_e(V) = 0$. Therefore, $eAv = eV$ by simplicity of eV. We conclude that $Av \supseteq AeAv = AeV = T_e(V)$. It follows that $T_e(V)$ is simple. $\qquad\square$

We now apply the proposition to induced and coinduced modules.

Corollary 4.21. *Let V be a semisimple eAe-module. Then*

$$\mathrm{Ind}_e(V)/\mathrm{rad}(\mathrm{Ind}_e(V)) \cong \mathrm{soc}(\mathrm{Coind}_e(V)).$$

Moreover, if V is simple, then $\mathrm{Ind}_e(V)/\mathrm{rad}(\mathrm{Ind}_e(V))$ and $\mathrm{soc}(\mathrm{Coind}_e(V))$ are isomorphic simple modules and there are isomorphisms

$$\mathrm{Res}_e(\mathrm{Ind}_e(V)/\mathrm{rad}(\mathrm{Ind}_e(V))) \cong V \cong \mathrm{Res}_e(\mathrm{soc}(\mathrm{Coind}_e(V)))$$

of eAe-modules.

Proof. By Propositions 4.5, 4.8, and 4.19 we have that $\mathrm{rad}(\mathrm{Ind}_e(V)) = N_e(V)$ and $\mathrm{soc}(\mathrm{Coind}_e(V)) = T_e(\mathrm{Coind}_e(V))$. The first statement then follows from Corollary 4.12. The second statement is immediate from Proposition 4.8, Proposition 4.20, the isomorphisms $e\,\mathrm{Ind}_e(V) \cong V \cong e\,\mathrm{Coind}_e(V)$ (see Proposition 4.5), and Corollary 4.7. $\square$

As a corollary, we can also describe when the natural map in Corollary 4.12 is an isomorphism.

Corollary 4.22. *Let V be a semisimple eAe-module and let*

$$\varphi\colon \mathrm{Ind}_e(V) \longrightarrow \mathrm{Coind}_e(V)$$

be given by $\varphi(a \otimes v)(b) = bav$ for $a \in Ae$, $b \in eA$, and $v \in V$. Then:

(i) φ is injective if and only if $\mathrm{Ind}_e(V)$ is semisimple;
(ii) φ is surjective if and only if $\mathrm{Coind}_e(V)$ is semisimple;
(iii) φ is an isomorphism if and only if $\mathrm{Ind}_e(V)$ and $\mathrm{Coind}_e(V)$ are both semisimple.

Proof. This is immediate from Proposition 4.5, Proposition 4.8, Corollary 4.12, and Proposition 4.19, which imply the equalities $\ker\varphi = \mathrm{rad}(\mathrm{Ind}_e(V))$ and $\varphi(\mathrm{Ind}_e(V)) = \mathrm{soc}(\mathrm{Coind}_e(V))$. $\square$

Now we set up a bijection between isomorphism classes of simple eAe-modules and isomorphism classes of simple A-modules not annihilated by e.

Theorem 4.23. *There is a bijection between isomorphism classes of simple eAe-modules and isomorphism classes of simple A-modules not annihilated by e induced by*

$$V \longmapsto \mathrm{Ind}_e(V)/N_e(\mathrm{Ind}_e(V)) = \mathrm{Ind}_e(V)/\mathrm{rad}(\mathrm{Ind}_e(V))$$
$$\cong \mathrm{soc}(\mathrm{Coind}_e(V)) = T_e(\mathrm{Coind}_e(V))$$
$$S \longmapsto \mathrm{Res}_e(S) = eS$$

for V a simple eAe-module and S a simple A-module with $eS \neq 0$.

Consequently, there is a bijection between the set of isomorphism classes of simple A-modules and the disjoint union of the sets of isomorphism classes of simple eAe-modules and of simple A/AeA-modules.

Proof. Applying Proposition 4.8, Lemma 4.15, Proposition 4.19, and Corollary 4.21 we have that the two maps are well defined and that

$$e\,\mathrm{soc}(\mathrm{Coind}_e(V)) \cong V.$$

It remains to show that if S is a simple A-module with $eS \neq 0$, then we have that $S \cong \operatorname{soc}(\operatorname{Coind}_e(eS))$. Indeed, we have by Proposition 4.1 isomorphisms

$$\operatorname{Hom}_A(S, \operatorname{soc}(\operatorname{Coind}_e(eS))) \cong \operatorname{Hom}_A(S, \operatorname{Coind}_e(eS)) \cong \operatorname{Hom}_{eAe}(eS, eS) \neq 0.$$

Because eS is simple by Lemma 4.15, and hence $\operatorname{soc}(\operatorname{Coind}_e(eS))$ is simple by Corollary 4.21, we conclude that $S \cong \operatorname{soc}(\operatorname{Coind}_e(eS))$ by Schur's lemma.

The final statement follows from what we have just proved and the observation that simple A-modules annihilated by e are exactly the same thing as simple A/AeA-modules. □

We shall need later the following lemma connecting primitive idempotents in A and eAe.

Lemma 4.24. *Suppose that S is a simple A-module such that $eS \neq 0$, and hence eS is a simple eAe-module. If $f \in eAe$ is a primitive idempotent with $eAef/e\operatorname{rad}(A)ef \cong eS$, then f is a primitive idempotent of A and $Af/\operatorname{rad}(A)f \cong S$.*

Proof. Let $f \in eAe$ be as above. Then $fAf = feAef$ and so $E(fAf) = E(feAef) = \{0, f\}$ by Proposition A.22. Thus f is primitive in A by another application of Proposition A.22. Because $0 \neq \operatorname{Hom}_{eAe}(eAef, eS) = feS = fS$, it follows that $\operatorname{Hom}_A(Af/\operatorname{rad}(A)f, S) \cong \operatorname{Hom}_A(Af, S) \cong fS \neq 0$. But $Af/\operatorname{rad}(A)f$ is simple, as is S, and so we conclude by Schur's lemma that $Af/\operatorname{rad}(A)f \cong S$, as required. □

4.3 Exercises

4.1. Give a detailed proof of Proposition 4.1.

4.2. Prove that if V is an A-module and W is an A/AeA-module, then there are natural isomorphisms

$$\operatorname{Hom}_A(W, V) \cong \operatorname{Hom}_{A/AeA}(W, N_e(V))$$
$$\operatorname{Hom}_A(V, W) \cong \operatorname{Hom}_{A/AeA}(V/T_e(V), W).$$

4.3. Let V be an eAe-module. Prove that the natural map

$$\varphi \colon \operatorname{Ind}_e(V) \longrightarrow \operatorname{Coind}_e(V)$$

from Corollary 4.12 is an isomorphism if and only if $\operatorname{Ind}_e(V) \cong \operatorname{Coind}_e(V)$.

4.4. Let A be a finite dimensional algebra and $e \in A$ an idempotent. Put $J = AeA$. Prove that eAe is semisimple if and only if $J\operatorname{rad}(A)J = 0$.

4.5. Let A be a finite dimensional algebra and $e \in A$ an idempotent. Let V be an A-module. Prove that

$$T_e(V) = AeV = \sum_{\varphi \in \mathrm{Hom}_A(Ae,V)} \varphi(Ae).$$

4.6. Let A be a finite dimensional algebra and $e \in A$ an idempotent. Prove that a finite dimensional A-module V is isomorphic to a module of the form $\mathrm{Ind}_e(W)$ with W a finite dimensional eAe-module if and only if there is an exact sequence

$$P_1 \longrightarrow P_0 \longrightarrow V \longrightarrow 0$$

where P_0 and P_1 are direct sums of summands of Ae.

4.7. Let A be a finite dimensional $\Bbbk$-algebra and let $e \in E(A)$. Prove that $A = AeA$ if and only if every projective indecomposable A-module is isomorphic to a direct summand in Ae.

4.8. Prove that $M_n(A)$ is Morita equivalent to A for all $n \geq 1$.

4.9. Prove that every finite dimensional $\Bbbk$-algebra A is Morita equivalent to a finite dimensional $\Bbbk$-algebra B such that $B/\mathrm{rad}(B)$ is isomorphic to a direct product of division algebras and that B is uniquely determined up to isomorphism. (Hint: let $e_1, \ldots, e_s$ form a complete set of orthogonal primitive idempotents for A and choose a subset $\{f_1, \ldots, f_r\}$ of $\{e_1, \ldots, e_s\}$ such that $Af_i \not\cong Af_j$ for $i \neq j$ and, for $i = 1, \ldots, s$, $Ae_i \cong Af_j$ for some j; put $e = f_1 + \cdots + f_r$ and let $B = eAe$; you will want to use that the $e_i + \mathrm{rad}(A)$ form a complete set of orthogonal primitive idempotents for $A/\mathrm{rad}(A)$ and that $\mathrm{rad}(B) = e\,\mathrm{rad}(A)e$.)

5

Irreducible Representations

Clifford-Munn-Ponizovskiĭ theory, developed in Clifford [Cli42], Munn [Mun55, Mun57b, Mun60], and Ponizovskiĭ [Pon58] (and in further detail in [LP69, RZ91]), gives a bijection between equivalence classes of irreducible representations of a finite monoid and equivalence classes of irreducible representations of its maximal subgroups (taken one per regular $\mathscr{J}$-class). We follow here the approach of [GMS09], using the techniques of Chapter 4. Let us commence by introducing formally the notion of a representation of a monoid.

5.1 Monoid algebras and representations

Let M be a monoid and $\Bbbk$ a field. A *representation* of M on a $\Bbbk$-vector space V is a homomorphism $\rho\colon M \longrightarrow \operatorname{End}_{\Bbbk}(V)$. We call $\dim V$ the *degree* of ρ. Two representations $\rho\colon M \longrightarrow \operatorname{End}_{\Bbbk}(V)$ and $\psi\colon M \longrightarrow \operatorname{End}_{\Bbbk}(W)$ are said to be *equivalent* if there is a vector space isomorphism $T\colon V \longrightarrow W$ such that $T^{-1}\psi(m)T = \rho(m)$ for all $m \in M$. A representation ρ is *faithful* if it is injective.

A *matrix representation* of M over $\Bbbk$ is a homomorphism $\rho\colon M \longrightarrow M_n(\Bbbk)$ for some $n \geq 0$. Notice that any matrix representation of a group takes values in the general linear group. Two matrix representations $\rho, \psi\colon M \longrightarrow M_n(\Bbbk)$ are *equivalent* if there is an invertible matrix $T \in M_n(\Bbbk)$ such that $T^{-1}\psi(m)T = \rho(m)$ for all $m \in M$. Of course, we can view each matrix representation as a representation by identifying $M_n(\Bbbk)$ with $\operatorname{End}_{\Bbbk}(\Bbbk^n)$ in the usual way. Conversely, if $\rho\colon M \longrightarrow \operatorname{End}_{\Bbbk}(V)$ is a representation of M on a finite dimensional vector space V, then we can obtain a matrix representation by choosing a basis. If we choose a different basis, this amounts to choosing an equivalent matrix representation. In fact, it is straightforward to verify that equivalence classes of degree n representations of M are in bijection with equivalence classes of matrix representations $\rho\colon M \longrightarrow M_n(\Bbbk)$. Since we only care about representations up to equivalence, we shall no longer distinguish in

© Springer International Publishing Switzerland 2016
B. Steinberg, *Representation Theory of Finite Monoids*,
Universitext, DOI 10.1007/978-3-319-43932-7_5

terminology between representations on finite dimensional vector spaces and matrix representations.

The *monoid algebra* $\Bbbk M$ of a monoid M over a field $\Bbbk$ is the $\Bbbk$-algebra constructed as follows. As a vector space, $\Bbbk M$ has basis M: so the elements of $\Bbbk M$ are formal sums $\sum_{m \in M} c_m m$ with $m \in M$, $c_m \in \Bbbk$ and with only finitely many $c_m \neq 0$. In practice, M will be finite and so this last constraint is unnecessary. The product is given by

$$\left(\sum_{m \in M} c_m m \right) \cdot \left(\sum_{m \in M} d_m m \right) = \sum_{m,n \in M} c_m d_n mn.$$

In other words, the product on M is extended to $\Bbbk M$ via the distributive law. When G is a group, $\Bbbk G$ is called the *group algebra*. Note that $\Bbbk M$ is finite dimensional if and only if M is finite.

The monoid algebra has the following universal property.

Proposition 5.1. *Let A be a $\Bbbk$-algebra and M a monoid. Then every monoid homomorphism $\varphi \colon M \longrightarrow A$ extends uniquely to a $\Bbbk$-algebra homomorphism $\Phi \colon \Bbbk M \longrightarrow A$.*

Proof. Let us define

$$\Phi \left(\sum_{m \in M} c_m m \right) = \sum_{m \in M} c_m \varphi(m).$$

The reader readily verifies that Φ is as required and is unique.   □

If A is a $\Bbbk$-algebra, then an A-module V is the same thing as a $\Bbbk$-vector space V together with a $\Bbbk$-algebra homomorphism $\rho \colon A \longrightarrow \mathrm{End}_{\Bbbk}(V)$. It then follows from Proposition 5.1 that a representation of a monoid M on a $\Bbbk$-vector space V is the same thing as a $\Bbbk M$-module structure on V and that two representations are equivalent if and only if the corresponding $\Bbbk M$-modules are isomorphic. More explicitly, if $\rho \colon M \longrightarrow \mathrm{End}_{\Bbbk}(V)$ is a representation, then the $\Bbbk M$-module structure on V is given by

$$\left(\sum_{m \in M} c_m m \right) \cdot v = \sum_{m \in M} c_m \rho(m) v$$

for $v \in V$. We shall say that the $\Bbbk M$-module V *affords* the corresponding representation $\rho \colon M \longrightarrow \mathrm{End}_{\Bbbk}(V)$ or, if we choose a basis for V, the corresponding matrix representation $\rho \colon M \longrightarrow M_n(\Bbbk)$.

Terminology for representations and modules tend to differ. For example, representations corresponding to simple $\Bbbk M$-modules are called *irreducible representations* and representations corresponding to semisimple $\Bbbk M$-modules are called *completely reducible*. We shall stick here principally to module theoretic terminology.

The set of isomorphism classes of simple $\Bbbk M$-modules will be denoted $\mathrm{Irr}_\Bbbk(M)$. The isomorphism class of a module V will typically be written $[V]$.

The following is an immediate consequence of the theorem of Frobenius and Schur (Theorem A.14), restricted to monoid algebras.

Corollary 5.2. *Let M be a finite monoid and $\Bbbk$ an algebraically closed field. Suppose that $S_1, \ldots, S_r$ form a complete set of representatives of the isomorphism classes of simple $\Bbbk M$-modules and that S_k affords the representation $\varphi^{(k)} \colon M \longrightarrow M_{n_k}(\Bbbk)$ with $\varphi^{(k)}(m) = (\varphi_{ij}^{(k)}(m))$. Then the mappings*

$$\varphi_{ij}^{(k)} \colon M \longrightarrow \Bbbk$$

with $1 \leq k \leq r$ and $1 \leq i, j \leq n_k$ are linearly independent in $\Bbbk^M$.

Proof. This follows immediately from Theorem A.14 and the observation that $\mathrm{Hom}_\Bbbk(\Bbbk M, \Bbbk) \cong \Bbbk^M$ via $\psi \mapsto \psi|_M$. $\square$

It is usual to identify $M_1(\Bbbk)$ with the multiplicative monoid of the field $\Bbbk$. The representation $\rho \colon M \longrightarrow \Bbbk$ given by $\rho(m) = 1$ for all $m \in M$ is called the *trivial representation*. We call $\Bbbk$, equipped with the corresponding module structure, the *trivial module*.

If V and W are $\Bbbk M$-modules, then their *tensor product* $V \otimes_\Bbbk W$ (usually written $V \otimes W$) becomes a $\Bbbk M$-module by defining $m(v \otimes w) = mv \otimes mw$ for $m \in M$, $v \in V$, and $w \in W$.

We shall also need the exterior power of a $\Bbbk M$-module. For this discussion, we assume that $\Bbbk$ is not of characteristic 2. If V is a $\Bbbk M$-module, then the tensor power $V^{\otimes r}$ is a $\Bbbk M$-$\Bbbk S_r$-bimodule, where the symmetric group S_r acts by permuting the tensor factors:

$$(v_1 \otimes \cdots \otimes v_r)f = v_{f(1)} \otimes \cdots \otimes v_{f(r)}$$

for $f \in S_r$ and $v_1, \ldots, v_r \in V$. The rth-*exterior power* of V is then

$$\Lambda^r(V) = V^{\otimes r} \otimes_{\Bbbk S_r} S_{(1^r)}$$

where we recall that $S_{(1^r)}$ is the sign representation of S_r. The equivalence class of a basic tensor $v_1 \otimes \cdots \otimes v_r$ is denoted by $v_1 \wedge \cdots \wedge v_r$ and one has the properties that $v_1 \wedge \cdots \wedge v_r = 0$ if $v_i = v_j$ for some $i \neq j$ and that

$$v_{f(1)} \wedge \cdots \wedge v_{f(r)} = \mathrm{sgn}(f)(v_1 \wedge \cdots \wedge v_r)$$

for $f \in S_r$. If $\dim V < r$, then $\Lambda^r(V) = 0$; otherwise, if $v_1, \ldots, v_n$ is a basis for V with $n \geq r$, then the $v_{i_1} \wedge \cdots \wedge v_{i_r}$ with $i_1 < i_2 < \cdots < i_r$ form a basis for $\Lambda^r(V)$. Hence $\dim \Lambda^r(V) = \binom{n}{r}$.

If $e \in E(M)$ is an idempotent and V is a $\Bbbk M$-module, then it is straightforward to verify that $e(V^{\otimes r}) = (eV)^{\otimes r}$ as a $\Bbbk[eMe]$-module because

$$e\left(\sum_{i=1}^{n} v_{i,1} \otimes \cdots \otimes v_{i,r}\right) = \sum_{i=1}^{n} ev_{i,1} \otimes \cdots \otimes ev_{i,r}.$$

It follows that $e\Lambda^r(V) \cong \Lambda^r(eV)$ as $\Bbbk[eMe]$-modules by Proposition 4.4.

If $\varphi: M \longrightarrow N$ is a homomorphism of monoids, then any $\Bbbk N$-module V can be viewed as a $\Bbbk M$-module via the action $mv = \varphi(m)v$ for $m \in M$ and $v \in V$. One sometimes says that the $\Bbbk M$-module V is the *inflation* of the $\Bbbk N$-module V to M. More generally, if $\psi: A \longrightarrow B$ is a homomorphism of finite dimensional algebras and V is a B-module, then we call the A-module structure on V given by $av = \psi(a)v$ for $a \in A$ and $v \in V$ the *inflation* of V along ψ.

Remark 5.3. If M is a monoid containing a zero element z, then $\Bbbk z$ is a one-dimensional central ideal of $\Bbbk M$ and the algebra $\Bbbk M/\Bbbk z$ is called the *contracted monoid algebra* of M. One can identify $\Bbbk M/\Bbbk z$-mod with the full subcategory of $\Bbbk M$-mod consisting of those modules V with $zV = 0$. One advantage of working with contracted monoid algebras is that if I is an ideal of M, then $\Bbbk M/\Bbbk I$ is isomorphic to the contracted monoid algebra of M/I. Notice that since z is a central idempotent, we can identify $\Bbbk M/\Bbbk z$ with $\Bbbk M(1 - z)$ and we have $\Bbbk M = \Bbbk z \times \Bbbk M(1 - z) \cong \Bbbk \times \Bbbk M/\Bbbk z$. Therefore, the representation theory of these two algebras is not very different. A disadvantage of contracted monoid algebras is that one no longer has the trivial module and so the Grothendieck ring will not have an identity. Essentially for this reason, we avoid working explicitly with contracted monoid algebras in this text.

5.2 Clifford-Munn-Ponizovskiĭ theory

We fix a finite monoid M and a field $\Bbbk$. If V is a $\Bbbk M$-module, $X \subseteq M$ and $Y \subseteq V$, then we put

$$XY = \left\{\sum_{i=1}^{n} k_i x_i y_i \mid k_i \in \Bbbk, x_i \in X, y_i \in Y\right\}.$$

Also, we let $\Bbbk X$ denote the $\Bbbk$-linear span of X in $\Bbbk M$. Note that if $I \subseteq M$ is a left (respectively, right or two-sided) ideal of M, then $\Bbbk I$ is a left (respectively, right or two-sided) ideal of $\Bbbk M$.

Let S be a simple $\Bbbk M$-module. We say that an idempotent $e \in E(M)$ is an *apex* for S if $eS \neq 0$ and $I_e S = 0$ where we recall that $I_e = eMe \setminus G_e$. Note that $I_e S = I_e eS$. Let us also recall the notation $I(e) = \{m \in M \mid e \notin MmM\}$ and that $eI(e)e = I_e$ (cf. Corollary 1.16).

We establish some basic properties of apexes.

Proposition 5.4. *Let S be a simple $\Bbbk M$-module with apex $e \in E(M)$.*

(i) $I(e) = \{m \in M \mid mS = 0\}$.
(ii) If $f \in E(M)$, then f is an apex for S if and only if $MeM = MfM$.

Proof. Since S is simple and $eS \neq 0$, we have $MeS = S$ and hence $I(e)S = I(e)MeS = I(e)eS$. Suppose that $I(e)S \neq 0$ (and in particular, $I(e) \neq \emptyset$). Note that $I(e)S$ is a submodule of S because $I(e)$ is an ideal of M. Thus, by simplicity of S, we obtain that $I(e)eS = I(e)S = S$ and so $eS = eI(e)eS = I_eS = 0$, a contradiction. We conclude that $I(e)S = 0$. If $e \in MmM$, then $0 \neq eS \subseteq MmMS = MmS$ shows that $mS \neq 0$, establishing (i).

If $f \in E(M)$ with $fS \neq 0$, then $f \notin I(e)$ by (i) and so $MeM \subseteq MfM$. It follows by symmetry that if f is another apex, then $MeM = MfM$. Conversely, if $f \in E(M)$ with $MfM = MeM$, then $fS \neq 0$ and $I_fS = 0$ by (i) because $f \notin I(e)$ and $I_f \subseteq I(e)$. Thus f is an apex for S. □

If M is a monoid and I is an ideal of M, then $\Bbbk M/\Bbbk I$ has basis the set $\{m + \Bbbk I \mid m \notin I\}$ (and all such cosets are distinct) with product

$$(m + \Bbbk I)(m' + \Bbbk I) = \begin{cases} mm' + \Bbbk I, & \text{if } mm' \notin I \\ 0, & \text{if } mm' \in I. \end{cases}$$

It is therefore natural to identify the coset $m + \Bbbk I$ with m for $m \notin I$ and so, for instance, we shall often write $(\Bbbk M/\Bbbk I)m$ instead of $(\Bbbk M/\Bbbk I)(m + \Bbbk I)$.

Fix an idempotent $e \in E(M)$ and put $A_e = \Bbbk M/\Bbbk I(e)$; notice that A_e-mod is the full subcategory of $\Bbbk M$-mod consisting of modules annihilated by $I(e)$. Observe that $eA_ee \cong \Bbbk[eMe]/\Bbbk I_e \cong \Bbbk G_e$ by Corollary 1.16. By Proposition 5.4, a simple $\Bbbk M$-module S has apex e if and only if it is annihilated by $I(e)$ but not by e; that is, the simple $\Bbbk M$-modules with apex e are precisely the inflations of simple A_e-modules S with $eS \neq 0$. We can then apply the theory of Chapter 4 to classify these modules.

Notice that, as $\Bbbk$-vector spaces, $A_ee \cong \Bbbk L_e$ and $eA_e \cong \Bbbk R_e$. Indeed, Theorem 1.13 shows that $J_e \cap Me = L_e$ and $J_e \cap eM = R_e$ and therefore $Me \setminus L_e \subseteq I(e)$ and $eM \setminus R_e \subseteq I(e)$. The corresponding left $\Bbbk M$-module structure on $\Bbbk L_e$ is defined by

$$m \odot \ell = \begin{cases} m\ell, & \text{if } m\ell \in L_e \\ 0, & \text{else} \end{cases}$$

for $m \in M$ and $\ell \in L_e$. From now on we will omit the symbol "$\odot$." The right $\Bbbk M$-module structure on $\Bbbk R_e$ is defined dually. Note that $\Bbbk L_e$ is also a free right $\Bbbk G_e$-module and $\Bbbk R_e$ is a free left $\Bbbk G_e$-module by Proposition 1.10 and Exercise 5.3. In the semigroup theory literature, $\Bbbk L_e$ and $\Bbbk R_e$ are known as the left and right *Schützenberger representations* associated with J_e due to the close connection with the paper of Schützenberger [Sch58].

If V is a $\mathbb{k}G_e$-module, then

$$\text{Hom}_{eA_ee}(eA_e, V) \cong \text{Hom}_{\mathbb{k}G_e}(\mathbb{k}R_e, V) \cong \text{Hom}_{G_e}(R_e, V)$$

where $\text{Hom}_{G_e}(R_e, V)$ denotes the vector space of G_e-equivariant mappings $\varphi\colon R_e \longrightarrow V$. The action of $m \in M$ on $\varphi \in \text{Hom}_{G_e}(R_e, V)$ is given by

$$(m\varphi)(r) = \begin{cases} \varphi(rm), & \text{if } rm \in R_e \\ 0, & \text{else} \end{cases}$$

for $r \in R_e$.

Using the induction and coinduction functors associated with A_e and eA_ee we define functors

$$\text{Ind}_{G_e} : \mathbb{k}G_e\text{-mod} \longrightarrow \mathbb{k}M\text{-mod}$$
$$\text{Coind}_{G_e} : \mathbb{k}G_e\text{-mod} \longrightarrow \mathbb{k}M\text{-mod}$$
$$\text{Res}_{G_e} : \mathbb{k}M\text{-mod} \longrightarrow \mathbb{k}G_e\text{-mod}$$
$$T_e : \mathbb{k}M\text{-mod} \longrightarrow \mathbb{k}M\text{-mod}$$
$$N_e : \mathbb{k}M\text{-mod} \longrightarrow \mathbb{k}M\text{-mod}$$

by putting

$$\text{Ind}_{G_e}(V) = A_ee \otimes_{eA_ee} V = \mathbb{k}L_e \otimes_{\mathbb{k}G_e} V$$
$$\text{Coind}_{G_e}(V) = \text{Hom}_{eA_ee}(eA_e, V) = \text{Hom}_{G_e}(R_e, V)$$
$$\text{Res}_{G_e}(V) = eV$$
$$T_e(V) = \mathbb{k}MeV = MeV$$
$$N_e(V) = \{v \in V \mid e\mathbb{k}Mv = 0\} = \{v \in V \mid eMv = 0\}.$$

Notice that the functors Ind_{G_e} and Coind_{G_e} are exact by Proposition 4.2 because $\mathbb{k}L_e$ and $\mathbb{k}R_e$ are free $\mathbb{k}G_e$-modules. Also, if V is an A_e-module, then $\mathbb{k}MeV = A_eeV$ and $e\mathbb{k}Mv = 0$ if and only if $eA_ev = 0$ for $v \in V$. We can now state the main theorem of this section, which describes the irreducible representations of a finite monoid in terms of the irreducible representations of its maximal subgroups. It is the fundamental theorem of Clifford-Munn-Ponizovskiĭ theory, as formulated in [GMS09].

Theorem 5.5. *Let M be a finite monoid and $\mathbb{k}$ a field.*

(i) *There is a bijection between isomorphism classes of simple $\mathbb{k}M$-modules with apex $e \in E(M)$ and isomorphism classes of simple $\mathbb{k}G_e$-modules induced by*

$$S \longmapsto \text{Res}_{G_e}(S) = eS$$
$$V \longmapsto V^\sharp = \text{Ind}_{G_e}(V)/N_e(\text{Ind}_{G_e}(V)) = \text{Ind}_{G_e}(V)/\text{rad}(\text{Ind}_{G_e}(V))$$
$$\cong \text{soc}(\text{Coind}_{G_e}(V)) = T_e(\text{Coind}_{G_e}(V))$$

for S a simple $\mathbb{k}M$-module with apex e and V a simple $\mathbb{k}G_e$-module.

(ii) Every simple $\Bbbk M$*-module has an apex (unique up to* $\mathscr{J}$*-equivalence).*

(iii) If V is a simple $\Bbbk G_e$*-module, then every composition factor of* $\mathrm{Ind}_{G_e}(V)$
and $\mathrm{Coind}_{G_e}(V)$ *has apex f with* $MeM \subseteq MfM$. *Moreover,* $V^{\sharp}$ *is the*
unique composition factor of either of these two modules with apex e and
it appears in both with multiplicity one.

Proof. Since simple $\Bbbk M$-modules with apex e are the same thing as inflations
of simple A_e-modules S with $eS \neq 0$, the first item follows from Theorem 4.23
applied to A_e and $eA_e e \cong \Bbbk G_e$.

To prove (ii) let S be a simple $\Bbbk M$-module and let I be minimal amongst
ideals of M satisfying $IS \neq 0$, that is, not annihilating S. As IS is a submodule, we have that $IS = S$ by simplicity of S and so $I^2 S = IS = S \neq 0$. Therefore, $I^2 = I$ by minimality. Choose $m \in I$ with $mS \neq 0$. Then $MmMS \neq 0$
and so $I = MmM$ by minimality. By Corollary 1.24 there is an idempotent $e \in E(M)$ with $I = MeM$. We claim that e is an apex for S. First
note that from $S = IS = MeMS = MeS$ we conclude that $eS \neq 0$. Also,
since $MI_e M = MeI(e)eM \subsetneq MeM = I$ by Corollary 1.16, it follows that
$MI_e MS = 0$ by minimality of I. Therefore, $I_e S = 0$. We conclude that e is
an apex for S. Uniqueness up to $\mathscr{J}$-equivalence follows from Proposition 5.4.

If $f \in E(M)$ with $f \in I(e)$, then from $f\,\mathrm{Ind}_{G_e}(V) = 0 = f\,\mathrm{Coind}_{G_e}(V)$
and Corollary 4.17, we conclude that f annihilates each composition factor of
$\mathrm{Ind}_{G_e}(V)$ and $\mathrm{Coind}_{G_e}(V)$ and hence none of their composition factors have
apex f. Also, from $e\,\mathrm{Ind}_{G_e}(V) = V = e\,\mathrm{Coind}_{G_e}(V)$ and Corollary 4.17, we
deduce that $V^{\sharp}$ is the only composition factor of either module with apex e
and it appears with multiplicity one in both.   □

As a corollary, we obtain the following parametrization of the irreducible
representations of M.

Corollary 5.6. *Let $e_1, \ldots, e_s$ be a complete set of idempotent representatives*
of the regular $\mathscr{J}$*-classes of M. Then there is a bijection between* $\mathrm{Irr}_{\Bbbk}(M)$ *and*
the disjoint union $\bigcup_{i=1}^{s} \mathrm{Irr}_{\Bbbk}(G_{e_i})$.

Proof. The corollary follows from Theorem 5.5 and the observation that each
simple $\Bbbk M$-module V has a unique apex amongst $e_1, \ldots, e_s$.   □

As an example, we compute the irreducible representations of an $\mathscr{R}$-trivial
monoid.

Corollary 5.7. *Let M be an* $\mathscr{R}$*-trivial monoid. The simple* $\Bbbk M$*-modules are*
in bijection with regular $\mathscr{J}$*-classes of M. More precisely, for each regular* $\mathscr{J}$*-*
class J_e with $e \in E(M)$, there is a one-dimensional simple $\Bbbk M$*-module S_{J_e}*
with corresponding representation $\chi_{J_e} : M \longrightarrow \Bbbk$ *given by*

$$\chi_{J_e}(m) = \begin{cases} 1, & \text{if } J_e \subseteq MmM \\ 0, & m \in I(e). \end{cases}$$

Proof. Since M is $\mathscr{R}$-trivial, it is immediate that $G_e = \{e\} = R_e$ for all $e \in E(M)$. Therefore, if $\Bbbk$ is the unique simple $\Bbbk G_e$-module and $S_{J_e} = \mathrm{Coind}_{G_e}(\Bbbk)$, then $S_{J_e} = \mathrm{Hom}_{G_e}(\Bbbk e, \Bbbk) \cong \Bbbk$ as a $\Bbbk$-vector space via the map $\varphi \mapsto \varphi(e)$. Trivially, S_{J_e} is a simple module, being one-dimensional. It is annihilated by $I(e)$ and if $m \notin I(e)$, then $(m\varphi)(e) = \varphi(em) = \varphi(e)$ because $em = e$ for all $m \notin I(e)$ by Corollary 2.7. Thus $m\varphi = \varphi$ and so the elements of $M \setminus I(e)$ act trivially on S_{J_e}. The result then follows from Theorem 5.5. $\square$

Example 5.8. Let P be a finite lattice. Then P is a $\mathscr{J}$-trivial monoid whose regular $\mathscr{J}$-classes are precisely the singletons of P. Hence, according to Corollary 5.7, all of the simple $\Bbbk P$-modules are one-dimensional. More precisely, for each $p \in P$, there is a degree one representation $\chi_p \colon P \longrightarrow \Bbbk$ given by

$$\chi_p(q) = \begin{cases} 1, & \text{if } q \geq p \\ 0, & \text{else} \end{cases}$$

for $q \in P$.

The simple modules for a number of $\mathscr{R}$-trivial monoids will be studied in detail in Chapter 14 in the context of analyzing Markov chains.

5.3 The irreducible representations of the full transformation monoid

As an example of the abstract theory we have just developed, we construct the irreducible representations of the full transformation monoid T_n of degree n. We recall that T_n is the monoid of all self-maps of the set $\{1, \ldots, n\}$. The representation theory of T_n has a very long history, beginning with the work of Hewitt and Zuckerman [HZ57]. In this section, we give explicit constructions of the simple $\Bbbk T_n$-modules over a field $\Bbbk$ of characteristic zero. In fact, the simple modules are all defined over $\mathbb{Q}$, as is the case with the symmetric group. These results are due to Putcha [Put96, Theorem 2.1], except that he did not make explicit the connection with exterior powers of which he only became aware later on (private communication). The proofs given here are module theoretic and are a significant simplification of Putcha's approach. We also give a construction of one of the families of simple modules in terms of polytabloids, similar to Grood's approach for the symmetric inverse monoid [Gro02], following a suggestion of Darij Grinberg. Let us fix a field $\Bbbk$ of characteristic zero for the entire section.

5.3.1 Construction of the simple modules

For $r \geq 0$, we put $[r] = \{1, \ldots, r\}$. The reader should consult Section B.4 for background and terminology concerning partitions and the representation theory of the symmetric group, as we shall use the notation from there. Let us quickly recall that $\lambda = (\lambda_1, \ldots, \lambda_m)$ is a *partition* of r if $\lambda_1 \geq \cdots \geq \lambda_m \geq 1$ and $\lambda_1 + \cdots + \lambda_m = r$. We put $|\lambda| = r$ to indicate that λ is a partition of r as we shall be dealing simultaneously with partitions of several integers. If λ is a partition, then S_λ will denote the Specht module associated with λ and c_λ will denote the corresponding Young symmetrizer.

Let $e_r \in T_n$, for $1 \leq r \leq n$, be the idempotent given by

$$e_r(i) = \begin{cases} i, & \text{if } i \leq r \\ 1, & \text{if } i > r \end{cases}$$

and note that $e_1, \ldots, e_n$ form a complete set of idempotent representatives of the $\mathscr{J}$-classes of T_n. Moreover, $e_r T_n e_r \cong T_r$ and hence $G_{e_r} \cong S_r$. Indeed, $e_r T_n e_r$ leaves $[r]$ invariant and the map $f \mapsto f|_{[r]}$ is an isomorphism of $e_r T_n e_r$ with T_r, as the reader can easily check. We will from now on identify G_{e_r} with S_r when convenient.

We begin by describing the simple modules $S_{(1^r)}^\sharp$ for $1 \leq r \leq n$ where we retain the notation of Theorem 5.5. First observe that $\Bbbk^n$ is a $\Bbbk T_n$-module, which we term the *natural module*, in the following way. Let $\{v_1, \ldots, v_n\}$ be the standard basis for $\Bbbk^n$ and define $f v_i = v_{f(i)}$ for $f \in T_n$. Notice that the action of f on a vector preserves the sum of its coefficients. Hence if we put

$$\mathrm{Aug}(\Bbbk^n) = \{(x_1, \ldots, x_n) \in \Bbbk^n \mid x_1 + \cdots + x_n = 0\}$$

then $\mathrm{Aug}(\Bbbk^n)$ is a $\Bbbk T_n$-submodule of $\Bbbk^n$ called the *augmentation submodule*. In fact, $\mathrm{Aug}(\Bbbk^n)$ is the kernel of the $\Bbbk T_n$-module homomorphism $\theta\colon \Bbbk^n \longrightarrow \Bbbk$ sending v_i to 1, for all $1 \leq i \leq n$, where $\Bbbk$ is given the trivial $\Bbbk T_n$-module structure. Note that $\dim \mathrm{Aug}(\Bbbk^n) = n - 1$.

Theorem 5.9. *Let $\Bbbk$ be a field of characteristic 0, $n \geq 1$ and $V = \mathrm{Aug}(\Bbbk^n)$. Then, for $1 \leq r \leq n$, the exterior power $\Lambda^{r-1}(V)$ is a simple $\Bbbk T_n$-module of dimension $\binom{n-1}{r-1}$ with apex e_r and with $e_r \Lambda^{r-1}(V)$ the sign representation $S_{(1^r)}$ of $G_{e_r} \cong S_r$, that is, $S_{(1^r)}^\sharp \cong \Lambda^{r-1}(V)$. More generally, if $r \leq m \leq n$, then $e_m \Lambda^{r-1}(V) \cong \Lambda^{r-1}(e_m V)$ is the simple $\Bbbk S_m$-module $S_{(m-r+1, 1^{r-1})}$.*

Proof. If $1 \leq m \leq n$, then clearly $e_m \Bbbk^n \cong \Bbbk^m$ and $e_m \mathrm{Aug}(\Bbbk^n) \cong \mathrm{Aug}(\Bbbk^m)$ under the identification of $e_m \Bbbk T_n e_m = \Bbbk[e_m T_n e_m]$ with $\Bbbk T_m$ induced by restricting an element of $e_m T_n e_m$ to $[m]$. In particular, as $e_m \Lambda^{r-1}(V) \cong \Lambda^{r-1}(e_m V)$ and $\dim e_m V = \dim \mathrm{Aug}(\Bbbk^m) = m - 1$, we conclude that $e_m \Lambda^{r-1}(V) = 0$ if $m < r$.

Assume now that $m \geq r$. By [FH91, Proposition 3.12], the exterior power $\Lambda^{r-1}(e_m V) \cong \Lambda^{r-1}(\mathrm{Aug}(\Bbbk^m))$ is a simple $\Bbbk S_m$-module of degree $\binom{m-1}{r-1}$ and by [FH91, Exercise 4.6] it is the Specht module $S_{(m-r+1,1^{r-1})}$. In particular, we have that $e_r \Lambda^{r-1}(V) \cong S_{(1^r)}$. It follows that $W = \Lambda^{r-1}(V)$ is a simple $\Bbbk T_n$-module with apex e_r and that $W \cong S_{(1^r)}^\sharp$. $\square$

To construct the remaining simple $\Bbbk T_n$-modules, we introduce some notation. For $1 \leq r \leq n$, let $T_{n,r}$ denote the set of all mappings $f \colon [r] \longrightarrow [n]$. Notice that T_n acts on the left of $T_{n,r}$ by post-composition and S_r acts on the right by precomposition. Moreover, these two actions commute and so $\Bbbk T_{n,r}$ is a $\Bbbk T_n$-$\Bbbk S_r$-bimodule. Let $I_{n,r} \subseteq T_{n,r}$ be the subset of injective mappings and let $L_{n,r} = T_{n,r} \setminus I_{n,r}$. Then $\Bbbk L_{n,r}$ is a sub-bimodule and so $\Bbbk T_{n,r}/\Bbbk L_{n,r}$ is a $\Bbbk T_n$-$\Bbbk S_r$-bimodule which can be identified as a $\Bbbk$-vector space with $\Bbbk I_{n,r}$. Under this identification, the right action of S_r is still via precomposition, but the left action of T_n is now given by

$$
f \odot g = \begin{cases} fg, & \text{if } fg \in I_{n,r} \\ 0, & \text{else} \end{cases}
$$

for $f \in T_n$ and $g \in I_{n,r}$. We drop from now on the notation "$\odot$."

The action of S_r on the right of $I_{n,r}$ is free because if $f \colon [r] \longrightarrow [n]$ is injective and $g \in S_r$, then $fg = f = f1_{[r]}$ implies $g = 1_{[r]}$. It is easy to see that two mappings are in the same S_r-orbit if and only if they have the same image. If $Y \subseteq [n]$ has cardinality r, then let $h_Y \colon [r] \longrightarrow Y$ be the unique order-preserving bijection. Viewing h_Y as an element of $I_{n,r}$, we have that the h_Y form a complete set of representatives of the S_r-orbits on $I_{n,r}$. Thus $\Bbbk I_{n,r}$ is a free right $\Bbbk S_r$-module with basis consisting of the h_Y with $Y \subseteq [n]$ and $|Y| = r$. Note that $h_{[r]} = 1_{[r]}$.

Theorem 5.10. *Let $\Bbbk$ be a field of characteristic 0 and λ a partition of r with $1 \leq r \leq n$ and $\lambda \neq (1^r)$. Then there is an isomorphism of $\Bbbk T_n$-modules*

$$
S_\lambda^\sharp \cong \Bbbk I_{n,r} \otimes_{\Bbbk S_r} S_\lambda \cong \Bbbk I_{n,r} c_\lambda
$$

where $I_{n,r}$ is the set of injective mappings from $[r]$ to $[n]$. Moreover,

$$
\dim S_\lambda^\sharp = \binom{n}{r} f_\lambda
$$

where $f_\lambda = \dim S_\lambda$ is the number of standard Young tableaux of shape λ.

Proof. Put $W_\lambda = \Bbbk I_{n,r} \otimes_{\Bbbk S_r} S_\lambda$. Note that $W_\lambda = \Bbbk I_{n,r} \otimes_{\Bbbk S_r} \Bbbk S_r c_\lambda \cong \Bbbk I_{n,r} c_\lambda$. Observe that since $\Bbbk I_{n,r}$ is a free right $\Bbbk S_r$-module with basis consisting of the h_Y with $Y \subseteq [n]$ and $|Y| = r$, we have the $\Bbbk$-vector space decomposition

$$
W_\lambda = \bigoplus_{Y \subseteq [n], |Y| = r} h_Y \otimes S_\lambda
$$

and so dim $W_\lambda = \binom{n}{r} f_\lambda$. The action of $f \in T_n$ on $h_Y \otimes v_Y$ is given by

$$f(h_Y \otimes v_Y) = \begin{cases} fh_Y \otimes v_Y, & \text{if } |f(Y)| = r \\ 0, & \text{else} \end{cases}$$

$$= \begin{cases} h_{f(Y)} \otimes (h_{f(Y)}^{-1} f h_Y) v_Y, & \text{if } |f(Y)| = r \\ 0, & \text{else.} \end{cases} \tag{5.1}$$

In particular, we have that W_λ is annihilated by all mappings of rank less than r and $e_r W_\lambda = 1_{[r]} \otimes S_\lambda \cong S_\lambda$ as a kS_r-module (recall that $h_{[r]} = 1_{[r]}$). Thus to prove the theorem, it suffices to show that W_λ is simple.

We prove the theorem by induction on the quantity $n - r$ (where we do not hold n fixed). If $n - r = 0$, then $I_{n,r} = S_n$ and $W_\lambda = S_\lambda$ made into a $\mathbb{C}T_n$-module by extending the S_n-action to $T_n \setminus S_n$ by having the non-permutations annihilate S_λ. Thus W_λ is simple in this case.

Assume that the result is true when $n - r = k$ and suppose that $n - r = k + 1$. If we view $I_{n-1,r}$ as a subset of $I_{n,r}$ in the natural way, then we have $e_{n-1} k I_{n,r} = k I_{n-1,r}$. Moreover, if we identify $e_{n-1} T_n e_{n-1}$ with T_{n-1} via the restriction to $[n-1]$, then $e_{n-1} k I_{n,r} \cong k I_{n-1,r}$ as a $k T_{n-1}$-$k S_r$-bimodule. Therefore, as $k T_{n-1}$-modules (under our identification), we have by Proposition 4.4 that

$$e_{n-1} W_\lambda \cong (e_{n-1} k I_{n,r}) \otimes_{kS_r} S_\lambda \cong k I_{n-1,r} \otimes_{kS_r} S_\lambda.$$

As $n - 1 - r = k$, it follows that $e_{n-1} W_\lambda$ is a simple $e_{n-1} k T_n e_{n-1}$-module by induction. Note that $1_{[r]} \otimes S_\lambda \subseteq e_{n-1} W_\lambda$.

For each $Y \subseteq [n]$ with $|Y| = r$, let $g_Y \in S_n$ be the unique permutation whose restriction to $[r]$ is h_Y and whose restriction to $[n] \setminus [r]$ is the unique order-preserving bijection $[n] \setminus [r] \longrightarrow [n] \setminus Y$. Then $g_Y(1_{[r]} \otimes S_\lambda) = h_Y \otimes S_\lambda$ from which we deduce that $k T_n e_{n-1} W_\lambda \supseteq k T_n(1_{[r]} \otimes S_\lambda) = W_\lambda$ and hence $k T_n e_{n-1} W_\lambda = W_\lambda$.

Let $v \in W_\lambda$ be nonzero. We must show that $k T_n v = W_\lambda$. Write

$$v = \sum_{Y \subseteq [n], |Y| = r} h_Y \otimes v_Y.$$

Suppose that $v_{Y'} \neq 0$. Then replacing v by $g_{Y'}^{-1} v$, which is nonzero because $g_{Y'}$ is invertible in T_n, we may assume that $v_{[r]} \neq 0$.

For $1 \leq i < j \leq n$, let $\eta_{i,j} \in T_n$ be the rank $n - 1$ idempotent defined by

$$\eta_{i,j}(x) = \begin{cases} x, & \text{if } x \neq j \\ i, & \text{if } x = j. \end{cases}$$

The image of $\eta_{i,j}$ is $[n] \setminus \{j\}$. Note that $\eta_{1,n} = e_{n-1}$ and that if $\sigma_{i,j} = (1\ i)(n\ j)$, then $\eta_{i,j} = \sigma_{i,j} e_{n-1} \sigma_{i,j}$ (where we interpret $(1\ 1)$ and $(n\ n)$ as the identity).

Suppose that $\eta_{i,j}v \neq 0$ for some $1 \leq i < j \leq n$. If we put $w = \sigma_{i,j}\eta_{i,j}v$, then $w = e_{n-1}\sigma_{i,j}v \in e_{n-1}W_\lambda$ is nonzero. Therefore, $e_{n-1}\Bbbk T_n e_{n-1}w = e_{n-1}W_\lambda$ by simplicity of $e_{n-1}W_\lambda$ over $e_{n-1}\Bbbk T_n e_{n-1}$. We conclude that $W_\lambda = \Bbbk T_n e_{n-1}W_\lambda \subseteq \Bbbk T_n w \subseteq \Bbbk T_n v$, as required.

So we are left with the case that $\eta_{i,j}v = 0$ for all $1 \leq i < j \leq n$. We will derive a contradiction to the assumption $\lambda \neq (1^r)$. Let $Y \subseteq [n]$ have cardinality r. Then the following computation is straightforward from (5.1):

$$\eta_{i,j}(h_Y \otimes v_Y) = \begin{cases} h_Y \otimes v_Y, & \text{if } j \notin Y \\ h_{(Y \cup \{i\}) \setminus \{j\}} \otimes (h_{(Y \cup \{i\}) \setminus \{j\}}^{-1} \eta_{i,j} h_Y) v_Y, & \text{if } j \in Y,\ i \notin Y \\ 0, & \text{if } \{i,j\} \subseteq Y. \end{cases}$$
(5.2)

Let us put $Y_m = [r+1] \setminus \{m\}$ for $1 \leq m \leq r+1$ and put

$$W_m = \bigoplus_{Y \neq Y_m} h_Y \otimes S_\lambda.$$

Note that $Y_{r+1} = [r]$. Let $1 \leq i < j \leq r+1$. Then it follows from (5.2) that

$$\eta_{i,j}(h_{Y_j} \otimes v_{Y_j}) = h_{Y_j} \otimes v_{Y_j}$$
$$\eta_{i,j}(h_{Y_i} \otimes v_{Y_i}) = h_{Y_j} \otimes (i\ i+1\ \cdots\ j-1)v_{Y_i}$$
$$\eta_{i,j}(h_Y \otimes v_Y) \in W_j, \text{ if } Y \notin \{Y_i, Y_j\}$$
(5.3)

since $\eta_{i,j}h_{Y_i}$ and $h_{Y_j}(i\ i+1\ \cdots\ j-1)$ are both the map $h\colon [r] \longrightarrow [n]$ given by

$$h(x) = \begin{cases} x, & \text{if } 1 \leq x < i \\ x+1, & \text{if } i \leq x < j-1 \text{ or } j \leq x \leq r \\ i, & \text{if } x = j-1. \end{cases}$$

In particular, if $2 \leq j \leq r+1$, then we have that

$$0 = \eta_{j-1,j}v = h_{Y_j} \otimes (v_{Y_{j-1}} + v_{Y_j}) + w$$

with $w \in W_j$ by (5.3). It follows that $v_{Y_{j-1}} = -v_{Y_j}$ and hence

$$v_{Y_m} = (-1)^{r+1-m}v_{Y_{r+1}} = (-1)^{r+1-m}v_{[r]}$$
(5.4)

for $1 \leq m \leq r+1$ because $Y_{r+1} = [r]$.

Next we compute that, for $1 \leq i \leq r-1$,

$$0 = \eta_{i,r+1}v = h_{Y_{r+1}} \otimes ((i\ i+1\ \cdots\ r)v_{Y_i} + v_{Y_{r+1}}) + w$$

with $w \in W_{r+1}$ by (5.3). Therefore, by (5.4), we have that

$$0 = (i\ i+1\ \cdots\ r)(-1)^{r+1-i}v_{[r]} + v_{[r]}$$

and so

$$(i \; i+1 \; \cdots \; r)v_{[r]} = (-1)^{r-i}v_{[r]} = \mathrm{sgn}((i \; i+1 \; \cdots \; r))v_{[r]} \qquad (5.5)$$

for all $1 \le i \le r-1$. Since $(r-1 \; r)$ and $(1 \; 2 \; \cdots \; r)$ generate S_r and $v_{[r]} \ne 0$, we conclude from (5.5) that $\Bbbk S_r v_{[r]} \cong S_{(1^r)}$ contradicting that $\lambda \ne (1^r)$. This contradiction completes the proof. $\square$

We next observe that the modules constructed above are induced modules.

Corollary 5.11. *Let $\Bbbk$ be a field of characteristic 0 and λ a partition of r with $\lambda \ne (1^r)$. Then $\mathrm{Ind}_{G_{e_r}}(S_\lambda) \cong S_\lambda^\sharp$ is a simple $\Bbbk T_n$-module.*

Proof. Note that L_{e_r} consists of all mappings $f \in T_n$ with $f|_{[r]}$ injective and $f(k) = f(1)$ for $r+1 \le k \le n$. Therefore, the bijection $\alpha \colon L_{e_r} \mapsto I_{n,r}$ given by $\alpha(f) = f|_{[r]}$ induces a $\Bbbk T_n$-$\Bbbk S_r$-bimodule isomorphism of $\Bbbk L_{e_r}$ and $\Bbbk I_{n,r}$, where we identify G_{e_r} with S_r via the restriction map. It follows that $\mathrm{Ind}\, G_{e_r}(S_\lambda) \cong \Bbbk I_{n,r} \otimes_{\Bbbk S_r} S_\lambda \cong S_\lambda^\sharp$ is simple by Theorem 5.10. $\square$

Let M be a monoid. If $\mathbb{F}$ is a subfield of $\Bbbk$, then we say that a $\Bbbk M$-module V is *defined over* $\mathbb{F}$ if there is an $\mathbb{F}M$-module W such that $V \cong \Bbbk \otimes_{\mathbb{F}} W$. Equivalently, V is defined over $\mathbb{F}$ if there is a basis for V such that the matrix representation $\rho \colon M \longrightarrow M_n(\Bbbk)$ afforded by V with respect to this basis takes values in $M_n(\mathbb{F})$.

Corollary 5.12. *Let $\Bbbk$ be a field of characteristic zero. Then each simple $\Bbbk T_n$-module is defined over $\mathbb{Q}$.*

Proof. Clearly, $\Lambda^r(\mathrm{Aug}(\Bbbk^n)) = \Bbbk \otimes_{\mathbb{Q}} \Lambda^r(\mathrm{Aug}(\mathbb{Q}^n))$ for $0 \le r \le n-1$. Since $c_\lambda \in \mathbb{Q}S_r$ for λ a partition of r, it follows that $\Bbbk I_{n,r}c_\lambda \cong \Bbbk \otimes_{\mathbb{Q}} \mathbb{Q}I_{n,r}c_\lambda$. We conclude that the simple $\Bbbk T_n$-modules are defined over $\mathbb{Q}$ by Theorems 5.9 and 5.10. $\square$

The following corollary computes the restriction of $e_m S_\lambda^\sharp$ to $\Bbbk G_{e_m}$ for the case $m \ge r$ and $\lambda \ne (1^r)$. It will also be used when we compute the character table of the symmetric inverse monoid. The reader is referred to Section B.4 for the definition of the outer product $V \boxtimes W$ of symmetric group representations.

Corollary 5.13. *Let $\Bbbk$ be a field of characteristic 0 and $n \ge 1$. Let λ be a partition of r with $0 \le r \le n$. Then $\Bbbk I_{n,r} \otimes_{\Bbbk S_r} S_\lambda \cong S_\lambda \boxtimes S_{(n-r)}$ as a $\Bbbk S_n$-module. Consequently, if $1 \le r \le m \le n$ and $\lambda \ne (1^r)$ is a partition of r, then $e_m S_\lambda^\sharp \cong S_\lambda \boxtimes S_{(m-r)}$ as a $\Bbbk S_m$-module.*

Proof. The case $r = 0$ is trivial as $\Bbbk I_{n,0} \cong S_{(n)}$ is the trivial $\Bbbk S_n$-module and S_λ is the trivial $\Bbbk S_0$-module. So assume that $r \ge 1$. We retain the notation h_Y and g_Y from the proof of Theorem 5.10 for an r-element subset Y of $[n]$.

The group S_n acts transitively on the set A of r-element subsets of $[n]$ and $S_r \times S_{n-r}$ is the stabilizer of $[r]$. It follows that the g_Y with $Y \in A$ form a

1	5	7
6	8	
9		

3	1	9
2	7	
4		

1	2	4
5	6	
7		

Fig. 5.1. Some 9-tableaux of shape $(3, 2, 1)$

set of left coset representatives of $S_r \times S_{n-r}$ in S_n. Thus $\Bbbk S_n$ is a free right $\Bbbk[S_r \times S_{n-r}]$-module with basis the g_Y and hence there is a $\Bbbk$-vector space decomposition

$$S_\lambda \boxtimes S_{(n-r)} = \bigoplus_{Y \in A} g_Y \otimes S_\lambda \otimes S_{(n-r)}.$$

Moreover, since $S_{(n-r)}$ is the trivial module, the S_n-action is given by

$$f(g_Y \otimes v \otimes 1) = f g_Y \otimes v \otimes 1$$
$$= g_{f(Y)} \otimes (g_{f(Y)}^{-1} f g_Y)(v \otimes 1)$$
$$= g_{f(Y)} \otimes (h_{f(Y)}^{-1} f h_Y) v \otimes 1$$

for $f \in S_n$. Comparing with (5.1), we conclude that $g_Y \otimes v \otimes 1 \mapsto h_Y \otimes v$, for $Y \in A$, provides a $\Bbbk S_n$-module isomorphism of $S_\lambda \boxtimes S_{(n-r)}$ with the restriction of $\Bbbk I_{n,r} \otimes_{\Bbbk S_r} S_\lambda$ to $\Bbbk S_n$. This completes the proof.

The final statement follows from Theorem 5.10 and the previous case via the observation $e_m \Bbbk I_{n,r} = \Bbbk I_{m,r}$, where we view $I_{m,r}$ as a subset of $I_{n,r}$, and by Proposition 4.4.   □

5.3.2 An approach via polytabloids

We now provide an alternative description of the simple $\Bbbk T_n$-module $S_\lambda^\sharp$, for $\lambda \neq (1^r)$, using polytabloids. This is analogous to the description of Specht modules in terms of polytabloids (cf. Section B.4) and a description by Grood [Gro02] of the irreducible representations of the symmetric inverse monoid. Our definition of polytabloids follows that of Grood [Gro02].

Let $1 \leq r \leq n$ and let λ be a partition of r. By an n-*tableau t of shape* λ, we mean a filling of the Young diagram of shape λ by r distinct integers from $[n]$. See Figure 5.1 for some examples. If t is an n-tableau of shape λ, then the *content* of t is the set $c(t)$ of entries of t. For example, if t is the first tableau in Figure 5.1, then $c(t) = \{1, 5, 6, 7, 8, 9\}$.

A *standard n-tableau* of shape λ is one whose entries are increasing along each row and column. For example, the first and last tableaux in Figure 5.1 are standard. The number of standard n-tableaux of shape λ is $\binom{n}{r} f_\lambda$ where f_λ is the number of standard Young tableaux of shape λ (in the usual sense).

If t is an n-tableau of shape λ and $f \in T_n$ is injective on $c(t)$, then we define ft to be the n-tableau of shape λ obtained from t by applying f to the entry of each of its boxes. For example, if t is the first n-tableau in Figure 5.1 and

$$f = \begin{pmatrix} 1 & 2 & 3 & 4 & 5 & 6 & 7 & 8 & 9 \\ 3 & 3 & 4 & 2 & 1 & 2 & 9 & 7 & 4 \end{pmatrix},$$

then ft is the second tableau in Figure 5.1.

If t, t' are n-tableaux of shape λ, we write $t \sim t'$ if they have the same entries in each row. The equivalence class of t is denoted by $[t]$ and is called an *n-tabloid of shape* λ. Notice that all tableaux in a tabloid have the same content and so $c([t]) = c(t)$ is well defined. Also, if $t \sim t'$ and $f \in T_n$ is injective on $c(t)$, then $ft \sim ft'$ and so we can define $f[t] = [ft]$.

Let $T^{\lambda,n}$ denote the set of n-tabloids of shape λ. Then there is a $\Bbbk T_n$-module structure on $M_{\lambda,n} = \Bbbk T^{\lambda,n}$ given by putting

$$f \cdot [t] = \begin{cases} [ft], & \text{if } f|_{c([t])} \text{ is injective} \\ 0, & \text{else} \end{cases}$$

for $f \in T_n$ and $[t] \in T^{\lambda,n}$. The reader is invited to check that $M_{\lambda,n} \cong \Bbbk I_{n,r} \otimes_{\Bbbk S_r} M_\lambda$ as a $\Bbbk T_n$-module where M_λ is the permutation module for $\Bbbk S_r$ constructed in Section B.4.

The following notation will be convenient. If $X \subseteq [n]$ and $f \in T_n$ with $f|_X$ injective, then $\widehat{f}_X$ will denote a fixed permutation in S_n that agrees with f on X. It will turn out that the choice of which permutation to take for $\widehat{f}_X$ is irrelevant for our purposes. The next lemma is then immediate from the definitions.

Lemma 5.14. *Let t be an n-tableau of shape λ with content X and $f \in T_n$ such that $f|_X$ is injective. Then $ft = \widehat{f}_X t$ and hence $f[t] = \widehat{f}_X[t]$.*

If t is an n-tableau of shape λ, we define C_t to be the set of all permutations in S_n that preserve the columns of t and fix $[n] \setminus c(t)$ pointwise. In other words, if $X_1, \ldots, X_s$ are the sets of entries of the s columns of t, then $C_t = S_{X_1} \times S_{X_2} \times \cdots \times S_{X_s}$, embedded in S_n in the usual way.

Lemma 5.15. *Let t be an n-tableau of shape λ and $g \in S_n$. Then $C_{gt} = gC_tg^{-1}$.*

Proof. This follows immediately from the observation that if $X_1, \ldots, X_s$ are the sets of entries of the columns of t, then $g(X_1), \ldots, g(X_s)$ are the entries of the columns of gt and that $c(gt) = gc(t)$. $\square$

If t is an n-tableau of shape λ, then the *n-polytabloid of shape* λ associated with t is

$$\varepsilon_t = \sum_{g \in C_t} \mathrm{sgn}(g)g[t] \in M_{\lambda,n}.$$

Notice that if $g \in C_t$, then $c(gt) = c(t)$ by definition. Hence each summand in ε_t has the same content and so we can define the content of a polytabloid by $c(\varepsilon_t) = c(t)$. Let $U_{\lambda,n}$ be the $\Bbbk$-subspace of $M_{\lambda,n}$ spanned by the n-polytabloids of shape λ. Then $U_{\lambda,n}$ is the direct sum over all r-element subsets X of $[n]$, where $r = |\lambda|$, of the subspaces $U_{\lambda,X}$ spanned by polytabloids of content X. Notice that $U_{\lambda,[r]} = S_\lambda$ as a vector space. This is the reason for our convention that elements of C_t must fix pointwise $[n] \setminus c(t)$.

If $f \in T_n$ and t is an n-tableau of shape λ such that f is injective on $X = c(t)$, then by Lemma 5.14 we have that $f\varepsilon_t = \widehat{f}_X \varepsilon_t$ as all tabloids appearing as a summand in ε_t have content X. This will allow us to show that $U_{\lambda,n}$ is a $\Bbbk T_n$-submodule.

Proposition 5.16. *Let λ be a partition of r with $1 \le r \le n$.*

(i) Suppose that t is an n-tableau of shape λ and $f \in T_n$. Then

$$f\varepsilon_t = \begin{cases} \varepsilon_{ft}, & \text{if } f|_{c(t)} \text{ is injective} \\ 0, & \text{else.} \end{cases}$$

(ii) $U_{\lambda,n}$ is a $\Bbbk T_n$-submodule.
(iii) If t_λ is the n-tableau of shape λ with j in the jth box, then $\Bbbk S_n \varepsilon_{t_\lambda} = U_{\lambda,n}$.
(iv) $U_{\lambda,n}$ has basis the set of ε_t where t is a standard n-tableau of shape λ and so $\dim U_{\lambda,n} = \binom{n}{r} f_\lambda$.

Proof. For the first item, as f annihilates all n-tabloids of content X whenever $f|_X$ is not injective, the only case to check is when f is injective on $X = c(t)$. In this case, using Lemma 5.14, Lemma 5.15, and our previous observation that $f\varepsilon_t = \widehat{f}_X \varepsilon_t$, we obtain

$$
\begin{aligned}
f\varepsilon_t &= \widehat{f}_X \varepsilon_t \\
&= \sum_{g \in C_t} \operatorname{sgn}(g) \widehat{f}_X g \widehat{f}_X^{-1} \widehat{f}_X [t] \\
&= \sum_{h \in C_{\widehat{f}_X t}} \operatorname{sgn}(\widehat{f}_X^{-1} h \widehat{f}_X) h [\widehat{f}_X t] \\
&= \varepsilon_{\widehat{f}_X t} \\
&= \varepsilon_{ft}
\end{aligned}
$$

as required.

The second item is immediate from the first. For the third item, let t be an n-tableau of shape λ and let $g \in S_n$ be any permutation with $g(j)$ the entry of the jth box of t. Then $gt_\lambda = t$ and so $g\varepsilon_{t_\lambda} = \varepsilon_{gt_\lambda} = \varepsilon_t$.

To prove the last item, observe that $U_{\lambda,[r]} = S^\lambda$ and hence has basis the set of ε_t with t a standard n-tableau of shape λ and content $[r]$. Let X be an r-element subset of $[n]$ and let $g_X \in S_n$ be any permutation extending the unique order-preserving bijection $h_X \colon [r] \longrightarrow X$. Then $U_{\lambda,X}$ is isomorphic to $U_{\lambda,[r]}$ as a vector space via the action of g_X. Moreover, a λ-tableau t (with content $[r]$) is standard if and only if $g_X t$ is standard. Therefore, the polytabloids ε_t with t a standard n-tableau of shape λ and content X form a basis for $U_{\lambda,X}$. As $U_{\lambda,n}$ is the direct sum of the $U_{\lambda,X}$ with X an r-element subset of $[n]$, the result follows. $\qquad\square$

We now prove that $U_{\lambda,n}$ is isomorphic to the simple $\Bbbk T_n$-module $S_\lambda^\sharp$ for $\lambda \ne (1^r)$.

Theorem 5.17. *Let* $\Bbbk$ *be a field of characteristic zero and* λ *a partition of* r *with* $1 \leq r \leq n$ *and* $\lambda \neq (1^r)$. *Then* $U_{\lambda,n}$ *is isomorphic to the simple* $\Bbbk T_n$-*module* $S_\lambda^\sharp$.

Proof. Let $e_r \in E(T_n)$ be defined as earlier in this section. Then clearly $e_r U_{\lambda,n} = U_{\lambda,[r]} \cong S_\lambda$ as a $\Bbbk G_{e_r}$-module (identifying G_{e_r} with S_r in the usual way) by construction. Also, each element of T_n of rank less than r annihilates $U_{\lambda,n}$ and so $U_{\lambda,n}$ is a $\Bbbk T_n/\Bbbk I(e_r)$-module. By Proposition 5.16(iii) we have that $U_{\lambda,n} = \Bbbk S_n \varepsilon_{t_\lambda} \subseteq \Bbbk T_n U_{\lambda,[r]} = \Bbbk T_n e_r U_{\lambda,n} \subseteq U_{\lambda,n}$ and so $U_{\lambda,n} = \Bbbk T_n e_r U_{\lambda,n}$. Therefore, there is a surjective homomorphism $\alpha \colon \mathrm{Ind}_{G_e}(S_\lambda) \longrightarrow U_{\lambda,n}$ of $\Bbbk T_n$-modules by Proposition 4.11 (with $A = \Bbbk T_n/\Bbbk I(e_r)$, $e = e_r$ and $V = U_{\lambda,n}$). Concretely, α maps a basic tensor $f \otimes \varepsilon_t$, with $f \in L_{e_r}$ and t a λ-tableau, to ε_{ft}. As $\mathrm{Ind}_{G_e}(S_\lambda)$ is isomorphic to the simple module $S_\lambda^\sharp$ by Corollary 5.11, α is an isomorphism and the result follows. $\qquad\square$

5.4 Semisimplicity

In this section, we characterize when $\Bbbk M$ is semisimple for a finite monoid M. The approach we pursue here differs from the classical approach in [CP61, Chapter 5] in that it is module theoretic and uses the machinery we have been developing, rather than the theory of semigroup algebras and Munn algebras.

If J_a is the $\mathscr{J}$-class of $a \in M$, then $I = MaM \setminus J_a$ is an ideal of M. Therefore, $\Bbbk[MaM]/\Bbbk I$ is an ideal of $\Bbbk M/\Bbbk I$. As a vector space it is $\Bbbk J_a$. The left $\Bbbk M$-module structure is given by

$$m \odot x = \begin{cases} mx, & \text{if } mx \in J_a \\ 0, & \text{if } mx \notin J_a \end{cases}$$

for $m \in M$ and $x \in J_a$. We omit the symbol "$\odot$" from now on. We shall repeatedly need the following lemma.

Lemma 5.18. *Let* $e \in E(M)$ *and let* n *be the number of* $\mathscr{L}$-*classes in* J_e. *Then* $\Bbbk J_e \cong n \cdot \Bbbk L_e$ *as a left* $\Bbbk M$-*module.*

Proof. By Theorem 1.13, if $f \in E(J_e)$, then $\Bbbk L_f$ is a submodule of $\Bbbk J_e$. Moreover, since J_e is the disjoint union of its $\mathscr{L}$-classes and each $\mathscr{L}$-class contains an idempotent (by Proposition 1.22), it suffices to show that $MeM = MfM$ implies that $\Bbbk L_e \cong \Bbbk L_f$. But $MeM = MfM$ implies that $I(e) = I(f)$ and that Me and Mf are isomorphic M-sets by Theorem 1.11. Moreover, stability (Theorem 1.13) implies that $Me \setminus I(e)e = L_e$ and $Mf \setminus I(f)f = L_f$. Since any isomorphism $Me \longrightarrow Mf$ of M-sets takes the set L_e of generators of Me to the set L_f of generators of Mf, we conclude that $\Bbbk L_e = \Bbbk Me/\Bbbk I(e)e \cong \Bbbk Mf/\Bbbk I(f)f = \Bbbk L_f$, as required. $\qquad\square$

Let $e \in E(M)$ and let W be a $\Bbbk G_e$-module. Then, putting $A_e = \Bbbk M/\Bbbk I(e)$, we have the natural homomorphism of A_e-modules

$$\varphi_W \colon \mathrm{Ind}_{G_e}(W) \longrightarrow \mathrm{Coind}_{G_e}(W)$$

from Corollary 4.12. For $\ell \in L_e$, $w \in W$ and $r \in R_e$, it is given by

$$\varphi_W(\ell \otimes w)(r) = (r \diamond \ell)w$$

where

$$r \diamond \ell = \begin{cases} r\ell, & \text{if } r\ell \in G_e \\ 0, & \text{else} \end{cases}$$

(note that $r\ell \in eMe$).

Theorem 5.19. *Let M be a finite monoid and $\Bbbk$ a field. Then $\Bbbk M$ is semisimple if and only if:*

(a) M is regular;
(b) the characteristic of $\Bbbk$ does not divide the order of G_e for any $e \in E(M)$;
(c) the natural homomorphism $\varphi_{\Bbbk G_e} \colon \mathrm{Ind}_{G_e}(\Bbbk G_e) \longrightarrow \mathrm{Coind}_{G_e}(\Bbbk G_e)$ is an isomorphism for all $e \in E(M)$.

Proof. Let us first assume that $\Bbbk M$ is semisimple. If M is not regular, then there is an ideal I with $I^2 \subsetneq I$ by Corollary 1.25. Then $\Bbbk I/\Bbbk I^2$ is a nonzero nilpotent ideal of $\Bbbk M/\Bbbk I^2$ and hence $\Bbbk M/\Bbbk I^2$ is not semisimple. As quotients of semisimple algebras are semisimple (Proposition A.8), we conclude that $\Bbbk M$ is not semisimple, a contradiction. It follows that M is regular. Let $e \in E(M)$. Then $A_e = \Bbbk M/\Bbbk I(e)$ is semisimple by Proposition A.8 and hence $eA_e e \cong \Bbbk G_e$ is semisimple by Proposition 4.18. Therefore, the characteristic of $\Bbbk$ does not divide $|G_e|$ by Maschke's theorem. Finally, since $\Bbbk G_e$ is a semisimple $eA_e e$-module and $\mathrm{Ind}_{G_e}(\Bbbk G_e)$, $\mathrm{Coind}_{G_e}(\Bbbk G_e)$ are semisimple A_e-modules, we have that $\varphi_{\Bbbk G_e}$ is an isomorphism by Corollary 4.22.

For the converse, Maschke's theorem yields that $\Bbbk G_e$ is semisimple for all $e \in E(M)$. Note that, for $e \in E(M)$, we have

$$\mathrm{Ind}_{G_e}(\Bbbk G_e) = \Bbbk L_e \otimes_{\Bbbk G_e} \Bbbk G_e \cong \Bbbk L_e.$$

Consider a principal series

$$\emptyset = I_0 \subsetneq I_1 \subsetneq \cdots \subsetneq I_s = M$$

for M. Let $J_k = I_k \setminus I_{k-1}$ for $k = 1, \ldots, s$. Then J_k is a regular $\mathscr{J}$-class (cf. Proposition 1.20) and $\Bbbk J_k \cong \Bbbk I_k/\Bbbk I_{k-1}$. Fix $e_k \in E(J_k)$ and let n_k be the number of $\mathscr{L}$-classes in J_k. Then by Lemma 5.18 we have an isomorphism

$$\Bbbk I_k/\Bbbk I_{k-1} \cong \Bbbk J_k \cong n_k \cdot \Bbbk L_{e_k} \tag{5.6}$$

of $\Bbbk M$-modules.

We claim that

$$\Bbbk M/\Bbbk I_{k-1} \cong \Bbbk J_k \oplus \Bbbk M/\Bbbk I_k \qquad (5.7)$$

as a $\Bbbk M$-module for $1 \leq k \leq s$. Indeed, let $A = \Bbbk M/\Bbbk I_{k-1}$; it is a finite dimensional $\Bbbk$-algebra. Then we have that $e_k A e_k \cong \Bbbk G_{e_k}$ (by Corollary 1.16), $\mathrm{Ind}_{e_k}(\Bbbk G_{e_k}) = \mathrm{Ind}_{G_{e_k}}(\Bbbk G_{e_k})$ and $\mathrm{Coind}_{e_k}(\Bbbk G_{e_k}) = \mathrm{Coind}_{G_{e_k}}(\Bbbk G_{e_k})$ (as $A e_k \cong \Bbbk L_{e_k}$ and $e_k A \cong \Bbbk R_{e_k}$ by Theorem 1.13). Because $\Bbbk G_{e_k}$ is semisimple, all of its modules are injective. Therefore, $\mathrm{Coind}_{G_{e_k}}(\Bbbk G_{e_k})$ is an injective A-module by Proposition 4.3. Since $\varphi_{\Bbbk G_{e_k}}$ is an isomorphism we conclude that $\mathrm{Ind}_{G_{e_k}}(\Bbbk G_{e_k}) \cong \Bbbk L_{e_k}$ is an injective A-module. From (5.6) we then conclude that $\Bbbk J_k$ is an injective A-module. Thus the exact sequence of A-modules

$$0 \longrightarrow \Bbbk J_k \longrightarrow \Bbbk A \longrightarrow \Bbbk M/\Bbbk I_k \longrightarrow 0$$

splits, establishing (5.7).

Applying (5.6) and (5.7) repeatedly, we obtain

$$\Bbbk M \cong \bigoplus_{k=1}^{s} \Bbbk J_k \cong \bigoplus_{k=1}^{s} n_k \cdot \Bbbk L_{e_k}. \qquad (5.8)$$

But $\Bbbk L_e = \mathrm{Ind}_{G_e}(\Bbbk G_e)$ is a semisimple $\Bbbk M/\Bbbk I(e)$-module, for $e \in E(M)$, by Corollary 4.22 because $\Bbbk G_e$ is semisimple and $\varphi_{\Bbbk G_e}$ is an isomorphism. Therefore, $\Bbbk L_e$ is a semisimple $\Bbbk M$-module and hence $\Bbbk M$ is a semisimple $\Bbbk M$-module by (5.8). We conclude that $\Bbbk M$ is a semisimple algebra, thereby completing the proof. $\qquad \square$

In the case that $\Bbbk M$ is semisimple, if V is a simple $\Bbbk G_e$-module, then $V^{\sharp} = \mathrm{Ind}_{G_e}(V)/\mathrm{rad}(\mathrm{Ind}_{G_e}(V)) = \mathrm{Ind}_{G_e}(V)$ and similarly, $V^{\sharp} = \mathrm{soc}(\mathrm{Coind}_{G_e}(V)) = \mathrm{Coind}_{G_e}(V)$.

Note that Exercise 4.3 shows that $\varphi_{\Bbbk G_e}$ is an isomorphism if and only if $\Bbbk L_e \cong \mathrm{Ind}_{G_e}(\Bbbk G_e) \cong \mathrm{Coind}_{G_e}(\Bbbk G_e) = \mathrm{Hom}_{G_e}(R_e, \Bbbk G_e)$. In Chapter 16 a homological proof of Theorem 5.19 will be given.

Classically, Theorem 5.19 is stated in the language of sandwich matrices. Let $e \in E(M)$. Then, by Proposition 1.10, L_e is a free right G_e-set and R_e is a free left G_e-set. Let $\lambda_1, \ldots, \lambda_\ell$ be a complete set of representatives for L_e/G_e and $\rho_1, \ldots, \rho_r$ be a complete set of representatives for $G_e \backslash R_e$. Note that ℓ is the number of $\mathscr{R}$-classes contained in J_e and r is the number of $\mathscr{L}$-classes contained in J_e by Corollary 1.17 and Exercise 1.18. As before, for $\lambda \in L_e$ and $\rho \in R_e$, put

$$\rho \diamond \lambda = \begin{cases} \rho\lambda, & \text{if } \rho\lambda \in G_e \\ 0, & \rho\lambda \in I_e. \end{cases}$$

Let $P(e)$ be the $r \times \ell$ matrix over $\Bbbk G_e$ with

$$P(e)_{ij} = \rho_i \diamond \lambda_j \in G_e \cup \{0\}.$$

One calls $P(e)$ a *sandwich matrix* for the $\mathscr{J}$-class J_e. One can prove that it depends on the choices that we have made only up to left and right multiplication by monomial matrices over G_e (cf. [CP61, KRT68]).

Observe that $\Bbbk L_e = \mathrm{Ind}_{G_e}(\Bbbk G_e)$ is a free right $\Bbbk G_e$-module with basis $\{\lambda_1, \ldots, \lambda_\ell\}$ via the right action of G_e on L_e. Also $\mathrm{Coind}_{G_e}(\Bbbk G_e) = \mathrm{Hom}_{G_e}(R_e, \Bbbk G_e)$ is a right $\Bbbk G_e$-module via $(\psi g)(x) = \psi(x)g$ for $g \in G_e$. Moreover, since R_e is a free left G_e-set with orbit representatives $\rho_1, \ldots, \rho_r$, it follows that $\mathrm{Coind}_{G_e}(\Bbbk G_e)$ is a free right $\Bbbk G_e$-module with basis $\{\rho_1^*, \ldots, \rho_r^*\}$ where

$$\rho_i^*(g\rho_j) = \begin{cases} g, & \text{if } i = j \\ 0, & \text{else} \end{cases}$$

for $g \in G_e$.

Lemma 5.20. *Let* $\varphi_{\Bbbk G_e} \colon \mathrm{Ind}_{G_e}(\Bbbk G_e) \longrightarrow \mathrm{Coind}_{G_e}(\Bbbk G_e)$ *be the natural homomorphism. Then the sandwich matrix* $P(e)$ *is the matrix for* $\varphi_{\Bbbk G_e}$*, as a homomorphism of free right* $\Bbbk G_e$*-modules, with respect to the bases* $\{\lambda_1, \ldots, \lambda_\ell\}$ *for* $\Bbbk L_e = \mathrm{Ind}_{G_e}(\Bbbk G_e)$ *and* $\{\rho_1^*, \ldots, \rho_r^*\}$ *for* $\mathrm{Hom}_{G_e}(R_e, \Bbbk G_e) = \mathrm{Coind}_{G_e}(\Bbbk G_e)$.

Proof. We compute

$$\varphi_{\Bbbk G_e}(\ell_j)(g\rho_k) = (g\rho_k) \diamond \ell_j = g(\rho_k \diamond \ell_j) = gP(e)_{kj} = \sum_{i=1}^{r} \rho_i^*(g\rho_k)P(e)_{ij}$$

for $g \in G$. Therefore, we conclude that

$$\varphi_{\Bbbk G_e}(\ell_j) = \sum_{i=1}^{r} \rho_i^* \cdot P(e)_{ij},$$

as was required. $\square$

Lemma 5.20 allows us to reformulate Theorem 5.19 in its classical form.

Theorem 5.21. *Let M be a finite monoid and $\Bbbk$ a field. Then $\Bbbk M$ is semisimple if and only if:*

(a) M is regular;
(b) the characteristic of $\Bbbk$ does not divide the order of G_e for any $e \in E(M)$;
(c) the sandwich matrix $P(e)$ is invertible over $\Bbbk G_e$ for all $e \in E(M)$.

In principle, this theorem reduces determining semisimplicity of a monoid algebra to checking invertibility of a matrix over a group algebra, but the latter problem is not a simple one.

Example 5.22. If $\Bbbk$ is any field and $n \geq 2$, then $\Bbbk T_n$ is not semisimple because the sandwich matrix $P(e_1)$ is $1 \times n$ where e_1 is the constant mapping to 1.

Example 5.23. Let $\Bbbk$ be a field of characteristic 0 and M a finite commutative monoid. Then because $\mathscr{L} = \mathscr{R} = \mathscr{J}$, each sandwich matrix $P(e)$ is 1×1 and hence invertible over $\Bbbk G_e$. We conclude from Theorem 5.21 that $\Bbbk M$ is semisimple if and only if M is regular (or, equivalently, an inverse monoid by Theorem 3.2).

Example 5.24. Let I_2 be the symmetric inverse monoid of degree 2 and $\Bbbk$ a field of characteristic 0. Then I_2 has three $\mathscr{J}$-classes J_0, J_1, J_2 where J_i consists of the rank i partial injective mappings. Let $e_i = 1_{[i]}$ for $i = 0, 1, 2$. Then the e_i are idempotent representatives of the $\mathscr{J}$-classes of I_2 and $G_{e_i} \cong S_i$. Clearly $J_0 = \{e_0\}$ and hence $P(e_0) = (e_0)$, which is invertible over $\Bbbk G_{e_0}$. Also, $J_2 = S_2 = G_{e_2}$ contains exactly one $\mathscr{R}$-class and one $\mathscr{L}$-class. If we take $\lambda_1 = e_2 = \rho_1$, then $P(e_2) = (e_2)$, which is invertible over $\Bbbk S_2$. Finally, if we let p_{ij} denote the rank one partial mapping taking j to i, then, for J_1, we may take $\lambda_1 = e_1, \lambda_2 = p_{21}, \lambda_3 = p_{31}, \rho_1 = e_1 = p_{11}, \rho_2 = p_{12}$, and $\rho_3 = p_{13}$. Then

$$\rho_i \diamond \lambda_j = \begin{cases} e_1, & \text{if } i = j \\ 0, & \text{else} \end{cases}$$

and so $P(e_1)$ is an identity matrix and hence invertible over $\Bbbk G_{e_1}$. We conclude that $\Bbbk I_2$ is semisimple by Theorem 5.21.

In Exercise 5.18, the reader will be asked to show that if M is an inverse monoid, then each sandwich matrix may be chosen to be an identity matrix and hence $\Bbbk M$ will be semisimple provided that the characteristic of $\Bbbk$ does not divide the order of any maximal subgroup of M. Chapter 9 will provide a different approach to the semisimplicity of inverse monoid algebras in good characteristic.

We end this section with a description of $\Bbbk M$ in the semisimple case.

Theorem 5.25. *Let M be a finite monoid and $\Bbbk$ a field such that $\Bbbk M$ is semisimple. Let $e_1, \ldots, e_s$ form a complete set of representatives of the $\mathscr{J}$-classes of idempotents of M and suppose that J_{e_i} contains n_i $\mathscr{L}$-classes. Then there is an isomorphism*

$$\Bbbk M \cong \prod_{i=1}^{s} M_{n_i}(\Bbbk G_{e_i})$$

of $\Bbbk$-algebras.

Proof. Under the hypotheses of the theorem, we have the isomorphism of $\Bbbk M$-modules $\Bbbk M \cong \bigoplus_{i=1}^{s} n_i \cdot \Bbbk L_{e_i}$ by (5.8). Let $V_1, \ldots, V_r$ form a complete set of representatives of the isomorphism classes of simple $\Bbbk G_{e_i}$-modules. From the decomposition $\Bbbk G_{e_i} = \bigoplus_{j=1}^{r} \dim V_j / \dim \operatorname{End}_{\Bbbk G_{e_i}}(V_j) \cdot V_j$ (coming from the semisimplicity of $\Bbbk G_{e_i}$ and Theorem A.7), we obtain that

$$\Bbbk L_{e_i} = \operatorname{Ind}_{G_{e_i}}(\Bbbk G_{e_i}) = \bigoplus_{j=1}^{r} \frac{\dim V_j}{\dim \operatorname{End}_{\Bbbk G_{e_i}}(V_j)} \cdot \operatorname{Ind}_{G_{e_i}}(V_j)$$

is the decomposition of $\Bbbk L_{e_i}$ into a direct sum of simple $\Bbbk M$-modules, all of which have apex e_i. It follows that $\mathrm{Hom}_{\Bbbk M}(\Bbbk L_{e_i}, \Bbbk L_{e_j}) = 0$ if $i \neq j$ by Schur's lemma. We conclude that

$$\Bbbk M^{op} \cong \mathrm{End}_{\Bbbk M}(\Bbbk M) \cong \prod_{i=1}^{s} M_{n_i}(\mathrm{End}_{\Bbbk M}(\Bbbk L_{e_i})).$$

But by Proposition 4.6, we have that

$$\mathrm{End}_{\Bbbk M}(\Bbbk L_{e_i}) = \mathrm{End}_{A_{e_i}}(\mathrm{Ind}_{e_i}(\Bbbk G_{e_i})) \cong \mathrm{End}_{\Bbbk G_{e_i}}(\Bbbk G_{e_i}) \cong \Bbbk G_{e_i}^{op}$$

(where $A_{e_i} = \Bbbk M / \Bbbk I(e_i)$) and the desired isomorphism follows. $\square$

The conclusion of Theorem 5.25 holds more generally if M is regular and the sandwich matrix $P(e)$ is invertible over $\Bbbk G_e$ for all $e \in E(M)$, as we shall see in Chapter 15.

5.5 Monomial representations

The induction and coinduction functors were first considered in semigroup theory in the matrix theoretic language of monomial representations and Schützenberger representations. The matrix form of the induction functors appeared independently in [LP69] and [RZ91], whereas coinduction first appeared in [RZ91]. Putcha works principally with the characters of these representations in [Put96, Put98]. In this section, we describe the monomial form of $\mathrm{Ind}_{G_e}(V)$ and $\mathrm{Coind}_{G_e}(V)$. We will work out in detail the case of $\mathrm{Ind}_{G_e}(V)$ and merely state the result for $\mathrm{Coind}_{G_e}(V)$, leaving the details to the reader in the exercises.

Fix an idempotent $e \in E(M)$. Let $\lambda_1, \ldots, \lambda_\ell$ form a complete set of representatives for L_e / G_e. Let $m \in M$. Then either $m\lambda_j \notin L_e$ or $m\lambda_j = \lambda_i g$ for a unique $g \in G_e$ and $i \in \{1, \ldots, \ell\}$. Let us define an $\ell \times \ell$ matrix $\Theta_e(m)$ over $G_e \cup \{0\} \subseteq \Bbbk G_e$ by

$$\Theta_e(m)_{ij} = \begin{cases} g, & \text{if } m\lambda_j = \lambda_i g \\ 0, & \text{else.} \end{cases}$$

Then $\Theta_e \colon M \longrightarrow M_\ell(\Bbbk G_e)$ is precisely the matrix representation afforded by $\Bbbk L_e$, viewed as a $\Bbbk M$-$\Bbbk G_e$-bimodule which is free over $\Bbbk G_e$. Classically, Θ_e is called the *left Schützenberger representation* of M associated with the $\mathscr{J}$-class J_e. Note that $\Theta_e(m)$ has at most one nonzero entry in each column. Such a matrix is called *column monomial*.

Suppose now that V is a finite dimensional $\Bbbk G_e$-module. Then as a $\Bbbk$-vector space, we have that

$$\mathrm{Ind}_{G_e}(V) = \Bbbk L_e \otimes_{\Bbbk G_e} V = \bigoplus_{j=1}^{\ell} \lambda_j \otimes V$$

because $\Bbbk L_e$ is a free right $\Bbbk G_e$-module with basis $\{\lambda_1, \dots, \lambda_\ell\}$. Moreover, if $\{v_1, \dots, v_n\}$ is a basis for V, then the $\lambda_j \otimes v_k$ with $1 \leq j \leq \ell$ and $1 \leq k \leq n$ form a basis for $\mathrm{Ind}_{G_e}(V)$.

Proposition 5.26. *If $m \in M$ and $v \in V$, then*

$$m(\lambda_j \otimes v) = \sum_{i=1}^{\ell} \lambda_i \otimes \Theta_e(m)_{ij} v.$$

Equivalently, the equality

$$m(\lambda_j \otimes v) = \begin{cases} \lambda_i \otimes gv, & \text{if } m\lambda_j = \lambda_i g \\ 0, & \text{if } m\lambda_j \notin L_e \end{cases}$$

holds.

Proof. Indeed, if $m\lambda_j \notin L_e$, then $m\lambda_j = 0$ in $\Bbbk L_e$ and so $m(\lambda_j \otimes v) = m\lambda_j \otimes v = 0$. If $m\lambda_j = \lambda_i g$ with $g \in G_e$, then $m(\lambda_j \otimes v) = m\lambda_j \otimes v = \lambda_i g \otimes v = \lambda_i \otimes gv$, as required. $\qquad \square$

As a consequence, we obtain the following block column monomial form for induced modules.

Theorem 5.27. *Let V be a $\Bbbk G_e$-module affording a matrix representation $\alpha \colon G_e \longrightarrow M_n(\Bbbk)$. Then $\mathrm{Ind}_{G_e}(V)$ affords the matrix representation*

$$\mathrm{Ind}_{G_e}(\alpha) \colon M \longrightarrow M_{\ell n}(\Bbbk)$$

given by the $\ell \times \ell$-block form

$$\mathrm{Ind}_{G_e}(\alpha)(m)_{ij} = \alpha(\Theta_e(m)_{ij})$$

where $\alpha(0)$ is the $n \times n$ zero matrix.

Proof. Let $\{v_1, \dots, v_n\}$ be a basis for V giving rise to α. Consider the basis

$$\{\lambda_j \otimes v_k \mid 1 \leq j \leq \ell, 1 \leq k \leq n\}$$

for $\mathrm{Ind}_{G_e}(V)$ with ordering $\lambda_1 \otimes v_1, \dots, \lambda_1 \otimes v_n, \dots, \lambda_\ell \otimes v_1, \dots, \lambda_\ell \otimes v_n$. Then by Proposition 5.26 we have that

$$m(\lambda_j \otimes v_k) = \sum_{i=1}^{\ell} \lambda_i \otimes \Theta_e(m)_{ij} v_k = \sum_{i=1}^{\ell} \sum_{t=1}^{n} \alpha(\Theta_e(m)_{ij})_{tk} (\lambda_i \otimes v_t).$$

The theorem follows. $\qquad \square$

The corresponding result for $\mathrm{Coind}_{G_e}(V)$ is as follows. Let $\rho_1, \ldots, \rho_r$ form a complete set of representatives for $G_e \backslash R_e$. Define $\Upsilon_e \colon M \longrightarrow M_r(\Bbbk G_e)$ by

$$\Upsilon_e(m)_{ij} = \begin{cases} g, & \text{if } \rho_i m = g\rho_j \\ 0, & \text{else.} \end{cases}$$

Then Υ is the representation afforded by viewing $\Bbbk R_e$ as a $\Bbbk G_e$-$\Bbbk M$-bimodule which is free as a left $\Bbbk G_e$-module. It is called the *right Schützenberger representation* of M associated with the $\mathscr{J}$-class J_e. Note that each row of $\Upsilon_e(m)$ contains at most one nonzero entry. Such a matrix is called *row monomial*.

If V is a $\Bbbk G_e$-module with basis $\{v_1, \ldots, v_n\}$, then $\mathrm{Coind}_{G_e}(V)$ has basis $\{\rho_{1,1}^*, \ldots, \rho_{1,n}^*, \ldots, \rho_{r,1}^*, \ldots, \rho_{r,n}^*\}$ where

$$\rho_{i,j}^*(g\rho_k) = \begin{cases} gv_j, & \text{if } i = k \\ 0, & \text{else} \end{cases}$$

for $g \in G_e$. The corresponding block row monomial matrix form for coinduced modules is encapsulated in the next theorem.

Theorem 5.28. *Let V be a $\Bbbk G_e$-module affording a matrix representation $\alpha \colon G_e \longrightarrow M_n(\Bbbk)$. Then $\mathrm{Coind}_{G_e}(V)$ affords the matrix representation*

$$\mathrm{Coind}_{G_e}(\alpha) \colon M \longrightarrow M_{rn}(\Bbbk)$$

given by the $r \times r$-block form

$$\mathrm{Coind}_{G_e}(\alpha)(m)_{ij} = \alpha(\Upsilon_e(m)_{ij})$$

where $\alpha(0)$ is the $n \times n$ zero matrix.

Theorem 5.29. *Let V be a $\Bbbk G_e$-module affording a matrix representation $\alpha \colon G_e \longrightarrow M_n(\Bbbk)$ with respect to the basis $\{v_1, \ldots, v_n\}$. Let $\lambda_1, \ldots, \lambda_\ell$ and $\rho_1, \ldots, \rho_r$ form complete sets of representatives for L_e/G_e and $G_e \backslash R_e$, respectively. Let $P(e)$ be the corresponding $r \times \ell$ sandwich matrix for J_e with $P(e)_{ij} = \rho_i \diamond \lambda_j$ and let $P(e) \otimes \alpha$ be the block $rn \times \ell n$ matrix given by*

$$[P(e) \otimes \alpha]_{ij} = \alpha(P(e)_{ij})$$

where $\alpha(0)$ is the $n \times n$ zero matrix.

Then $P(e) \otimes \alpha$ is the matrix for the natural map

$$\varphi_V \colon \mathrm{Ind}_{G_e}(V) \longrightarrow \mathrm{Coind}_{G_e}(V)$$

with respect to the bases $\{\lambda_i \otimes v_j \mid 1 \le i \le \ell, 1 \le j \le n\}$ for $\mathrm{Ind}_{G_e}(V)$ and $\{\rho_{i,j}^* \mid 1 \le i \le r, 1 \le j \le n\}$ for $\mathrm{Coind}_{G_e}(V)$. Therefore, $N_e(\mathrm{Ind}_{G_e}(V))$ can be identified with the null space of $P(e) \otimes \alpha$ and $T_e(\mathrm{Coind}_{G_e}(V))$ can be identified with the column space of $P(e) \otimes \alpha$.

Proof. Let $g \in G_e$. Then we compute

$$\varphi_V(\lambda_j \otimes v_k)(g\rho_i) = g(\rho_i \diamond \lambda_j)v_k = gP(e)_{ij}v_k = \sum_{m=1}^{n} \alpha(P(e)_{ij})_{mk}gv_m.$$

It follows that

$$\varphi_V(\lambda_j \otimes v_k) = \sum_{i=1}^{r} \sum_{m=1}^{n} \alpha(P(e)_{ij})_{mk}\rho_{i,m}^*,$$

and so $P(e) \otimes \alpha$ is the matrix of φ_V, as was required. The final statement then follows from Corollary 4.12. $\qquad\square$

By considering the particular case where V is simple, Theorem 5.29 provides a matrix theoretic description of the irreducible representations of M in light of Theorem 5.5.

Corollary 5.30. *Let V be a simple $\Bbbk G_e$-module affording an irreducible matrix representation $\alpha\colon G_e \longrightarrow M_n(\Bbbk)$ with respect to the basis $\{v_1, \ldots, v_n\}$. Let $\lambda_1, \ldots, \lambda_\ell$ and $\rho_1, \ldots, \rho_r$ form complete sets of representatives for L_e/G_e and $G_e\backslash R_e$, respectively. Let $P(e)$ be the corresponding $r \times \ell$ sandwich matrix for J_e with $P(e)_{ij} = \rho_i \diamond \lambda_j$ and let $P(e) \otimes \alpha$ be the block $rn \times \ell n$ matrix given by*

$$[P(e) \otimes \alpha]_{ij} = \alpha(P(e)_{ij})$$

where $\alpha(0)$ is the $n \times n$ zero matrix. Then the simple module $V^\sharp$ can be identified with the quotient of $\Bbbk^{\ell n}$ by the null space of $P(e) \otimes \alpha$ and with the column space of $P(e)\otimes\alpha$ as subrepresentations of the representations in Theorems 5.27 and 5.28, respectively.

5.6 Semisimplicity of the algebra of $M_n(\mathbb{F}_q)$

Let $\mathbb{F}_q$ be a q-element field of characteristic $p > 0$ and let $\Bbbk$ be a field. Kovács proved [Kov92] that if the characteristic of $\Bbbk$ is different from p and does not divide the order of the general linear group $GL_n(\mathbb{F}_q)$, then $\Bbbk M_n(\mathbb{F}_q)$ is semisimple. The special case of $\Bbbk = \mathbb{C}$ was proved independently, and at about the same time, by Okniński and Putcha [OP91] who considered more generally finite monoids of Lie type. Kovács, in fact, proved the following result.

Theorem 5.31 (Kovács). *Let $\mathbb{F}$ be a finite field of characteristic p and let $\Bbbk$ be a field of characteristic different from p. Then there is an isomorphism*

$$\Bbbk M_n(\mathbb{F}) \cong \prod_{r=0}^{n} M_{k_r}(\Bbbk GL_r(\mathbb{F}))$$

where k_r is the number of r-dimensional subspaces of $\mathbb{F}^n$ and $GL_0(\mathbb{F})$ is interpreted as the trivial group.

An immediate consequence of the theorem is the following semisimplicity result.

Corollary 5.32. *Let $\mathbb{F}$ be a finite field of characteristic p and let $\Bbbk$ be a field of characteristic different from p. Then $\Bbbk M_n(\mathbb{F})$ is semisimple if and only if the characteristic of $\Bbbk$ does not divide the order of $GL_n(\mathbb{F})$. In particular, $\Bbbk M_n(\mathbb{F})$ is semisimple whenever $\Bbbk$ is of characteristic zero.*

Theorem 5.31 has recently been put to good effect by Kuhn in his study of generic representation theory [Kuh15]. We present here a rendering of Kovács's proof of Theorem 5.31. Our approach is less linear algebraic and is instead based on our study of the symmetric inverse monoid.

Let us begin with a well-known ring theoretic lemma.

Lemma 5.33. *Let A be a $\Bbbk$-algebra and I an ideal of A which is a unital algebra. Then $A \cong I \times A/I$ as a $\Bbbk$-algebra.*

Proof. Let e be the identity of I. Observe that e is a central idempotent. Indeed, if $a \in A$, then $ae, ea \in I$ and so $ae = e(ae) = (ea)e = ea$. It now follows that $I = Ae$, $A/I \cong A(1-e)$ and $A \cong Ae \times A(1-e) \cong I \times A/I$. $\square$

Put $A = \Bbbk M_n(\mathbb{F})$. For $0 \le r \le n$, let I_r be the ideal of $M_n(\mathbb{F})$ consisting of matrices of rank at most r and put $A_r = \Bbbk I_r/\Bbbk I_{r-1}$ for $0 \le r \le n$ where we set $I_{-1} = \emptyset$. Note that the $I_r \setminus I_{r-1}$ are the $\mathscr{J}$-classes of $M_n(\mathbb{F})$. The crucial step in Kovács's proof is to show that A_r is a unital algebra for each $0 \le r \le n$. Let us show how this implies the theorem. Note that if $e \in M_n(\mathbb{F})$ is a rank r idempotent, then $eM_n(\mathbb{F})e \cong M_r(\mathbb{F})$, and hence $G_e \cong GL_r(\mathbb{F})$, as is easily verified by considering the vector space decomposition $\mathbb{F}^n = e\mathbb{F}^n \oplus (1-e)\mathbb{F}^n$.

Proposition 5.34. *Suppose that A_r has an identity for all $0 \le r \le n$. Then Theorem 5.31 holds.*

Proof. We claim that $A/\Bbbk I_{r-1} \cong A_r \times A/\Bbbk I_r$ for $0 \le r \le n$. Indeed, we have that A_r is an ideal of $A/\Bbbk I_{r-1}$ containing an identity element and so the isomorphism follows from Lemma 5.33. Applying the claim iteratively, we obtain that $A \cong A_0 \times A_1 \times \cdots \times A_n$. Let e_r be a rank r idempotent of $M_n(\mathbb{F})$. Then $A_r e_r A_r = A_r$, $e_r A_r e_r \cong \Bbbk GL_r(\mathbb{F})$ and $e_r A_r = e_r A/\Bbbk I_{r-1} \cong \Bbbk R_{e_r}$ as a left $\Bbbk GL_r(\mathbb{F})$-module by Corollary 1.16 and Theorem 1.13.

Recall that G_{e_r} acts freely on the left of R_{e_r} by Proposition 1.10 with orbits in bijection with the set of $\mathscr{L}$-classes of the $\mathscr{J}$-class of rank r matrices by Corollary 1.17 and Exercise 1.18. Two rank r matrices are $\mathscr{L}$-equivalent if and only if they are row equivalent, i.e., they have the same row space. Therefore, G_{e_r} has k_r orbits on R_{e_r}. Thus $\Bbbk R_{e_r}$ is a free $\Bbbk GL_r(\mathbb{F})$-module of rank k_r by Exercise 5.3. Corollary 4.14 then yields

$$A_r^{op} \cong \mathrm{End}_{e_r A_r e_r}(e_r A_r) \cong \mathrm{End}_{\Bbbk GL_r(\mathbb{F})}(\Bbbk R_{e_r}) \cong M_{k_r}(\Bbbk GL_r(\mathbb{F})^{op})$$

and so, since $M_{k_r}(\Bbbk GL_r(\mathbb{F})^{op})^{op} \cong M_{k_r}(\Bbbk GL_r(\mathbb{F}))$,

$$\Bbbk M_n(\mathbb{F}) \cong \prod_{r=0}^{n} A_r \cong \prod_{r=0}^{n} M_{k_r}(\Bbbk GL_r(\mathbb{F}))$$

as required.   □

It remains to construct the identity element of A_r. A crucial insight of Kovács is that the identity is a linear combination of semi-idempotent matrices. To define the notion, it will be convenient to turn to the setting of linear maps and partial linear maps on a vector space in order to be able to work in a more coordinate-free manner. Most of what we shall do works over an arbitrary field.

Let V be a finite dimensional vector space over a field $\mathbb{F}$. We write $\mathrm{End}(V)$ instead of $\mathrm{End}_{\mathbb{F}}(V)$ for the rest of this section. By a *partial linear mapping* on V, we mean a linear mapping $T: W \longrightarrow V$ where W is a subspace of V. The partial linear maps form a submonoid $\mathrm{PEnd}(V)$ of the monoid of all partial mappings on V. The partial injective linear mappings form a submonoid $\mathrm{PAut}(V)$. If $T \in \mathrm{PEnd}(V)$, we write $\mathrm{dom}(T)$ for the domain of T and put $\mathrm{rk}\, T = \dim \mathrm{im}\, T$. The monoids $\mathrm{PEnd}(V)$, $\mathrm{PAut}(V)$ and $\mathrm{End}(V)$ all have group of units the general linear group $GL(V)$ and hence being *conjugate* in each of these monoids shall mean being conjugate by an element of $GL(V)$. If W is a subspace of V, then the inclusion mapping of W into V is a partial linear mapping that we think of as the partial identity map on W and denote by 1_W, as was done in Chapter 3.

If T is a partial linear map, then we have a decreasing chain of subspaces

$$\mathrm{im}\, T^0 \supseteq \mathrm{im}\, T^1 \supseteq \mathrm{im}\, T^2 \supseteq \cdots$$

which must stabilize by finite dimensionality. If $\mathrm{im}\, T^m = \mathrm{im}\, T^{m+1}$, then $\mathrm{im}\, T^m = \bigcap_{k \geq 0} \mathrm{im}\, T^k$. We put

$$\mathrm{im}^{\infty} T = \bigcap_{k \geq 0} \mathrm{im}\, T^k$$

and call it the *eventual range* of T. Since the equalities

$$T(\mathrm{im}^{\infty} T) = T(\mathrm{im}\, T^m) = \mathrm{im}\, T^{m+1} = \mathrm{im}^{\infty} T \qquad (5.9)$$

hold (where m is as above), it follows that $T|_{\mathrm{im}^{\infty} T}: \mathrm{im}^{\infty} T \longrightarrow \mathrm{im}^{\infty} T$ is a surjective total linear mapping and hence a vector space automorphism by finite dimensionality.

Define the *rank sequence* of T to be

$$\vec{r}(T) = (\mathrm{rk}\, T^0, \mathrm{rk}\, T^1, \mathrm{rk}\, T^2, \ldots).$$

This is a weakly decreasing sequence of nonnegative integers with first entry $\dim V$ that is constant as soon as two consecutive values are equal and whose entries are eventually $\dim \mathrm{im}^{\infty} T$. In particular, there are only finitely many possible rank sequences of partial linear maps on V. If μ, ν are rank sequences, then we write $\mu \leq \nu$ if $\mu_i \leq \nu_i$ for all $i \geq 0$. We say that T is *semi-idempotent* if T fixes $\mathrm{im}^{\infty} T$ pointwise.

Proposition 5.35. *Let* $S, T \in \mathrm{PEnd}(V)$ *and suppose that* S *is a restriction of* T. *Then* $\vec{r}(S) \leq \vec{r}(T)$. *Moreover, if* T *is semi-idempotent, then so is* S.

Proof. Clearly, $\mathrm{im}\, S^k \subseteq \mathrm{im}\, T^k$ for all $k \geq 0$ and so $\vec{r}(S) \leq \vec{r}(T)$. Furthermore, we have that
$$\mathrm{im}^\infty S = \bigcap_{k \geq 0} \mathrm{im}\, S^k \subseteq \bigcap_{k \geq 0} \mathrm{im}\, T^k = \mathrm{im}^\infty T.$$
Therefore, if T is semi-idempotent and $v \in \mathrm{im}^\infty S$, then $v \in \mathrm{im}^\infty T$ and so $Sv = Tv = v$. We conclude that S is semi-idempotent. □

For the remainder of this section, we shall use the notation of Section 3.2, which the reader should review. Suppose that B is a basis for V. Then we can identify the symmetric inverse monoid I_B with a submonoid of $\mathrm{PAut}(V)$ by sending a partial injective map $f \in I_B$ to the unique partial injective linear map T_f with domain spanned by $\mathrm{dom}(f)$ that agrees with f on $\mathrm{dom}(f)$ (where $\mathrm{dom}(f) \subseteq B$ denotes the domain of f). Note that T_f has image the linear span of $f(B)$ and hence $\mathrm{im}^\infty T_f$ is the linear span of $\mathrm{rec}(f)$, $\vec{r}(f) = \vec{r}(T_f)$ and f is semi-idempotent if and only if T_f is semi-idempotent. Observe that if f, g are conjugate in I_B, say $f = hgh^{-1}$ with $h \in S_B$, then $T_h \in GL(V)$ and $T_f = T_h T_g T_h^{-1}$ and so T_f, T_g are conjugate.

There is a representation $\rho \colon I_B \longrightarrow \mathrm{End}(V)$ defined by $\rho(f) = S_f$ where
$$S_f b = \begin{cases} f(b), & \text{if } b \in \mathrm{dom}(f) \\ 0, & \text{else} \end{cases}$$
for $b \in B$. The matrix representation corresponding to ρ using the basis B is the faithful representation of I_B by partial permutation matrices alluded to in Chapter 3. It is clear that if $f \in I_B$, then $\mathrm{im}\, S_f$ is the linear span of $f(B)$ and hence $\mathrm{im}\, S_f = \mathrm{im}\, T_f$. Therefore, $\mathrm{im}\, T_f^k = \mathrm{im}\, S_f^k$ for all $k \geq 0$ and, consequently, $\mathrm{im}^\infty T_f = \mathrm{im}^\infty S_f$ is the linear span of $\mathrm{rec}(f)$. It also follows that $\vec{r}(T_f) = \vec{r}(f) = \vec{r}(S_f)$. Moreover, T_f is a restriction of S_f and T_f is semi-idempotent if and only if S_f is semi-idempotent, if and only if f is semi-idempotent. Summarizing the above discussion, we obtain the following proposition.

Proposition 5.36. *Let* V *be a finite dimensional vector space over a field* $\mathbb{F}$ *with basis* B. *Then the following are equivalent for* $f \in I_B$.

  (i) f is semi-idempotent.
  (ii) T_f is semi-idempotent.
  (iii) S_f is semi-idempotent.

Moreover, S_f *extends* T_f *and* $\vec{r}(T_f) = \vec{r}(f) = \vec{r}(S_f)$.

We now count the number of extensions of T_f to total endomorphisms with the same rank sequence in the case that f is semi-idempotent.

Lemma 5.37. *Let V be a finite dimensional vector space over a finite field $\mathbb{F}$ with basis B and let $f \in I_B$ be semi-idempotent. The number of extensions $S \in \text{End}(V)$ of T_f with $\vec{r}(S) = \vec{r}(f) = \vec{r}(T_f)$ is a power of the characteristic of $\mathbb{F}$. Moreover, each such extension S is semi-idempotent.*

Proof. We prove the final statement first. If $S \in \text{End}(V)$ is an extension of T_f, then $\text{im}\, T_f^k \subseteq \text{im}\, S^k$ for all $k \geq 0$. If $\vec{r}(S) = \vec{r}(T_f)$, then we conclude that $\dim \text{im}\, S^k = \dim \text{im}\, T_f^k$ for all $k \geq 0$ and so $\text{im}\, S^k = \text{im}\, T_f^k$ for all $k \geq 0$. Consequently, we have $\text{im}^\infty T_f = \text{im}^\infty S$ and hence $S|_{\text{im}^\infty S} = T_f|_{\text{im}^\infty T_f}$ is the identity mapping. Therefore, S is semi-idempotent.

Let us put $B' = \text{dom}(f)$ and $B'' = B \setminus B'$. If $S \in \text{End}(V)$ is an extension of T_f, then $Sb = f(b)$ for all $b \in B'$. Thus it remains to describe how S can be defined on basis elements $b \in B''$. For each $b \in B''$, let m_b be the size of the orbit of b under f. Note that each element of B'' is transient. We claim that an extension S of T_f satisfies $\vec{r}(S) = \vec{r}(f)$ if and only if $Sb \in \text{im}\, T_f^{m_b}$ for all $b \in B''$. The crucial point is that, if $b \in B''$ and x_b is the source of the orbit of b, then $b = f^{m_b-1}(x_b)$ and $b \notin f^{m_b}(B)$ because $b \notin \text{dom}(f)$. In particular, $m_b - 1$ is the largest integer k such that $b \in f^k(B)$. Therefore, if $S \in \text{End}(V)$ is an extension of T_f with $\vec{r}(S) = \vec{r}(T_f)$, then $Sb \in S(\text{im}\, T_f^{m_b-1}) \subseteq \text{im}\, S^{m_b} = \text{im}\, T_f^{m_b}$ and so the condition is necessary.

For sufficiency, suppose that $Sb \in \text{im}\, T_f^{m_b}$ for all $b \in B''$ and $Sb = f(b)$ for all $b \in B'$ (and so, in particular, S extends T_f). We prove by induction on k that $\text{im}\, S^k \subseteq \text{im}\, T_f^k$. As the reverse inclusion is clear, it will follow that $\vec{r}(S) = \vec{r}(T_f) = \vec{r}(f)$. Trivially, $\text{im}\, S^0 = V = \text{im}\, T_f^0$. Assume, by induction, that $\text{im}\, S^r \subseteq \text{im}\, T_f^r$. Then $\text{im}\, S^r = \text{im}\, T_f^r$ has basis $f^r(B)$. If $b \in f^r(B) \cap B'$, then $Sb = f(b) = T_f(b) \in \text{im}\, T_f^{r+1}$, as required. If $b \in f^r(B) \cap B''$, then $r \leq m_b - 1$. Therefore, we have $Sb \in \text{im}\, T_f^{m_b} \subseteq \text{im}\, T_f^{r+1}$ by construction. This proves that $\text{im}\, S^{r+1} \subseteq \text{im}\, T_f^{r+1}$, thereby establishing the claim.

The lemma follows from the claim because, for each $b \in B''$, there are $|\text{im}\, T_f^{m_b}| = |\mathbb{F}|^{\dim \text{im}\, T_f^{m_b}}$ choices for Sb and this cardinality is a power of the characteristic of $\mathbb{F}$. Thus the number of choices for S is the product $\prod_{b \in B''} |\text{im}\, T_f^{m_b}|$, which is a power of the characteristic of $\mathbb{F}$. $\square$

The crux of Kovács's argument involves classifying semi-idempotent (total) linear mappings and semi-idempotent partial injective linear mappings up to conjugacy. Of critical importance is that in both cases conjugacy is determined by the rank sequence and that the same rank sequences arise. Let us begin with the case of (total) linear maps. It will be convenient to recall the *Fitting decomposition* of a linear operator.

Let $T: V \longrightarrow V$ be a linear map on a finite dimensional vector space. Then there is an increasing sequence of subspaces

$$0 \subseteq \ker T \subseteq \ker T^2 \subseteq \cdots$$

which must stabilize at some subspace, that is,

$$\ker^\infty T = \bigcup_{k \geq 0} \ker T^k$$

coincides with $\ker T^m$ for some $m > 0$.

Theorem 5.38 (Fitting decomposition). *Let V be a finite dimensional vector space over a field $\mathbb{F}$ and let $T\colon V \longrightarrow V$ be a linear map.*

(i) $\ker^\infty T$ is the largest T-invariant subspace W of V such that $T|_W$ is nilpotent.

(ii) $\operatorname{im}^\infty T$ is the largest T-invariant subspace W of V such that $T|_W$ is invertible.

(iii) $V = \ker^\infty T \oplus \operatorname{im}^\infty T$.

(iv) If $V = U \oplus W$ where U, W are T-invariant subspaces such that $T|_U$ is nilpotent and $T|_W$ is invertible, then $U = \ker^\infty T$ and $W = \operatorname{im}^\infty T$.

Proof. Clearly $\ker^\infty T$ is T-invariant. If $\ker^\infty T = \ker T^m$, then T^m annihilates $\ker^\infty T$ and so $T|_{\ker^\infty T}$ is nilpotent. If W is T-invariant and $T|_W$ is nilpotent, then $W \subseteq \ker T^k$ for some $k \geq 0$. But then $W \subseteq \ker^\infty T$. This establishes the first item.

From (5.9) and the observation immediately following it, we have that $\operatorname{im}^\infty T$ is a T-invariant subspace and $T|_{\operatorname{im}^\infty T}$ is invertible. If $W \subseteq V$ is T-invariant and $T|_W$ is invertible, then $W = T^k(W)$ for all $k \geq 0$ and hence $W \subseteq \operatorname{im}^\infty T$. This proves the second item.

If $v \in \ker^\infty T \cap \operatorname{im}^\infty T$, then $T^k v = 0$ for some $k > 0$. But T^k is invertible on $\operatorname{im}^\infty T$ and so we conclude that $v = 0$. Now let $v \in V$ be arbitrary and suppose that $\operatorname{im} T^m = \operatorname{im}^\infty T$ with $m \geq 0$. Then $T^m v \in \operatorname{im}^\infty T$. As $T^m|_{\operatorname{im}^\infty T}$ is invertible, there exists $w \in \operatorname{im}^\infty T$ such that $T^m w = T^m v$. Then $T^m(v - w) = 0$ and so $v - w \in \ker^\infty T$. Therefore, $v = (v - w) + w \in \ker^\infty T + \operatorname{im}^\infty T$. This yields the third item.

Item (iv) is an immediate consequence of (i)–(iii). □

If $T \in \operatorname{End}(V)$ is semi-idempotent, then $T|_{\operatorname{im}^\infty T}$ is the identity. It follows from Theorem 5.38 that if $S, T \in \operatorname{End}(V)$ are semi-idempotent, then S and T are conjugate if and only if $\dim \operatorname{im}^\infty S = \dim \operatorname{im}^\infty T$ and the nilpotent operators $S|_{\ker^\infty S}$ and $T|_{\ker^\infty T}$ have the same Jordan canonical form.

Lemma 5.39. *Let S, T be semi-idempotent linear operators on a finite dimensional vector space V over a field $\mathbb{F}$. Then S and T are conjugate if and only if $\vec{r}(S) = \vec{r}(T)$.*

Proof. Clearly, if S and T are conjugate, then $\vec{r}(S) = \vec{r}(T)$. For the converse, the equality $\vec{r}(S) = \vec{r}(T)$ implies that $\dim \operatorname{im}^\infty S = \dim \operatorname{im}^\infty T$. By the previous discussion it then suffices to show that if T is a semi-idempotent operator, then the Jordan form of the nilpotent operator $N = T|_{\ker^\infty T}$ is determined by $\vec{r}(T)$. But $\dim \operatorname{im} T^i - \dim \operatorname{im} T^{i+1} = \dim \operatorname{im} N^i - \dim \operatorname{im} N^{i+1}$ is the number of Jordan blocks of N of degree greater than i for $i \geq 0$. Thus the Jordan canonical form of N is determined by $\vec{r}(T)$. This completes the proof. □

Our next aim is to establish that if $S \in \mathrm{PAut}(V)$ is semi-idempotent, then there is a basis B for V such that $S = T_f$ for some semi-idempotent $f \in I_B$. This should be thought of as the analogue of Jordan canonical form in this context and it will allow us to connect tightly semi-idempotent linear operators and semi-idempotent partial injective linear maps.

Lemma 5.40. *Let V be a finite dimensional vector space over a field $\mathbb{F}$ and let S be a semi-idempotent partial injective linear map on V. Then there exist a basis B for V and semi-idempotent $f \in I_B$ such that $S = T_f$.*

Proof. Assume that $\mathrm{im}\, S^k = \mathrm{im}^\infty S$ with $k > 0$. We proceed to prove by downward induction that, for each $0 \le i \le k$, there is a basis $B_i = B_i' \uplus B_i''$ (disjoint union) for $\mathrm{im}\, S^i$ such that

(i) B_i' is a basis for $\mathrm{dom}(S) \cap \mathrm{im}\, S^i$;
(ii) $S(B_i') \subseteq B_i$.

Choose a basis B_k for $\mathrm{im}\, S^k = \mathrm{dom}(S) \cap \mathrm{im}\, S^k$ and put $B_k' = B_k$, $B_k'' = \emptyset$. Since $S|_{\mathrm{im}\, S^k}$ is the identity mapping, $S(B_k') = B_k' = B_k$. Assume inductively that $B_i = B_i' \uplus B_i''$ has been constructed satisfying (i) and (ii) where $1 \le i \le k$. As S is a partial injective linear mapping, we conclude that

$$S \colon \mathrm{dom}(S) \cap \mathrm{im}\, S^{i-1} \longrightarrow \mathrm{im}\, S^i$$

is an isomorphism of vector spaces. Therefore, the set $S^{-1}(B_i)$ is a basis for $\mathrm{dom}(S) \cap \mathrm{im}\, S^{i-1}$. Thus $B_i' \cup S^{-1}(B_i)$ spans $\mathrm{dom}(S) \cap \mathrm{im}\, S^{i-1}$ with B_i' a linearly independent set and so we can find a basis B_{i-1}' of $\mathrm{dom}(S) \cap \mathrm{im}\, S^{i-1}$ such that

$$B_i' \subseteq B_{i-1}' \subseteq B_i' \cup S^{-1}(B_i). \tag{5.10}$$

As the linear span of B_i'' intersects $\mathrm{dom}(S)$ trivially, $B_{i-1}' \uplus B_i''$ is a linearly independent subset of $\mathrm{im}\, S^{i-1}$. Therefore, we can find a set of vectors B_{i-1}'' such that $B_i'' \subseteq B_{i-1}''$ and $B_{i-1} = B_{i-1}' \uplus B_{i-1}''$ is a basis for $\mathrm{im}\, S^{i-1}$ satisfying (i). By construction, $B_i = B_i' \uplus B_i'' \subseteq B_{i-1}$. Hence, (5.10) yields

$$S(B_{i-1}') \subseteq S(B_i') \cup S(S^{-1}(B_i)) \subseteq B_i \subseteq B_{i-1}$$

by (ii) of the inductive hypothesis. This completes the induction.

Let $B = B_0$. Then B is a basis for $\mathrm{im}\, S^0 = V$ and $B = B_0' \uplus B_0''$ where B_0' is a basis for $\mathrm{dom}(S)$ and $S(B_0') \subseteq B$. Thus we can define $f \in I_B$ by $f = S|_B$. Clearly, $S = T_f$ as they both have domain the linear span of B_0' and agree on its basis B_0'. Note that f is semi-idempotent by Proposition 5.36 as $T_f = S$ is semi-idempotent. $\qquad\square$

As a corollary of the previous lemma and our description of conjugacy in symmetric inverse monoids, we obtain the desired characterization of conjugacy for semi-idempotent partial injective linear maps.

Corollary 5.41. *Let S, T be semi-idempotent partial injective linear maps on a finite dimensional vector space V over a field $\mathbb{F}$. Then S and T are conjugate if and only if $\vec{r}(S) = \vec{r}(T)$.*

Proof. If S and T are conjugate, then the equality $\vec{r}(S) = \vec{r}(T)$ is evident. So assume that $\vec{r}(S) = \vec{r}(T)$. By Lemma 5.40, there exist bases B and B' for V and $f \in I_B$, $g \in I_{B'}$ semi-idempotent partial bijections such that S is the partial injective linear map associated with f via the embedding of I_B in $\mathrm{PAut}(V)$ coming from the basis B and T is the partial injective linear map associated with g via the embedding of $I_{B'}$ in $\mathrm{PAut}(V)$ coming from the basis B'. Let $h \colon B' \longrightarrow B$ be a bijection and put $k = hgh^{-1}$. Then $k \in I_B$ is semi-idempotent and if $U \in GL(V)$ is defined by $Ub' = h(b')$ for $b \in B'$, then $T_k = UTU^{-1}$ (where T_k is defined with respect to the basis B) and hence T_k is conjugate to T. We then have that

$$\vec{r}(f) = \vec{r}(S) = \vec{r}(T) = \vec{r}(T_k) = \vec{r}(k)$$

and so f, k are conjugate in I_B by Corollary 3.21, whence $S = T_f$ is conjugate to T_k in $\mathrm{PAut}(V)$. We conclude that S and T are conjugate, as required. $\quad\square$

The next lemma now makes precise the connection between semi-idempotent linear operators and semi-idempotent partial injective linear maps.

Lemma 5.42. *Let V be a finite dimensional vector space over a field $\mathbb{F}$.*

(i) *If $S \in \mathrm{PAut}(V)$ is semi-idempotent and $T \in \mathrm{End}(V)$ is a semi-idempotent extension of S, then $\vec{r}(S) \leq \vec{r}(T)$.*

(ii) *Suppose that $S, T \in \mathrm{PAut}(V)$ are semi-idempotent with the same rank sequence μ. Then, for any rank sequence $\nu \geq \mu$, the number of semi-idempotent operators in $\mathrm{End}(V)$ with rank sequence ν extending S is equal to the number of semi-idempotent operators in $\mathrm{End}(V)$ with rank sequence ν extending T.*

(iii) *If $S \in \mathrm{PAut}(V)$ is semi-idempotent, then there exists $T \in \mathrm{End}(V)$ a semi-idempotent operator extending S with $\vec{r}(S) = \vec{r}(T)$. Moreover, if $\mathbb{F}$ is finite, then the number of such T is a power of the characteristic of $\mathbb{F}$.*

(iv) *If $S \in \mathrm{End}(V)$ is semi-idempotent, then there exists a semi-idempotent element $T \in \mathrm{PAut}(V)$ such that T is a restriction of S and $\vec{r}(S) = \vec{r}(T)$.*

Proof. The first item is a special case of Proposition 5.35. To prove (ii), observe that there is an invertible operator U such that $S = UTU^{-1}$ by Corollary 5.41. Conjugation by U then provides a bijection between semi-idempotent extensions of T with rank sequence ν and semi-idempotent extensions of S with rank sequence ν.

To prove the third item, let S be a semi-idempotent partial injective linear map. Then by Lemma 5.40 there is a basis B for V and a semi-idempotent $f \in I_B$ such that $S = T_f$. Then Proposition 5.36 shows that S_f is a semi-idempotent operator extending T_f with $\vec{r}(S) = \vec{r}(T_f) = \vec{r}(S_f)$ and

Lemma 5.37 shows that if $\mathbb{F}$ is finite, then the number of extensions is a power of the characteristic of $\mathbb{F}$. This establishes (iii).

Finally, suppose that $S \in \text{End}(V)$ is semi-idempotent. Then we can choose an ordered basis B for V such that the matrix of S with respect to B is in Jordan canonical form. Since the matrix of S is a direct sum of nilpotent Jordan blocks and an identity matrix, we have that $S(B) \subseteq B \cup \{0\}$ and if we define a partial mapping f on B by

$$f(b) = \begin{cases} Sb, & \text{if } Sb \neq 0 \\ \text{undefined}, & \text{if } Sb = 0, \end{cases}$$

then $f \in I_B$ is semi-idempotent and $S = S_f$. Thus $T = T_f$ is a semi-idempotent partial injective linear map, which is a restriction of S, that satisfies $\vec{r}(S) = \vec{r}(T)$ by Proposition 5.36.   ⊔

Example 5.43. To illustrate the proof of Lemma 5.42(iv), consider the semi-idempotent matrix

$$S = \begin{bmatrix} 0 & 1 & 0 & 0 & 0 & 0 & 0 & 0 \\ 0 & 0 & 1 & 0 & 0 & 0 & 0 & 0 \\ 0 & 0 & 0 & 0 & 0 & 0 & 0 & 0 \\ 0 & 0 & 0 & 0 & 1 & 0 & 0 & 0 \\ 0 & 0 & 0 & 0 & 0 & 0 & 0 & 0 \\ 0 & 0 & 0 & 0 & 0 & 1 & 0 & 0 \\ 0 & 0 & 0 & 0 & 0 & 0 & 1 & 0 \\ 0 & 0 & 0 & 0 & 0 & 0 & 0 & 1 \end{bmatrix}$$

in Jordan canonical form. Then $S = S_f$ where

$$f = \begin{pmatrix} 1 & 2 & 3 & 4 & 5 & 6 & 7 & 8 \\ 1 & 2 & & 4 & 6 & 7 & 8 \end{pmatrix}$$

in I_8 is semi-idempotent.

We are now ready to prove Theorem 5.31.

Proof (of Theorem 5.31). By Proposition 5.34, it suffices to show that A_r is a unital algebra for each $0 \leq r \leq n$, where we retain the notation of that proposition. We view A_r as having basis the rank r matrices in $M_n(\mathbb{F})$ with product extending linearly the product

$$S \odot T = \begin{cases} ST, & \text{if } \text{rk}(ST) = r \\ 0, & \text{else} \end{cases}$$

on rank r matrices $S, T \in M_n(\mathbb{F})$. From now on we drop the notation "$\odot$."

The transpose operation on $M_n(\mathbb{F})$, which preserves ranks, induces an involution on A_r and hence to show that A_r has an identity, it suffices to show that it has a left identity. The monoid $M_n(\mathbb{F})$ is regular, so if T is a rank r matrix, then we can find an idempotent E with $EM_n(\mathbb{F}) = TM_n(\mathbb{F})$. In particular, this means that E has rank r and $ET = T$. Thus to prove that $e \in A_r$ is a left identity, it suffices to prove that $eE = E$ for all rank r idempotent matrices E.

Put $V = \mathbb{F}^n$ and let $\{b_1, \ldots, b_n\}$ be the standard basis for V. Let W be the span of the first r vectors $b_1, \ldots, b_r$ and let e_r be the rank r idempotent matrix fixing W and annihilating the span of $b_{r+1}, \ldots, b_n$. Let us define a sequence $\mu \in \mathbb{N}^{\mathbb{N}}$ to be *admissible* if $\mu = \vec{r}(T)$ for some semi-idempotent matrix $T \in M_n(\mathbb{F})$ of rank r and denote by $\mathfrak{A}$ the set of admissible sequences; of course, $\mathfrak{A}$ is finite. Note that $\mu \in \mathfrak{A}$ if and only if μ is the rank sequence of some semi-idempotent partial injective linear map of rank r by Lemma 5.42. Also, since every rank r semi-idempotent partial injective linear map is conjugate to one with domain W, it follows that $\mathfrak{A}$ is precisely the set of rank sequences of semi-idempotent partial injective linear maps with domain W.

Note that a partial linear mapping on V with domain W is the same thing as a linear mapping $W \longrightarrow V$. Using our given bases $\{b_1, \ldots, b_r\}$ for W and $\{b_1, \ldots, b_n\}$ for V, we can identify partial linear maps on V with domain W with $n \times r$-matrices T over $\mathbb{F}$. The corresponding partial linear map is injective if and only if T has full rank, that is, rank r. The $n \times r$-matrix corresponding to the partial identity 1_W is the matrix whose first r rows form the $r \times r$ identity matrix and whose remaining rows form an $(n - r) \times r$ zero matrix.

If $\mu \in \mathfrak{A}$, let $a_\mu \in A_r$ be the sum of all semi-idempotent matrices T in $M_n(\mathbb{F})$ such that $\vec{r}(T) = \mu$. The conjugation action of $GL_n(\mathbb{F})$ on $M_n(\mathbb{F})$ preserves rank and hence induces an action of $GL_n(\mathbb{F})$ on A_r by automorphisms that we also denote by conjugation. Notice that $Ua_\mu U^{-1} = a_\mu$ for all $U \in GL_n(\mathbb{F})$ as a consequence of Lemma 5.39. Our aim is to show that we can choose $k_\mu \in \Bbbk$, for each $\mu \in \mathfrak{A}$, such that

$$e = \sum_{\mu \in \mathfrak{A}} k_\mu a_\mu$$

is a left identity for A_r. Note that $UeU^{-1} = e$ for all $U \in GL_n(\mathbb{F})$.

Observe that if E is a rank r idempotent matrix, then $E = Ue_rU^{-1}$ for some invertible matrix U (this follows from elementary linear algebra, or by observing that E and e_r are both semi-idempotent with rank sequence $(n, r, r, r, \ldots)$). Let us assume for the moment that $ee_r = e_r$. Then $eE = eUe_rU^{-1} = U(U^{-1}eU)e_rU^{-1} = Uee_rU^{-1} = Ue_rU^{-1} = E$. Thus to prove that e is a left identity for A_r, it suffices to prove that $ee_r = e_r$.

If S is a rank r semi-idempotent matrix and $Se_r \neq 0$ in A_r, then

$$Se_r = [T \mid 0]$$

where $S|_W = T \colon W \longrightarrow V$ has rank r. Thus T can be viewed as a semi-idempotent partial injective linear mapping with domain W. It follows that ee_r is a linear combination of matrices of the form $[T \mid 0]$ with T a semi-idempotent partial injective linear map with domain W.

Fix a semi-idempotent partial injective linear mapping T with domain W and rank sequence $\mu \in \mathfrak{A}$. If $\nu \in \mathfrak{A}$, then the number of semi-idempotent matrices S with $\bar{r}(S) = \nu$ and $S|_W = T$ is a nonnegative integer $c_{\mu,\nu}$ that depends only on μ and ν and, moreover, $c_{\mu,\nu} = 0$ unless $\nu \geq \mu$ by Lemma 5.42. Thus the coefficient of $[T \mid 0]$ in ee_r is $\sum_{\nu \geq \mu} c_{\mu,\nu} k_\nu$.

Also note that 1_W is the unique semi-idempotent partial injective linear mapping with domain W and rank sequence $\bar{r} = (n, r, r, r, \ldots)$. Indeed, if $T \in \mathrm{PAut}(V)$ is semi-idempotent with rank sequence $\bar{r}$ and domain W, then $\mathrm{im}^\infty T \subseteq W$ and $\dim \mathrm{im}^\infty T = r = \dim W$, whence $T = T|_{\mathrm{im}^\infty T} = 1_W$ by semi-idempotence. Furthermore, $\bar{r}$ is the unique maximal element of $\mathfrak{A}$ and $e_r = [1_W \mid 0]$ with our notational conventions.

It follows that $ee_r = e_r$ if and only if the system of equations

$$1 = c_{\bar{r},\bar{r}} k_{\bar{r}}$$
$$0 = \sum_{\nu \geq \mu} c_{\mu,\nu} k_\nu \qquad (\mu \in \mathfrak{A} \setminus \{\bar{r}\}) \tag{5.11}$$

has a solution. By Lemma 5.42(iii), $c_{\mu,\mu}$ is a power of the characteristic of $\mathbb{F}$ for all $\mu \in \mathfrak{A}$ and hence if we order $\mathfrak{A}$ lexicographically, then the system of equations (5.11) is upper triangular with diagonal entries invertible in $\Bbbk$ by the hypothesis on $\Bbbk$. It follows that the system has a solution and so A_r has an identity element. This completes the proof of Theorem 5.31 and its corollary, Corollary 5.32. $\qquad\square$

Remark 5.44. It is not difficult to see from the proof that Theorem 5.31 remains valid so long as $\Bbbk$ is a commutative ring with unit such that the characteristic of $\mathbb{F}$ is invertible in $\Bbbk$ [Kov92]. In fact, the solution to system (5.11) is over $\mathbb{Z}[1/p]$ where p is the characteristic of $\mathbb{F}$.

Remark 5.45. Theorem 5.21 and Corollary 5.32 imply that each sandwich matrix of $M_n(\mathbb{F})$, where $\mathbb{F}$ is a finite field, is invertible over the complex group algebra of the corresponding maximal subgroup. In fact, it will follow from Theorems 15.6 and 5.31 that each sandwich matrix is invertible over the group algebra of the maximal subgroup for any field of characteristic different from that of $\mathbb{F}$.

The following corollary makes explicit how to go from simple modules over the general linear group to simple modules over the full matrix monoid using Theorem 5.31.

Corollary 5.46. *Let $\mathbb{F}$ be a finite field of characteristic p and let $\Bbbk$ be a field whose characteristic is different from p. Let e_r be the rank r idempotent matrix in $M_n(\mathbb{F})$ that fixes the first r standard basis vectors and annihilates the remaining ones and identify the maximal subgroup G_{e_r} with $GL_r(\mathbb{F})$. Let k_r be the number of r-dimensional subspaces of $\mathbb{F}^n$. If V is a simple $\Bbbk GL_r(\mathbb{F})$-module, then $V^\sharp = \mathrm{Ind}_{G_{e_r}}(V)$ and hence has dimension $k_r \cdot \dim V$.*

Proof. We continue to denote by I_k the ideal of $M_n(\mathbb{F})$ consisting of matrices of rank at most k, with $I_{-1} = \emptyset$. Consider again the algebra $A_r = \Bbbk I_r / \Bbbk I_{r-1}$ and recall from the proof of Proposition 5.34 that $e_r A_r e_r \cong \Bbbk GL_r(\mathbb{F})$ and that $A_r e_r \cong \Bbbk L_{e_r}$ is a free right $\Bbbk GL_r(\mathbb{F})$-module of rank k_r (actually, the dual of the latter fact was shown, but the same proof applies using column equivalence instead of row equivalence). As A_r is a unital algebra (by the proof of Theorem 5.31) and $A_r = A_r e_r A_r$, it follows that

$$\mathrm{Ind}_{e_r} : \Bbbk GL_r(\mathbb{F})\text{-mod} \longrightarrow A_r\text{-mod}$$

is an equivalence of categories by Theorem 4.13 and hence sends simple modules to simple modules. But then we have that $\mathrm{Ind}_{G_{e_r}}(V) = \mathrm{Ind}_{e_r}(V) = A_r e_r \otimes_{\Bbbk GL_r(\mathbb{F})} V$ is a simple $\Bbbk M_n(\mathbb{F})$-module of dimension $k_r \cdot \dim V$ and hence $V^\sharp = \mathrm{Ind}_{G_{e_r}}(V)$ by Theorem 5.5. This completes the proof. $\qquad\square$

The irreducible characters of the general linear group $GL_n(\mathbb{F}_q)$ over the field of complex numbers were determined by Green [Gre55] and so one can use Corollary 5.46 to describe, in principle, the irreducible characters of $M_n(\mathbb{F}_q)$ over the complex numbers; the reader should consult Chapter 7 for the definition of characters.

5.7 Exercises

5.1. Let M be a monoid and $\Bbbk$ a field. Let I be an ideal of M and let $\Bbbk I$ be the $\Bbbk$-subspace of $\Bbbk M$ spanned by I. Prove that $\Bbbk I$ is an ideal.

5.2. Verify that if V, W are $\Bbbk M$-modules, then so is $V \otimes W$.

5.3. Let G be a group acting freely on a set X and let $\Bbbk$ be a field. Let $T \subseteq X$ be a complete set of representatives of the orbits of G on X. Prove that $\Bbbk X$ is a free $\Bbbk G$-module with basis T.

5.4. Let M be a monoid and $\Bbbk$ a field. Prove that M is finitely generated as a monoid if and only if $\Bbbk M$ is finitely generated as a $\Bbbk$-algebra.

5.5. Let M be a finite monoid and $\Bbbk$ a field. Suppose that $G_e = \{e\}$ and J_e is a subsemigroup of M. Prove that the unique simple $\Bbbk M$-module with apex e affords the one-dimensional representation $\chi_{J_e} : M \longrightarrow \Bbbk$ given by

$$\chi_{J_e}(m) = \begin{cases} 1, & \text{if } e \in MmM \\ 0, & \text{if } m \in I(e). \end{cases}$$

Deduce that if each maximal subgroup of M is trivial and each regular $\mathscr{J}$-class of M is a subsemigroup, then each simple $\Bbbk M$-module is one-dimensional.

5.6. Compute the dimension of $\mathrm{rad}(\mathbb{C}T_n)$.

5.7. Let $\rho\colon M \longrightarrow M_n(\Bbbk)$ be an irreducible representation with apex $e \in E(M)$. Extending ρ to $\Bbbk M$, prove that the identity matrix belongs to $\rho(\Bbbk I) = \rho(\Bbbk J_e)$ where $I = MeM$. (Hint: $A = \rho(\Bbbk M)$ is simple by Proposition A.12.)

5.8 (Rhodes). Suppose that M is a nontrivial finite monoid with a faithful irreducible representation $\rho\colon M \longrightarrow M_n(\Bbbk)$ with apex $e \in E(M)$.

(a) Prove $I = MeM$ is the unique minimal nonzero ideal of M.
(b) Prove that M acts faithfully on both the left and right of I. (Hint: use Exercise 5.7.)

5.9. Let M be a finite monoid, $\Bbbk$ a field and $e \in E(M)$. Recall that D denotes the standard duality; see Appendix A.

(a) Prove that $\mathrm{Coind}_{G_e}(\Bbbk G_e) \cong D(\Bbbk R_e) = \mathrm{Hom}_{\Bbbk}(\Bbbk R_e, \Bbbk) \cong \Bbbk^{R_e}$ where if $\varphi\colon R_e \longrightarrow \Bbbk$ is a mapping and $m \in M$, then

$$(m\varphi)(r) = \begin{cases} \varphi(rm), & \text{if } rm \in R_e \\ 0, & \text{else} \end{cases}$$

for $r \in R_e$. (Hint: let $\tau\colon \Bbbk G_e \longrightarrow \Bbbk$ be the functional defined on $g \in G_e$ by

$$\tau(g) = \begin{cases} 1, & \text{if } g \neq e \\ 0, & \text{else}; \end{cases}$$

map $\varphi \in \mathrm{Hom}_{G_e}(R_e, \Bbbk G_e)$ to $\tau \circ \varphi\colon R_e \longrightarrow \Bbbk$.)
(b) Prove that if V is a finite dimensional $\Bbbk G_e$-module, then $\mathrm{Coind}_{G_e}(V) \cong D(D(V) \otimes_{\Bbbk G_e} \Bbbk R_e)$.

5.10. Let $r, n \geq 1$. Prove that $\Bbbk T_{n,r} \cong (\Bbbk^n)^{\otimes r}$ as $\Bbbk T_n$-$\Bbbk S_r$-bimodules where S_r acts on the right of $(\Bbbk^n)^{\otimes r}$ by permuting the tensor factors.

5.11. Let $n \geq 1$ and let e_r be the idempotent from Section 5.3 for $1 \leq r \leq n$. Prove that $\mathrm{Ind}_{G_{e_r}}(V) \cong \Bbbk I_{n,r} \otimes_{\Bbbk S_r} V$ for any $\Bbbk S_r$-module V.

5.12. Let $1 \leq r \leq n$ and let $\Bbbk$ be a field of characteristic 0. We retain the notation of Section 5.3.

(a) Prove that $\Bbbk I_{n,r} \otimes_{\Bbbk S_r} S_{(1^r)} \cong \Lambda^r(\Bbbk^n)$ as $\Bbbk T_n$-modules.
(b) Prove that $\Lambda^r(\Bbbk^n)/\Lambda^r(\mathrm{Aug}(\Bbbk^n)) \cong \Lambda^{r-1}(\mathrm{Aug}(\Bbbk^n))$ as $\Bbbk T_n$-modules.
(c) Prove that $\Lambda^r(\Bbbk^n) \cong \Bbbk T_n \varepsilon_r$ where

$$\varepsilon_r = \frac{1}{r!} \sum_{g \in G_{e_r}} \mathrm{sgn}(g|_{[r]})g$$

is a primitive idempotent. (Hint: Exercise 5.10 may be useful.)
(d) Deduce that $\Lambda^r(\Bbbk^n)$ is a projective indecomposable module with simple top $\Lambda^{r-1}(\mathrm{Aug}(\Bbbk^n))$ and $\mathrm{rad}(\Lambda^r(\Bbbk^n)) = \Lambda^r(\mathrm{Aug}(\Bbbk^n))$.

(e) Prove that there is an exact sequence

$$0 \longrightarrow \mathrm{Coind}_{G_{e_r}}(S_{(1^r)}) \longrightarrow D(\varepsilon_r \Bbbk T_n) \longrightarrow \mathrm{Coind}_{G_{e_{r-1}}}(S_{(1^{r-1})}) \longrightarrow 0$$

where D denotes the standard duality.

5.13. Let λ be a partition of r with $1 \leq r \leq n$ and $\lambda \neq (1^r)$. Show that there is an isomorphism $\Bbbk I_{n,r} \otimes_{\Bbbk S_r} S_\lambda \longrightarrow U_{\lambda,n}$ sending $f \otimes \varepsilon_t$ to ε_{ft}, for $f \in I_{n,r}$ and t a λ-tableau, where ft is the n-tableau of shape λ obtained by applying f to the entries of t.

5.14. Let $\Bbbk$ be a field of characteristic 0. Adapt the proof of Theorem 5.10 to show that if e_r is a rank r idempotent of PT_n (respectively, I_n) and S_λ is a simple $\Bbbk S_r$-module, then $\mathrm{Ind}_{G_{e_r}}(S_\lambda)$ is a simple $\Bbbk PT_n$-module (respectively, $\Bbbk I_n$-module). Also, give a description of $\mathrm{Ind}_{G_{e_r}}(S_\lambda)$ in terms of n-polytabloids of shape λ.

5.15. Let M be a finite monoid and $\mathbb{F} \subseteq \Bbbk$ be fields. Suppose that, for each $e \in E(M)$, every simple $\Bbbk G_e$-module is defined over $\mathbb{F}$. Prove that each simple $\Bbbk M$-module is defined over $\mathbb{F}$.

5.16. Let M be a finite monoid and $e \in E(M)$.

(a) Prove that J_e is a subsemigroup if and only if the sandwich matrix $P(e)$ has no zeroes.

(b) Use Corollary 5.30 to give another proof of Exercise 5.5.

5.17. Prove Theorem 5.28.

5.18. Let M be finite inverse monoid and $e \in E(M)$. Prove that sandwich matrix $P(e)$ can be chosen to be an identity matrix. Deduce that $\Bbbk M$ is semisimple as long as the characteristic of $\Bbbk$ does not divide the order of any maximal subgroup of M. (Hint: use Corollary 1.17 and Exercise 3.5.)

5.19. Compute sandwich matrices for each $\mathscr{J}$-class of T_2, T_3, and T_4.

5.20. Compute matrix representations afforded by the simple $\mathbb{C}T_2$ and $\mathbb{C}T_3$-modules.

5.21. Let $\mathbb{F}_q$ be the finite field of order q and let $T \in M_n(\mathbb{F}_q)$. Prove that $\ker^\infty T = \ker T^\omega$ and $\mathrm{im}^\infty T = \mathrm{im}\, T^\omega$.

5.22. Let $\mathbb{F}$ be a field and let $e \in M_n(\mathbb{F})$ be a rank r idempotent. Prove that $eM_n(\mathbb{F})e \cong M_r(\mathbb{F})$.

5.23. Compute explicitly the identities of A_0, A_1, A_2 for $\mathbb{C}M_2(\mathbb{F}_2)$.

5.24 (Putcha). An ideal I of a finite dimensional algebra A is called a *heredity ideal* [CPS88] if the following hold.

(i) $I = AeA$ for some idempotent $e \in A$.

(ii) eAe is semisimple.

(iii) I is a projective (left) A-module.

A finite dimensional algebra A is *quasi-hereditary* [CPS88] if there is a filtration

$$\{0\} = I_0 \subsetneq I_1 \subsetneq \cdots \subsetneq I_n = A \tag{5.12}$$

of A by ideals such that I_k/I_{k-1} is a heredity ideal of A/I_{k-1} for $1 \leq k \leq n$. One calls (5.12) a *heredity chain*.

Prove that if M is a regular monoid and $\Bbbk$ is a field of characteristic 0, then $\Bbbk M$ is quasi-hereditary. (Hint: use a principal series for M to construct a heredity chain for $\Bbbk M$; can you find a shorter chain?)

Part III

Character Theory

6

The Grothendieck Ring

In this chapter, we introduce the Grothendieck ring of a finite monoid M over a field $\Bbbk$. The main result is that the Grothendieck ring of M is isomorphic to the direct product of the Grothendieck rings of its maximal subgroups (one per regular $\mathscr{J}$-class). This result was first proved by McAlister for $\Bbbk = \mathbb{C}$ in the language of virtual characters [McA72]. In Chapter 7, the results of this chapter will be used to study the ring of characters and the character table of a finite monoid. Throughout this chapter we hold fixed a finite monoid M and a field $\Bbbk$.

6.1 The Grothendieck ring

If V is a $\Bbbk M$-module, then we denote by $[V]$ the isomorphism class of V. The additive group of the *Grothendieck ring* $G_0(\Bbbk M)$ of M with respect to $\Bbbk$ is the abelian group generated by the set of isomorphism classes of finite dimensional $\Bbbk M$-modules subject to the relations $[V] = [U] + [W]$ whenever there is an exact sequence of the form

$$0 \longrightarrow U \longrightarrow V \longrightarrow W \longrightarrow 0.$$

Notice that if V, W are $\Bbbk M$-modules, then there is an exact sequence

$$0 \longrightarrow V \longrightarrow V \oplus W \longrightarrow W \longrightarrow 0$$

and so $[V] + [W] = [V \oplus W]$. It follows that every element of $G_0(\Bbbk M)$ can be written in the form $[V] - [W]$ with V, W finite dimensional $\Bbbk M$-modules.

A ring structure is defined on $G_0(\Bbbk M)$ by putting $[V][W] = [V \otimes W]$ and extending linearly, where all unlabeled tensor products are over the ground field $\Bbbk$. The identity is the trivial module $\Bbbk$. We must check that the multiplication is well defined.

© Springer International Publishing Switzerland 2016
B. Steinberg, *Representation Theory of Finite Monoids*,
Universitext, DOI 10.1007/978-3-319-43932-7_6

Proposition 6.1. *Let M be a monoid and $\Bbbk$ a field. Then $G_0(\Bbbk M)$ is a commutative ring with unit.*

Proof. The tensor product of modules is commutative and associative up to isomorphism. Also, the natural map $\Bbbk \otimes V \longrightarrow V$ given by $c \otimes v \mapsto cv$ is a $\Bbbk M$-module isomorphism if $\Bbbk$ is given the structure of the trivial module. Therefore, the free abelian group on the set of isomorphism classes of finite dimensional $\Bbbk M$-modules can be made a unital commutative ring by putting $[V][W] = [V \otimes W]$ and extending linearly, where the trivial module is the multiplicative identity.

We then just need to show that the subgroup generated by the set of elements of the form $[V] - ([U] + [W])$ for which there is an exact sequence of the form

$$0 \longrightarrow U \longrightarrow V \longrightarrow W \longrightarrow 0$$

is an ideal. But this is clear because tensoring over a field is exact. It follows that $G_0(\Bbbk M)$ is a commutative ring with unit. □

As an abelian group, $G_0(\Bbbk M)$ turns out to be free with basis the simple modules. To prove this, let us first establish some notation and a basic lemma. If V is a finite dimensional $\Bbbk M$-module and S is a simple $\Bbbk M$-module, we denote by $[V : S]$ the multiplicity of S as a composition factor of V.

Lemma 6.2. *Suppose that*

$$0 \longrightarrow U \longrightarrow V \longrightarrow W \longrightarrow 0$$

is an exact sequence of finite dimensional $\Bbbk M$-modules and that S is a simple $\Bbbk M$-module. Then $[V : S] = [U : S] + [W : S]$.

Proof. Without loss of generality, we may assume that U is contained in V and $W = V/U$. Let $0 = U_0 \subseteq \cdots \subseteq U_r = U$ and $0 = W_0/U \subseteq \cdots \subseteq W_s/U = W$ be composition series with $U \subseteq W_i$. Then we have that

$$0 = U_0 \subseteq \cdots \subseteq U_r \subseteq W_1 \subseteq \cdots \subseteq W_s = V$$

is a composition series for V. From the isomorphism

$$W_{i+1}/W_i \cong (W_{i+1}/U)/(W_i/U)$$

and the Jordan-Hölder theorem, we conclude $[V : S] = [U : S] + [W : S]$. □

Now we prove that the isomorphism classes of simple $\Bbbk M$-modules form a basis for $G_0(\Bbbk M)$.

Proposition 6.3. *The additive group of $G_0(\Bbbk M)$ is free abelian with basis the set of isomorphism classes of simple $\Bbbk M$-modules. Moreover, if V is a finite dimensional $\Bbbk M$-module, then the decomposition*

$$[V] = \sum_{[S] \in \mathrm{Irr}_{\Bbbk}(M)} [V : S][S] \tag{6.1}$$

holds.

Proof. We prove that (6.1) holds by induction on the number of composition factors (i.e., the length) of V. If $V = 0$, there is nothing to prove. Otherwise, let U be a maximal proper submodule of V and consider the exact sequence

$$0 \longrightarrow U \longrightarrow V \longrightarrow V/U \longrightarrow 0.$$

Then V/U is simple and by induction $[U] = \sum_{[S] \in \mathrm{Irr}_{\Bbbk}(M)} [U : S][S]$. Applying Lemma 6.2, we conclude that

$$
\begin{aligned}
[V] &= [U] + [V/U] \\
&= ([U : V/U] + 1)[V/U] + \sum_{[S] \in \mathrm{Irr}_{\Bbbk}(M) \setminus \{[V/U]\}} [U : S] \\
&= \sum_{[S] \in \mathrm{Irr}_{\Bbbk}(M)} [V : S][S]
\end{aligned}
$$

as required. It follows that $G_0(\Bbbk M)$ is generated by $\mathrm{Irr}_{\Bbbk}(M)$.

Fix a simple $\Bbbk M$-module S. By Lemma 6.2 the mapping $[V] \mapsto [V : S]$ extends to a homomorphism $f_S \colon G_0(\Bbbk M) \longrightarrow \mathbb{Z}$. Moreover, if S, S' are simple $\Bbbk M$-modules, then

$$
f_{S'}([S]) = \begin{cases} 1, & \text{if } [S] = [S'] \\ 0, & \text{else.} \end{cases}
$$

Suppose that $0 = \sum_{[S] \in \mathrm{Irr}_{\Bbbk}(M)} n_{[S]} \cdot [S]$ with the $n_{[S]} \in \mathbb{Z}$. If S' is a simple $\Bbbk M$-module, then

$$
0 = f_{S'}\left(\sum_{[S] \in \mathrm{Irr}_{\Bbbk}(M)} n_{[S]} \cdot [S] \right) = n_{[S']}.
$$

Thus the set $\mathrm{Irr}_{\Bbbk}(M)$ is linearly independent. This completes the proof. $\qquad \square$

It is worth mentioning that there is another important ring associated with the representation theory of M over $\Bbbk$, called the *representation ring* of M, denoted $\mathcal{R}_{\Bbbk}(M)$. Namely, one can take the free abelian group $\mathcal{R}_{\Bbbk}(M)$ on the isomorphism classes of finite dimensional $\Bbbk M$-modules modulo the relations $[V] + [W] = [V \oplus W]$. One again defines multiplication by $[V][W] = [V \otimes W]$. Notice that there is a natural surjective ring homomorphism

$$\mathcal{R}_{\Bbbk}(M) \longrightarrow G_0(\Bbbk M).$$

As an abelian group, $\mathcal{R}_{\Bbbk}(M)$ is easily checked (using the Krull-Schmidt theorem) to be a free abelian group on the set of isomorphism classes of finite dimensional indecomposable $\Bbbk M$-modules. This may very well be an infinite set. Since many natural finite monoids are of so-called 'wild representation type,' meaning that classifying their indecomposable modules is as hard as classifying the indecomposable modules of any finitely generated $\Bbbk$-algebra, it seems unlikely that we will be able to understand $\mathcal{R}_{\Bbbk}(M)$ in much generality.

6.2 The restriction isomorphism

Now let $e \in E(M)$ be an idempotent. The restriction functor

$$\mathrm{Res}_{G_e} : \Bbbk M\text{-mod} \longrightarrow \Bbbk G_e\text{-mod}$$

given by $V \mapsto eV$ is an exact functor and hence induces a homomorphism $\mathrm{Res}_{G_e} : G_0(\Bbbk M) \longrightarrow G_0(\Bbbk G_e)$ of abelian groups defined by $\mathrm{Res}_{G_e}([V]) = [eV]$.

Proposition 6.4. *Let $e \in E(M)$. Then $\mathrm{Res}_{G_e} : G_0(\Bbbk M) \longrightarrow G_0(\Bbbk G_e)$ is a ring homomorphism.*

Proof. Clearly, if $\Bbbk$ denotes the trivial module, then we have that $\mathrm{Res}_{G_e}([\Bbbk]) = [e\Bbbk] = [\Bbbk]$. Let V, W be $\Bbbk M$-modules and $v_1, \ldots, v_m \in V$ and $w_1, \ldots, w_m \in W$. We compute that

$$e \cdot \left(\sum_{i=1}^{m} v_i \otimes w_i \right) = \sum_{i=1}^{m} e v_i \otimes e w_i$$

from which it follows that $e(V \otimes W) = eV \otimes eW$. Thus Res_{G_e} is a ring homomorphism. □

Fix now a complete set of idempotent representatives $e_1, \ldots, e_s$ of the regular $\mathscr{J}$-classes of M. Assume that we have chosen the ordering so that $Me_iM \subseteq Me_jM$ implies $i \leq j$. In particular, $e_s = 1$ and e_1 belongs to the minimal ideal of M. We fix a total ordering $\preceq$ on $\mathrm{Irr}_\Bbbk(M)$ such that if e_i is the apex of S and e_j is the apex of S' with $i < j$, then $[S] \prec [S']$. We put a corresponding total ordering, also denoted $\preceq$, on the disjoint union $\bigcup_{i=1}^{s} \mathrm{Irr}_\Bbbk(G_e)$ by putting $[S_1] \preceq [S_2]$ if and only if $[S_1^\sharp] \preceq [S_2^\sharp]$, where we retain the notation from Theorem 5.5. The main theorem of this chapter is the following.

Theorem 6.5. *Let $e_1, \ldots, e_s$ be a complete set of idempotent representatives of the regular $\mathscr{J}$-classes of M. Then the mapping*

$$\mathrm{Res}: G_0(\Bbbk M) \longrightarrow \prod_{i=1}^{s} G_0(\Bbbk G_{e_i})$$

given by

$$\mathrm{Res}([V]) = ([e_1 V], \ldots, [e_s V])$$

is a ring isomorphism. Moreover, if we take as bases $\mathrm{Irr}_\Bbbk(M)$ for $G_0(\Bbbk M)$ and $\bigcup_{i=1}^{s} \mathrm{Irr}_\Bbbk(G_{e_i})$ for $\prod_{i=1}^{s} G_0(\Bbbk G_{e_i})$, ordered as above, then the matrix L for Res is lower triangular with ones along the diagonal (i.e., unipotent lower triangular).

Proof. It follows from Proposition 6.4 that Res is a ring homomorphism. Let $W^\sharp$ be a simple $\Bbbk M$-module with apex e_j corresponding to a simple $\Bbbk G_{e_j}$-module W under Theorem 5.5. Then $e_j W^\sharp \cong W$ and $e_i W^\sharp = 0$ unless

$Me_jM \subseteq Me_iM$ by Proposition 5.4. Hence $e_iW^\sharp = 0$ if $i < j$. Therefore, we have

$$\mathrm{Res}([W^\sharp]) = \sum_{i=1}^s [e_iW^\sharp] = \sum_{i=1}^s \sum_{[V]\in\mathrm{Irr}_k(G_{e_i})} [e_iW^\sharp : V][V]$$

$$= [W] + \sum_{i=j+1}^s \sum_{[V]\in\mathrm{Irr}_k(G_{e_i})} [e_iW^\sharp : V][V].$$

This establishes that L is lower triangular with ones along the diagonal. In particular, L is invertible. $\square$

We call the matrix L from Theorem 6.5 the *decomposition matrix* of M over k.

Let us compute the decomposition matrix for the full transformation monoid T_n. We use the idempotents $e_1,\ldots,e_n$ from Section 5.3. For $1 \leq r \leq n$, denote by $\mathcal{P}_r$ the set of all partitions of r and put $\mathcal{P}(n) = \bigcup_{r=1}^n \mathcal{P}_r$. We can view L as a $\mathcal{P}(n) \times \mathcal{P}(n)$ matrix where $L_{\lambda,\mu} = [e_m S_\mu^\sharp : S_\lambda]$ for μ a partition of r and λ a partition of m with $1 \leq m,r \leq n$. It is convenient to order $\mathcal{P}(n)$ so that $\alpha \prec \beta$ if $|\alpha| < |\beta|$, or $|\alpha| = |\beta|$ and there exists j such that $\alpha_i = \beta_i$ for $i < j$ and $\alpha_j < \beta_j$ where $\alpha = (\alpha_1,\ldots,\alpha_r)$ and $\beta = (\beta_1,\ldots,\beta_s)$.

Proposition 6.6. *Let k be a field of characteristic 0. The decomposition matrix L for the full transformation monoid T_n over k is given by*

$$L_{\lambda,\mu} = \begin{cases} 1, & \text{if } \mu = (1^r) \text{ and } \lambda = (m-r+1,1^{r-1}) \text{ with } m \geq r \\ 1, & \text{if } \mu \neq (1^r), \mu \subseteq \lambda \text{ and } \lambda \setminus \mu \text{ is a horizontal strip} \\ 0, & \text{else} \end{cases}$$

for $|\mu| = r$ and $|\lambda| = m$.

Proof. The result for $\mu = (1^r)$ follows directly from Theorem 5.9; the remaining cases follow from Corollary 5.13 and Pieri's rule (Theorem B.13). $\square$

Example 6.7. We compute here the decomposition matrix L for T_3 using Proposition 6.6. In this case $\mathcal{P}(3) = \{(1),(1^2),(2),(1^3),(2,1),(3)\}$. Then $(3)\setminus(2)$ and $(2,1)\setminus(2)$ are horizontal strips and so we have that

$$L = \begin{bmatrix} 1&0&0&0&0&0 \\ 0&1&0&0&0&0 \\ 1&0&1&0&0&0 \\ 0&0&0&1&0&0 \\ 0&1&1&0&1&0 \\ 1&0&1&0&0&1 \end{bmatrix}.$$

6.3 The triangular Grothendieck ring

If S, T are one-dimensional simple $\Bbbk M$-modules, then $S \otimes T$ is also one-dimensional. Hence the one-dimensional simple $\Bbbk M$-modules form a basis for a unital subring $G_0^\triangledown(\Bbbk M)$ of $G_0(\Bbbk M)$, which we call the *triangular Grothendieck ring* of M over $\Bbbk$. The reason for this terminology is based on the following lemma.

Lemma 6.8. *Let V be a finite dimensional $\Bbbk M$-module. Then V affords a representation $\varphi \colon M \longrightarrow M_n(\Bbbk)$ by upper triangular matrices if and only if each composition factor of V is one-dimensional, i.e., $[V] \in G_0^\triangledown(\Bbbk M)$.*

Proof. By Proposition 6.3 we have that $[V] \in G_0^\triangledown(\Bbbk M)$ if and only if each composition factor of V is one-dimensional. Suppose first that V affords a representation $\varphi \colon M \longrightarrow M_n(\Bbbk)$ by upper triangular matrices. Let $e_1, \ldots, e_n$ denote the standard basis vectors and let V_i be the subspace of $\Bbbk^n$ spanned by $e_1, \ldots, e_i$ for $i = 0, \ldots, n$. Then because φ is a representation by upper triangular matrices, we have that each V_i is a $\Bbbk M$-submodule. Moreover, V_i/V_{i-1} is one-dimensional for $1 \le i \le n$. Thus

$$0 = V_0 \subseteq V_1 \subseteq \cdots \subseteq V_n = V \tag{6.2}$$

is a composition series with V_i/V_{i-1} one-dimensional for $i = 1, \ldots, n$.

Conversely, suppose that (6.2) is a composition series for V with one-dimensional composition factors. Let $v_1 \in V_1$ be a nonzero vector. Assume inductively that we have found vectors $v_1, \ldots, v_i$ such that $\{v_1, \ldots, v_j\}$ is a basis for V_j for all $0 \le j \le i$. Choose $v_{i+1} \in V_{i+1}$ such that $v_{i+1} + V_i \ne V_i$. Then $\{v_1, \ldots, v_j\}$ is a basis for V_j for $0 \le j \le i+1$. It follows that we have a basis $\{v_1, \ldots, v_n\}$ for V such that $\{v_1, \ldots, v_j\}$ is a basis for V_j for $0 \le j \le n$. The matrix representation $\varphi \colon M \longrightarrow M_n(\Bbbk)$ afforded by V with respect to the ordered basis $v_1, \ldots, v_n$ is clearly by upper triangular matrices.   □

We shall call a $\Bbbk M$-module that satisfies the equivalent conditions of Lemma 6.8 a *triangularizable module*.

Corollary 6.9. *The tensor product of triangularizable modules is again triangularizable and the trivial module is triangularizable.*

Proof. This follows from Lemma 6.8 and Proposition 6.3 because $G_0^\triangledown(\Bbbk M)$ is a subring of $G_0(\Bbbk M)$.   □

6.4 The Grothendieck group of projective modules

We end this chapter by considering the *Grothendieck group* $K_0(\Bbbk M)$. By the Krull-Schmidt theorem, the set of isomorphism classes of finite dimensional projective $\Bbbk M$-modules is a free commutative monoid on the isomorphism classes of projective indecomposable $\Bbbk M$-modules with respect to the binary operation of direct sum. The group $K_0(\Bbbk M)$ is then the corresponding group of fractions. It consists of all formal differences $[P] - [Q]$ of finite dimensional projective $\Bbbk M$-modules with the operation

$$[P] - [Q] + [P'] - [Q'] = [P \oplus P'] - [Q \oplus Q'].$$

It is a free abelian group with basis the set of isomorphism classes of projective indecomposable modules. Of course, if $\Bbbk M$ is semisimple, then $K_0(\Bbbk M) = G_0(\Bbbk M)$.

There is a natural abelian group homomorphism

$$C \colon K_0(\Bbbk M) \longrightarrow G_0(\Bbbk M)$$

given by

$$C([P]) = [P] = \sum_{[S] \in \mathrm{Irr}_\Bbbk(M)} [P : S][S]$$

for a finite dimensional projective module P. If $P_1, \ldots, P_r$ form a complete set of representatives of the isomorphism classes of projective indecomposable $\Bbbk M$-modules and $S_i = P_i / \mathrm{rad}(P_i)$ is the corresponding simple module, for $1 \leq i \leq r$, then the matrix for C with respect to the bases $\{[P_1], \ldots, [P_r]\}$ and $\{[S_1], \ldots, [S_r]\}$ of $K_0(\Bbbk M)$ and $G_0(\Bbbk M)$, respectively, is called the *Cartan matrix* of $\Bbbk M$. We shall return to the Cartan matrix, and how to compute it, later in the text.

Notice that if P is a finite dimensional projective module, then because the functor $\mathrm{Hom}_{\Bbbk M}(P, -)$ is exact, there is a well-defined functional $f_P \colon G_0(\Bbbk M) \longrightarrow \mathbb{Z}$ given by $f_P([V]) = \dim \mathrm{Hom}_{\Bbbk M}(P, V)$. Moreover, one has $f_{P \oplus Q}([V]) = f_P([V]) + f_Q([V])$. Thus we have a bilinear pairing

$$\langle -, - \rangle \colon K_0(\Bbbk M) \times G_0(\Bbbk M) \longrightarrow \mathbb{Z}$$

given by

$$\langle [P], [V] \rangle = \dim \mathrm{Hom}_{\Bbbk M}(P, V).$$

If $\Bbbk$ is an algebraically closed field and $P_i = \Bbbk M f_i$ is a projective indecomposable with corresponding primitive idempotent f_i and with simple quotient $P_i / \mathrm{rad}(P_i) = S_i$, then

$$\langle [P_i], [V] \rangle = \dim \mathrm{Hom}_{\Bbbk M}(P_i, V) = \dim f_i V = [V : S_i] \tag{6.3}$$

(cf. Proposition A.24). It follows that, in the setting of an algebraically closed field, $\{[P_1], \ldots, [P_r]\}$ and $\{[S_1], \ldots, [S_r]\}$ are dual bases with respect to the pairing.

The reader will verify in Exercise 17.15 that the projective $\Bbbk M$-modules are not in general closed under tensor product and so $K_0(\Bbbk M)$ is not usually a ring.

6.5 Exercises

6.1. Let M be a finite monoid and $\Bbbk$ a field. Let $e_1, \ldots, e_s$ be a complete set of idempotent representatives of the regular $\mathscr{J}$-classes of M. Prove that elements of the form $[\mathrm{Ind}_{G_{e_i}}(V)]$ with $[V] \in \mathrm{Irr}_{\Bbbk}(G_{e_i})$ form a basis for $G_0(\Bbbk M)$.

6.2. Let M be a finite $\mathscr{R}$-trivial monoid. Prove that the submonoid of $G_0(\mathbb{C}M)$ consisting of the isomorphism classes of simple $\mathbb{C}M$-modules is isomorphic to the lattice $\Lambda(M)$ made a monoid via the join operation.

6.3. Compute the decomposition matrix for the symmetric inverse monoid I_3 over $\mathbb{C}$.

6.4. Explicitly compute the decomposition matrix for the full transformation monoid T_4 over $\mathbb{C}$.

6.5. Let M be a finite monoid, $\equiv$ a congruence on M, and $\Bbbk$ a field. Prove that $G_0(\Bbbk[M/\equiv])$ can be identified with a subring of $G_0(\Bbbk M)$.

6.6. Let M be a finite monoid and $\Bbbk$ an algebraically closed field. Let $\rho_1, \ldots, \rho_k$ be all the degree one representations of M. Let $\equiv$ be the congruence on M defined by $m \equiv n$ if and only if $\rho_i(m) = \rho_i(n)$ for all $i = 1, \ldots, k$. Prove that $G_0^{\vee}(\Bbbk M) \cong G_0(\Bbbk[M/\equiv])$.

7

Characters and Class Functions

In this chapter, we work exclusively over $\mathbb{C}$, although most of the results hold in greater generality (cf. [MQS15], where the theory is worked out over an arbitrary field). We study the ring $\mathrm{Cl}(M)$ of class functions on a finite monoid M. It turns out that $\mathrm{Cl}(M) \cong \mathbb{C} \otimes_{\mathbb{Z}} G_0(\mathbb{C}M)$. The character table of a monoid is defined and shown to be invertible. In fact, it is block upper triangular with group character tables on the diagonal blocks. Inverting the character table allows us to determine, in principle, the composition factors of a representation directly from its character. The fundamental results of this chapter are due to McAlister [McA72] and, independently, to Rhodes and Zalcstein [RZ91].

7.1 Class functions and generalized conjugacy classes

A *class function* on a finite monoid M is a mapping $f \colon M \longrightarrow \mathbb{C}$ such that the following two properties hold for all $m, n \in M$:

(a) $f(mn) = f(nm)$;
(b) $f(m^{\omega+1}) = f(m)$.

We recall that if m has index c and period d, then $m^{\omega+1} = m^k$ where $k \geq c$ and $k \equiv 1 \bmod d$.

It is clear that the set $\mathrm{Cl}(M)$ of class functions on M is a $\mathbb{C}$-algebra with respect to pointwise operations. We call $\mathrm{Cl}(M)$ the *ring of class functions* on M. The next proposition verifies that our notion of class function coincides with the usual notion for finite groups.

Proposition 7.1. *If G is a finite group, then $f \colon G \longrightarrow \mathbb{C}$ is a class function if and only if f is constant on conjugacy classes.*

© Springer International Publishing Switzerland 2016
B. Steinberg, *Representation Theory of Finite Monoids*,
Universitext, DOI 10.1007/978-3-319-43932-7_7

Proof. If f is a class function on G, then $f(gxg^{-1}) = f(xg^{-1}g) = f(x)$ and so f is constant on conjugacy classes. Conversely, if $f: G \longrightarrow \mathbb{C}$ is constant on conjugacy classes, then $f(gh) = f(g^{-1}(gh)g) = f(hg)$. Also, since $g^{\omega+1} = g$, we have that $f(g^{\omega+1}) = f(g)$. Thus f is a class function. $\square$

It will be convenient to have an analogue of conjugacy for monoids such that class functions are precisely those functions which are constant on conjugacy classes. We define an equivalence relation $\sim$ on M by $m \sim n$ if and only if there exist $x, x' \in M$ such that $xx'x = x$, $x'xx' = x'$, $x'x = m^{\omega}$, $xx' = n^{\omega}$ and $xm^{\omega+1}x' = n^{\omega+1}$. One should think of x' as a generalized inverse of x and therefore think of $m^{\omega+1}$ and $n^{\omega+1}$ as being conjugate.

Proposition 7.2. *The relation $\sim$ is an equivalence relation on M.*

Proof. To see that $m \sim m$, take $x = m^{\omega} = x'$. Suppose that $m \sim n$ and that $xm^{\omega+1}x' = n^{\omega+1}$ with $xx'x = x$, $x'xx' = x'$, $x'x = m^{\omega}$, $xx' = n^{\omega}$. Then $x'n^{\omega+1}x = x'xm^{\omega+1}x'x = m^{\omega}m^{\omega+1}m^{\omega} = m^{\omega+1}$ and so $n \sim m$. To verify transitivity, suppose that $a \sim b$ and $b \sim c$. Then we can find $x, x', y, y' \in M$ such that $xx'x = x$, $x'xx' = x'$, $x'x = a^{\omega}$, $xx' = b^{\omega} = y'y$, $yy'y = y$, $y'yy' = y'$, $yy' = c^{\omega}$, $xa^{\omega+1}x' = b^{\omega+1}$ and $yb^{\omega+1}y' = c^{\omega+1}$. Let $z = yx$ and $z' = x'y'$. Then we compute that $zz'z = yxx'y'yx = yy'yy'yx = yx = z$, $z'zz' = x'y'yxx'y' = x'xx'xx'y' = x'y' = z'$, $z'z = x'y'yx = x'xx'x = a^{\omega}$ and $zz' = yxx'y' = yy'yy' = c^{\omega}$. Also, we have $za^{\omega+1}z' = yxa^{\omega+1}x'y' = yb^{\omega+1}y' = c^{\omega+1}$. This establishes that $a \sim c$, completing the proof. $\square$

We shall call $\sim$-classes by the name *generalized conjugacy classes* and denote the $\sim$-class of an element m by $[m]_{\sim}$. The next proposition shows that the class functions are exactly the functions which are constant on generalized conjugacy classes.

Proposition 7.3. *Let $f: M \longrightarrow \mathbb{C}$ be a mapping. Then f is a class function if and only if it is constant on generalized conjugacy classes.*

Proof. Suppose first that f is a class function and that $m \sim n$. Let $x, x' \in M$ with $xx'x = x$, $x'xx' = x'$, $x'x = m^{\omega}$, $xx' = n^{\omega}$ and $xm^{\omega+1}x' = n^{\omega+1}$. Then we have that $f(n) = f(n^{\omega+1}) = f(xm^{\omega+1}x') = f(x'xm^{\omega+1}) = f(m^{\omega}m^{\omega+1}) = f(m^{\omega+1}) = f(m)$ and so f is constant on generalized conjugacy classes.

Conversely, suppose that f is constant on generalized conjugacy classes. Since $m \sim m^{\omega+1}$ (by taking $x = m^{\omega} = x'$), it follows that $f(m^{\omega+1}) = f(m)$. Next, let $m, n \in M$. Choose $k > 0$ such that $x^k = x^{\omega}$ for all $x \in M$ (see Remark 1.3). Note that $x^{\omega}x^r = x^r$ for any $r \geq k$ as a consequence of Corollary 1.2. Let $x = n(mn)^{2k-1} = (nm)^{2k-1}n$ and $x' = (mn)^k m = m(nm)^k$. Then we compute

$$x'x = (mn)^k mn(mn)^{2k-1} = (mn)^{3k} = (mn)^{\omega}$$

$$xx' = (nm)^{2k-1}nm(nm)^k = (nm)^{3k} = (nm)^{\omega}$$

$$xx'x = (nm)^{\omega}(nm)^{2k-1}n = (nm)^{2k-1}n = x$$

$$x'xx' = (mn)^{\omega}(mn)^k m = (mn)^k m = x'$$

and $x(mn)^{\omega+1}x' = n(mn)^{2k-1}(mn)^{k+1}(mn)^k m = n(mn)^{4k}m = nm(nm)^{4k} = (nm)^{\omega+1}$. Therefore, $mn \sim nm$ and so $f(mn) = f(nm)$, completing the proof that f is a class function. □

Proposition 7.3 essentially says that $\sim$ is the smallest equivalence relation on M such that $mn \sim nm$ and $m^{\omega+1} \sim m$ for all $m, n \in M$.

Let $e_1, \ldots, e_s$ form a complete set of idempotent representatives of the regular $\mathcal{J}$-classes of M. Then it turns out that each generalized conjugacy class of M intersects exactly one maximal subgroup G_{e_i}, and in one of its conjugacy classes. If G is a group, then $\mathrm{Conj}(G)$ will denote the set of conjugacy classes of elements of G. The conjugacy class of an element $g \in G$ will be denoted by $\mathrm{cl}(g)$.

Proposition 7.4. *Let $e_1, \ldots, e_s$ form a complete set of idempotent representatives of the regular $\mathcal{J}$-classes of M. Then the restriction of $\sim$ to G_{e_i} is conjugacy and the natural map*

$$\psi: \bigcup_{i=1}^{s} \mathrm{Conj}(G_{e_i}) \longrightarrow M/\!\sim$$

given by $\psi(\mathrm{cl}(g)) = [g]_\sim$ is a bijection.

Proof. If $g, h \in G_{e_i}$ and $h = xgx^{-1}$ with $x \in G_{e_i}$, then putting $x' = x^{-1}$, we have $xx'x = x$, $x'xx' = x'$, $g^\omega = e_i = h^\omega = x'x = xx'$ and $h^{\omega+1} = h = xgx^{-1} = xg^{\omega+1}x'$. Therefore, $g \sim h$. Conversely, suppose that $g \sim h$ and that x, x' satisfy $xx'x = x$, $x'xx' = x'$, $x'x = g^\omega = e_i = h^\omega = xx'$ and $xg^{\omega+1}x' = h^{\omega+1}$. Then $x, x' \in e_i M e_i$ and are inverse units, hence $x, x' \in G_{e_i}$ and $x' = x^{-1}$. As $g = g^{\omega+1}$ and $h = h^{\omega+1}$, it follows that $h = xgx^{-1}$ and so h, g are conjugate in G_{e_i}. Therefore, $\sim$ restricts to G_{e_i} as conjugacy and hence ψ is a well-defined mapping, injective on each $\mathrm{Conj}(G_{e_i})$. If $g \in G_{e_i}$ and $h \in G_{e_j}$ with $g \sim h$, then there exist $x, x' \in M$ with $x'x = g^\omega = e_i$ and $xx' = h^\omega = e_j$. But then $Me_iM = Me_jM$ by Theorem 1.11 and so $e_i = e_j$. Thus ψ is injective. It remains to show that it is surjective.

Let $m \in M$ and suppose that $Mm^\omega M = Me_iM$. By Theorem 1.11, there exist $x, x' \in M$ such that $xx'x = x$, $x'xx' = x'$, $x'x = m^\omega$ and $xx' = e_i$. Let $g = xm^{\omega+1}x'$. Then $e_ige_i = xx'xm^{\omega+1}x'xx' = xm^{\omega+1}x' = g$ and so $g \in e_iMe_i$. Also, we have that $MgM = Mm^\omega M = Me_iM$ because $x'gx = m^\omega m^{\omega+1}m^\omega = m^{\omega+1}$ and $Mm^{\omega+1}M = Mm^\omega M$. Therefore, $g \in G_{e_i}$ by Corollary 1.16. But $g \sim m$. Therefore, $[m]_\sim = \psi(\mathrm{cl}(g))$, as required. □

As an immediate consequence, we can identify the ring of class functions on M with the product of the rings of class functions on the G_{e_i} with $i = 1, \ldots, s$.

Corollary 7.5. *Let $e_1, \ldots, e_s$ form a complete set of idempotent representatives of the regular $\mathcal{J}$-classes of M. Then the restriction map*

$$\mathrm{res}: \mathrm{Cl}(M) \longrightarrow \prod_{i=1}^{s} \mathrm{Cl}(G_{e_i})$$

given by

$$\mathrm{res}(f) = (f|_{G_{e_1}}, \ldots, f|_{G_{e_s}})$$

is an isomorphism of $\mathbb{C}$-algebras.

Proof. Identifying $\mathrm{Cl}(M)$ with the algebra A of mappings $f: M/\!\!\sim \longrightarrow \mathbb{C}$ and identifying $\prod_{i=1}^{s} \mathrm{Cl}(G_{e_i})$ with the algebra B of mappings

$$f: \bigcup_{i=1}^{s} \mathrm{Conj}(G_{e_i}) \longrightarrow \mathbb{C}$$

(using Proposition 7.1), we see that

$$\mathrm{res}: \mathrm{Cl}(M) \longrightarrow \prod_{i=1}^{s} \mathrm{Cl}(G_{e_i})$$

corresponds to the algebra homomorphism $\psi^*: A \longrightarrow B$ given by $f \mapsto f \circ \psi$, where ψ is the bijection from Proposition 7.4. Thus the mapping res is an isomorphism. □

As a corollary, we may compute the dimension of $\mathrm{Cl}(M)$.

Corollary 7.6. *Let M be a finite monoid and let $e_1, \ldots, e_s$ be a complete set of idempotent representatives of the regular $\mathscr{J}$-classes of M. Then the equality*

$$\dim \mathrm{Cl}(M) = \kappa_1 + \cdots + \kappa_s$$

holds where $\kappa_i = |\mathrm{Conj}(G_{e_i})|$.

Proof. This is immediate from Proposition 7.1 and Corollary 7.5. □

A natural basis for $\mathrm{Cl}(M)$ consists of the indicator functions $\delta_C: M \longrightarrow \mathbb{C}$ of the generalized conjugacy classes where, for a class C,

$$\delta_C(m) = \begin{cases} 1, & \text{if } m \in C \\ 0, & \text{else.} \end{cases} \tag{7.1}$$

In the next section we shall see that the irreducible characters of M form another basis for $\mathrm{Cl}(M)$. This in turn will allow us to identify $\mathrm{Cl}(M)$ with $\mathbb{C} \otimes_{\mathbb{Z}} G_0(\mathbb{C}M)$.

7.2 Character theory

Let $\rho: M \longrightarrow M_n(\mathbb{C})$ be a representation of a monoid M. Then the *character* of ρ is the mapping $\chi_\rho: M \longrightarrow \mathbb{C}$ defined by

$$\chi_\rho(m) = \mathrm{Tr}(\rho(m))$$

where $\mathrm{Tr}(A)$ is the trace of a matrix A. If ρ is equivalent to φ, then $\chi_\rho = \chi_\varphi$. Indeed, if T is invertible with $\rho(m) = T\varphi(m)T^{-1}$ for all $m \in M$, then

$$\chi_\rho(m) = \mathrm{Tr}(\rho(m)) = \mathrm{Tr}(T\varphi(m)T^{-1}) = \mathrm{Tr}(\varphi(m)) = \chi_\varphi(m).$$

Hence if V is a finite dimensional $\mathbb{C}M$-module, then we can put $\chi_V = \chi_\rho$ where ρ is the representation afforded by V with respect to some basis. If $V \cong W$, then clearly $\chi_V = \chi_W$ because they afford equivalent representations. We sometimes say that V *affords* the character χ_V. The character of an irreducible representation will be called an *irreducible character*. The zero mapping will be considered the character of the zero module. A mapping $f \colon M \longrightarrow \mathbb{C}$ is called a *character* if $f = \chi_V$ for some finite dimensional $\mathbb{C}M$-module V. One says that f is a *virtual character* if $f = \chi_V - \chi_W$ for some $\mathbb{C}M$-modules V and W, that is, if f is a difference of two characters.

The following is an immediate consequence of the theorem of Frobenius and Schur (Corollary 5.2).

Theorem 7.7. *Let $V_1, \ldots, V_r$ be a set of pairwise non-isomorphic simple $\mathbb{C}M$-modules. Then the characters $\chi_{V_1}, \ldots, \chi_{V_r}$ are linearly independent as complex-valued functions on M.*

Proof. Suppose that $\varphi^{(k)} \colon M \longrightarrow M_{d_k}(\mathbb{C})$ is a representation afforded by V_k for $k = 1, \ldots, r$. If $c_1 \chi_{V_1} + \cdots + c_r \chi_{V_r} = 0$ with $c_1, \ldots, c_r \in \mathbb{C}$, then

$$0 = \sum_{k=1}^{r} \sum_{i=1}^{d_k} c_k \varphi_{ii}^{(k)}$$

and so $c_1 = \cdots = c_r = 0$ by Corollary 5.2. $\qquad\square$

We shall need the following algebraic properties of character values in the sequel.

Lemma 7.8. *Let $\rho \colon M \longrightarrow M_n(\mathbb{C})$ be a representation and let $m \in M$ have index c and period d.*

(i) The minimal polynomial of $\rho(m)$ divides $x^c(x^d - 1)$.
(ii) Every nonzero eigenvalue of $\rho(m)$ is a dth-root of unity.
(iii) $\chi_\rho(m)$ is an algebraic integer.
(iv) $\chi_\rho(m) = n$ if and only if $\rho(m) = I$.

Proof. From $m^c = m^{c+d}$, we obtain that $\rho(m)^c = \rho(m)^{c+d}$ and hence (i) follows. Item (ii) is an immediate consequence of (i). Since $\mathrm{Tr}(\rho(m))$ is the sum of the eigenvalues with multiplicities, (iii) is a consequence of (ii). For the final statement, let $\lambda_1, \ldots, \lambda_n$ be the eigenvalues of $\rho(m)$ with multiplicities. By (ii) and the Cauchy-Schwarz inequality we have that

$$\left| \sum_{i=1}^{n} \lambda_i \right| \le \sum_{i=1}^{n} |\lambda_i| \le n$$

with equality if and only if $\lambda_1 = \cdots = \lambda_n \neq 0$. Therefore, if $n = \chi_\rho(m) = \mathrm{Tr}(\rho(m)) = \sum_{i=1}^n \lambda_i$, then $\lambda_1 = \cdots = \lambda_n = 1$ and so the minimal polynomial of $\rho(m)$ divides both $(x-1)^n$ and $x^c(x^d-1)$. But the greatest common divisor of these two polynomials is $x-1$, whence $\rho(m) = I$, completing the proof. $\square$

As a consequence, we deduce that characters are class functions.

Proposition 7.9. *Characters of representations are class functions.*

Proof. Let $\rho\colon M \longrightarrow M_n(\mathbb{C})$ be a representation. Then we compute that

$$\chi_\rho(m_1 m_2) = \mathrm{Tr}(\rho(m_1)\rho(m_2)) = \mathrm{Tr}(\rho(m_2)\rho(m_1)) = \chi_\rho(m_2 m_1).$$

Next let $m \in M$ have index c and period d. Then $m^{\omega+1} = m^k$ where $k \geq c$ and $k \equiv 1 \bmod d$. Let $\lambda_1, \ldots, \lambda_n$ be the eigenvalues of $\rho(m)$ with multiplicity. Then $\lambda_1^k, \ldots, \lambda_n^k$ are the eigenvalues of $\rho(m)^k = \rho(m^k) = \rho(m^{\omega+1})$. By Lemma 7.8 each nonzero eigenvalue λ of $\rho(m)$ is a dth-root of unity and hence satisfies $\lambda^k = \lambda$. Since this latter equality is obviously true for $\lambda = 0$, we conclude that $\lambda_i = \lambda_i^k$ for all $i = 1, \ldots, n$. Thus we have

$$\chi_\rho(m) = \mathrm{Tr}(\rho(m)) = \sum_{i=1}^n \lambda_i = \sum_{i=1}^n \lambda_i^k = \mathrm{Tr}(\rho(m)^k) = \chi_\rho(m^{\omega+1}).$$

This completes the proof that χ_ρ is a class function. $\square$

We may now deduce that the irreducible characters form a basis for $\mathrm{Cl}(M)$.

Theorem 7.10. *Let $V_1, \ldots, V_r$ be a complete set of representatives of the isomorphism classes of simple $\mathbb{C}M$-modules. Then the characters $\chi_{V_1}, \ldots, \chi_{V_r}$ form a basis for $\mathrm{Cl}(M)$. In particular, the number of isomorphism classes of simple $\mathbb{C}M$-modules coincides with the number of generalized conjugacy classes of M.*

Proof. The characters $\chi_{V_1}, \ldots, \chi_{V_r}$ are linearly independent by Theorem 7.7. Also, we know that $\dim \mathrm{Cl}(M)$ is the number of generalized conjugacy classes. Fix a complete set of idempotent representatives $e_1, \ldots, e_s$ of the regular $\mathscr{J}$-classes of M. Put $\kappa_i = |\mathrm{Conj}(G_{e_i})|$. Then $\dim \mathrm{Cl}(M) = \kappa_1 + \cdots + \kappa_s$ by Corollary 7.6. But $\kappa_i = |\mathrm{Irr}_{\mathbb{C}}(G_{e_i})|$ by Corollary B.5 and hence $\kappa_1 + \cdots + \kappa_s = r$ by Corollary 5.6. Therefore, $\chi_{V_1}, \ldots, \chi_{V_r}$ is a basis for $\mathrm{Cl}(M)$. $\square$

We define the *character table* $X(M)$ to be the transpose of the change of basis matrix between the basis of irreducible characters and the basis of indicator functions $\{\delta_C \mid C \in M/\sim\}$, cf. (7.1), for $\mathrm{Cl}(M)$. In other words,

$$X(M)_{\chi,C} = \chi(C)$$

where $\chi(C)$ is the value taken by the character χ on the generalized conjugacy class C. Notice that $X(M)$ is invertible by Theorem 7.10. It seems unfortunate to use the transpose of the change of basis matrix, but this tradition is too deeply entrenched to change. From the definition of $X(M)$ as the transpose of the change of basis matrix, the following proposition is immediate.

Proposition 7.11. *If* $f\colon M \longrightarrow \mathbb{C}$ *is a class function and* $\chi_1, \ldots, \chi_r$ *are the irreducible characters of* M, *then*

$$f = \sum_{i=1}^{r} \sum_{C \in M/\sim} f(C) X(M)_{C,\chi_i}^{-1} \cdot \chi_i$$

where $f(C)$ *denotes the value of* f *on the generalized conjugacy class* C.

To further enhance our understanding of the ring of class functions and the character table, it will be invaluable to connect $\mathrm{Cl}(M)$ with $G_0(\mathbb{C}M)$.

Proposition 7.12. *The assignment* $V \mapsto \chi_V$ *enjoys the following properties.*

(i) The character of the trivial representation is identically 1.
(ii) Given an exact sequence of $\mathbb{C}M$*-modules*

$$0 \longrightarrow U \longrightarrow V \longrightarrow W \longrightarrow 0$$

one has that $\chi_V = \chi_U + \chi_W$.
(iii) $\chi_{V \otimes W} = \chi_V \cdot \chi_W$.

Proof. The first item is clear. To prove (ii), without loss of generality assume that $U \leq V$ and $W = V/U$. Let $\{u_1, \ldots, u_k\}$ be a basis for U and extend it to a basis $\{u_1, \ldots, u_k, v_1, \ldots, v_n\}$ for V so that $\{v_1 + U, \ldots, v_n + U\}$ is a basis for V/U. Let $\rho\colon M \longrightarrow M_k(\mathbb{C})$ be the representation afforded by U with respect to the basis $\{u_1, \ldots, u_k\}$ and let $\psi\colon M \longrightarrow M_n(\mathbb{C})$ be the representation afforded by V/U with respect to the basis $\{v_1 + U, \ldots, v_n + U\}$. Then the representation $\varphi\colon M \longrightarrow M_{k+n}(\mathbb{C})$ afforded by V has the block form

$$\varphi(m) = \begin{bmatrix} \rho(m) & * \\ 0 & \psi(m) \end{bmatrix}$$

and so

$$\chi_V(m) = \chi_\varphi(m) = \mathrm{Tr}(\varphi(m)) = \mathrm{Tr}(\rho(m)) + \mathrm{Tr}(\psi(m)) = \chi_U(m) + \chi_W(m),$$

as required.

Let us turn to (iii). Let $\{v_1, \ldots, v_r\}$ and $\{w_1, \ldots, w_s\}$ be bases for V and W, respectively. Then $\{v_i \otimes w_j \mid 1 \leq i \leq r, 1 \leq j \leq s\}$ is a basis for $V \otimes W$. Let $\rho\colon M \longrightarrow M_r(\mathbb{C})$ and $\psi\colon M \longrightarrow M_s(\mathbb{C})$ be the representations afforded by V and W, respectively, using these bases. Clearly, we have

$$m(v_i \otimes w_j) = mv_i \otimes mw_j = \sum_{k=1}^{r} \rho(m)_{ki}(v_k \otimes mw_j)$$

$$= \sum_{k=1}^{r} \sum_{\ell=1}^{s} \rho(m)_{ki} \psi(m)_{\ell j}(v_k \otimes w_\ell).$$

In particular, the coefficient of $v_i \otimes w_j$ is $\rho(m)_{ii}\psi(m)_{jj}$ and so

$$\chi_{V\otimes W}(m) = \sum_{i=1}^{r}\sum_{j=1}^{s} \rho(m)_{ii}\psi(m)_{jj} = \left(\sum_{i=1}^{r}\rho(m)_{ii}\right)\left(\sum_{j=1}^{s}\psi(m)_{jj}\right)$$

$$= \chi_V(m)\chi_W(m),$$

as required. □

It follows from the proposition that the mapping sending $[V]$ to χ_V induces a ring homomorphism $G_0(\mathbb{C}M) \longrightarrow \mathrm{Cl}(M)$.

Corollary 7.13. *The assignment $[V] \mapsto \chi_V$ induces an injective ring homomorphism $\Delta_M \colon G_0(\mathbb{C}M) \longrightarrow \mathrm{Cl}(M)$. The image of Δ_M is the subring of virtual characters. Moreover, Δ induces an isomorphism $\mathbb{C}\otimes_\mathbb{Z} G_0(\mathbb{C}M) \cong \mathrm{Cl}(M)$.*

Proof. Clearly, Δ is a ring homomorphism with image the subring of virtual characters. Injectivity of Δ_M and the final statement follow because if $\mathrm{Irr}_\mathbb{C}(M) = \{[V_1], \ldots, [V_r]\}$, then $\{[V_1], \ldots, [V_r]\}$ is a basis for $G_0(\mathbb{C}M)$ as a free abelian group and $\{\chi_{V_1}, \ldots, \chi_{V_r}\}$ is a basis for $\mathrm{Cl}(M)$ as a $\mathbb{C}$-vector space by Theorem 7.10. □

To complete the picture relating the Grothendieck ring to the ring of class functions, we will show that the isomorphisms Res and res from Theorem 6.5 and Corollary 7.5 are intertwined by the maps sending a representation to its character.

Proposition 7.14. *Suppose that the $\mathbb{C}M$-module V affords the character χ_V and $e \in E(M)$. Then $\chi_{eV} = (\chi_V)|_{eMe}$ as a character of eMe and $\chi_{eV} = (\chi_V)|_{G_e}$ as a character of G_e.*

Proof. The direct sum decomposition $V = eV \oplus (1-e)V$ is into eMe-invariant subspaces. Moreover, eMe annihilates $(1-e)V$. Choose a basis B for eV and B' for $(1-e)V$. Let $\psi \colon eMe \longrightarrow M_r(\mathbb{C})$ be the representation afforded by eV with respect to the basis B. Choosing the basis $B \cup B'$ for V affords a matrix representation $\rho \colon M \longrightarrow M_n(\mathbb{C})$ with

$$\rho(m) = \begin{bmatrix} \psi(m) & 0 \\ 0 & 0 \end{bmatrix}$$

for $m \in eMe$. Therefore, $\chi_V(m) = \mathrm{Tr}(\rho(m)) = \mathrm{Tr}(\psi(m)) = \chi_{eV}(m)$ for $m \in eMe$, as was required. □

We now present the fundamental theorem of this section.

Theorem 7.15. *Let M be a finite monoid. Let $e_1, \ldots, e_s$ form a complete set of idempotent representatives of the regular $\mathscr{J}$-classes of M. Then there is a commutative diagram*

$$G_0(\mathbb{C}M) \xrightarrow{\text{Res}} \prod_{i=1}^{s} G_0(\mathbb{C}G_{e_i})$$

$$\Delta_M \downarrow \qquad\qquad \downarrow \prod_{i=1}^{s} \Delta_{G_{e_i}}$$

$$\mathrm{Cl}(M) \xrightarrow[\text{res}]{} \prod_{i=1}^{s} \mathrm{Cl}(G_{e_i}) \tag{7.2}$$

where

$$\mathrm{Res}([V]) = ([e_1 V], \dots, [e_s V]),$$
$$\mathrm{res}(f) = (f|_{G_{e_1}}, \dots, f|_{G_{e_s}})$$

and where Δ_M, $\Delta_{G_{e_i}}$ *are given by* $[V] \mapsto \chi_V$. *The horizontal maps are isomorphisms and the vertical maps are monomorphisms inducing a commutative diagram*

$$\mathbb{C} \otimes_{\mathbb{Z}} G_0(\mathbb{C}M) \xrightarrow{\text{Res}} \prod_{i=1}^{s} \mathbb{C} \otimes_{\mathbb{Z}} G_0(\mathbb{C}G_{e_i})$$

$$\downarrow \qquad\qquad\qquad \downarrow$$

$$\mathrm{Cl}(M) \xrightarrow[\text{res}]{} \prod_{i=1}^{s} \mathrm{Cl}(G_{e_i}). \tag{7.3}$$

of isomorphisms of $\mathbb{C}$*-algebras.*

Proof. Proposition 7.14 implies that the diagram commutes. The remainder of the theorem follows directly from Theorem 6.5, Corollary 7.5, and Corollary 7.13. $\qquad\square$

An important corollary of Theorem 7.15 is a characterization of the virtual characters of a finite monoid in terms of those of its maximal subgroups.

Corollary 7.16. *Let M be a finite monoid and let $e_1, \dots, e_s$ form a complete set of idempotent representatives of the regular $\mathscr{J}$-classes of M. Then the following are equivalent for a class function $f \colon M \longrightarrow \mathbb{C}$.*

(i) f is a virtual character.
(ii) $f|_{G_e}$ is a virtual character of G_e for each $e \in E(M)$
(iii) $f|_{G_{e_i}}$ is a virtual character of G_{e_i} for $i = 1, \dots, s$.

Proof. The restriction of a virtual character to a maximal subgroup is again a virtual character by Proposition 7.14 and so (i) implies (ii). Obviously, (ii) implies (iii). Assume that (iii) holds. Suppose that $f|_{G_{e_i}} = \chi_{V_i} - \chi_{W_i}$ with V_i, W_i finite dimensional $\Bbbk G_{e_i}$-modules for $i = 1, \dots, s$. Then by Theorem 6.5, there exist finite dimensional $\Bbbk M$-modules V, W with

$$\mathrm{Res}([V] - [W]) = ([V_1] - [W_1], \dots, [V_s] - [W_s]).$$

By commutativity of the diagram (7.2) from Theorem 7.15, it follows that

$$\mathrm{res}(\chi_V - \chi_W) = (\chi_{V_1} - \chi_{W_1}, \ldots, \chi_{V_s} - \chi_{W_s}) = \mathrm{res}(f)$$

and so $f = \chi_V - \chi_W$ because res is an isomorphism. This establishes that (iii) implies (i) and completes the proof of the theorem.   □

As in Section 6.2, we fix a complete set of idempotent representatives $e_1, \ldots, e_s$ of the regular $\mathscr{J}$-classes of M and assume that we have chosen the ordering so that $Me_iM \subseteq Me_jM$ implies $i \leq j$. Again, we fix a total ordering $\preceq$ on $\mathrm{Irr}_\Bbbk(M)$ such that if e_i is the apex of S and e_j is the apex of S' with $i < j$, then $[S] \prec [S']$. We put a corresponding total ordering, also denoted $\preceq$, on the disjoint union $\bigcup_{i=1}^s \mathrm{Irr}_\Bbbk(G_e)$ by putting $[S_1] \preceq [S_2]$ if and only if $[S_1^\sharp] \preceq [S_2^\sharp]$, where we retain the notation from Theorem 5.5. Let us choose a total ordering $\leq$ on the generalized conjugacy classes of M such that if $C \cap G_{e_i} \neq \emptyset$ and $C' \cap G_{e_j} \neq \emptyset$ with $i < j$, then $C < C'$. We totally order $\bigcup_{i=1}^s \mathrm{Conj}(G_{e_i})$ by transporting $\leq$ along ψ^{-1} where ψ is the bijection in Proposition 7.4.

Retaining all this notation, we take $\mathrm{Irr}_\mathbb{C}(M)$ as a basis for $\mathbb{C} \otimes_\mathbb{Z} G_0(\mathbb{C}M)$, $\bigcup_{i=1}^s \mathrm{Irr}_\mathbb{C}(G_{e_i})$ as a basis for $\prod_{i=1}^s \mathbb{C} \otimes_\mathbb{Z} G_0(\mathbb{C}G_{e_i})$, the set of indicator functions of generalized conjugacy classes of M as a basis for $\mathrm{Cl}(M)$ and the union of the sets of indicator functions of the conjugacy classes of the G_{e_i} as a basis for $\prod_{i=1}^s \mathrm{Cl}(G_{e_i})$. These bases are ordered using the orderings $\preceq$ in the first two cases and using the orderings coming from $\leq$ in the latter two. Then $X(M)^T$ is the matrix of left hand vertical arrow in (7.3), the matrix of res is the identity matrix, the matrix of Res is the decomposition matrix (cf. Theorem 6.5), and the matrix of the right hand vertical map is

$$\begin{bmatrix} X(G_{e_1})^T & 0 & \cdots & 0 \\ 0 & X(G_{e_2})^T & 0 & \vdots \\ \vdots & 0 & \ddots & 0 \\ 0 & \cdots & 0 & X(G_{e_s})^T \end{bmatrix}.$$

Taking transposes, we have thus proved the following result.

Corollary 7.17. *With respect to the above ordering of simple modules and generalized conjugacy classes, we have that $X(M) = UX$ where*

$$X = \begin{bmatrix} X(G_{e_1}) & 0 & \cdots & 0 \\ 0 & X(G_{e_2}) & 0 & \vdots \\ \vdots & 0 & \ddots & 0 \\ 0 & \cdots & 0 & X(G_{e_s}) \end{bmatrix}.$$

and U is a unipotent upper triangular matrix (the transpose of the decomposition matrix). Hence $X(M)$ is block upper triangular with the diagonal blocks the character tables of the maximal subgroups of M, one per regular $\mathscr{J}$-class.

We shall work out the character table of the full transformation monoid in the next section.

Recall that if $f, h\colon G \longrightarrow \mathbb{C}$ are class functions on a group G, then their inner product is given by

$$\langle f, h \rangle_G = \frac{1}{|G|} \sum_{g \in G} f(g)\overline{h(g)}.$$

Also, recall that the irreducible characters form an orthonormal basis for $\mathrm{Cl}(G)$ with respect to this inner product (cf. Theorem B.4).

Corollary 7.18. *Let V be a finite dimensional $\Bbbk M$-module. Then the equality*

$$\chi_V = \sum_{i=1}^{s} \sum_{S \in \mathrm{Irr}_{\mathbb{C}}(G_{e_i})} \langle \chi_V, \chi_S \rangle_{G_{e_i}} \left(\sum_{S' \in \mathrm{Irr}_{\mathbb{C}}(M)} L^{-1}_{[S'],[S]} \chi_{S'} \right)$$

holds, where L is the decomposition matrix and χ_V is viewed as a character of G_{e_i} by restriction.

Proof. This follows because we have

$$[V] = \mathrm{Res}^{-1}\left(\sum_{i=1}^{s} [e_i V] \right)$$

$$= \mathrm{Res}^{-1}\left(\sum_{i=1}^{s} \sum_{S \in \mathrm{Irr}_{\mathbb{C}}(G_{e_i})} \langle \chi_V, \chi_S \rangle_{G_{e_i}} [S] \right)$$

$$= \sum_{i=1}^{s} \sum_{S \in \mathrm{Irr}_{\mathbb{C}}(G_{e_i})} \langle \chi_V, \chi_S \rangle_{G_{e_i}} \cdot \mathrm{Res}^{-1}([S])$$

as the irreducible characters of a finite group form an orthonormal basis for the ring of class functions. $\qquad\square$

Corollary 7.18 provides one method to compute the composition factors of a finite dimensional $\mathbb{C}M$-module V from its character χ_V. Another possible way to do this is via primitive idempotents, although computing primitive idempotents for monoid algebras does not seem to be an easy task.

Proposition 7.19. *Let V be a finite dimensional $\mathbb{C}M$-module with character χ_V and S a simple $\mathbb{C}M$-module. Suppose that $S \cong \mathbb{C}Mf/\mathrm{rad}(\mathbb{C}M)f$ where*

$$f = \sum_{m \in M} c_m \cdot m$$

is a primitive idempotent. Then the equality

$$[V : S] = \sum_{m \in M} c_m \chi_V(m)$$

holds.

Proof. By Proposition A.24, we have $[V : S] = \dim fV$. If $\rho\colon M \longrightarrow M_n(\mathbb{C})$ is the representation afforded by V, and we extend ρ linearly to $\mathbb{C}M$, then we have

$$\dim fV = \operatorname{Tr}(\rho(f)) = \sum_{m \in M} c_m \operatorname{Tr}(\rho(m)) = \sum_{m \in M} c_m \chi_V(m)$$

because $\rho(f)$ is idempotent. This completes the proof. $\square$

Let us compute the character table of an $\mathscr{R}$-trivial monoid. Recall from Corollary 5.7 that if M is $\mathscr{R}$-trivial, then there is a one-dimensional simple module S_J for each regular $\mathscr{J}$-class J of M and the corresponding character is given by

$$\chi_J(m) = \begin{cases} 1, & \text{if } J \subseteq MmM \\ 0, & \text{else.} \end{cases} \tag{7.4}$$

Corollary 2.7 shows that there is a homomorphism $\sigma\colon M \longrightarrow \Lambda(M)$, where

$$\Lambda(M) = \{MeM \mid e \in E(M)\},$$

given by $\sigma(m) = Mm^\omega M$. Moreover, the fibers of σ are exactly the generalized conjugacy classes of M. Indeed, each regular $\mathscr{J}$-class of M consists of idempotents by Corollary 2.7 and hence $m^{\omega+1} = m^\omega$ for $m \in M$. Moreover, two idempotents are generalized conjugates if and only if they generate the same two-sided ideal by Exercise 7.7. Thus we can identify both the set of simple modules and the set of generalized conjugacy classes of M with the lattice $\Lambda(M)$.

We can easily deduce from (7.4) that $X(M)$ is the zeta function of $\Lambda(M)$ (see Appendix C for the definition of the zeta function of a poset). Indeed, if J is the $\mathscr{J}$-class of an idempotent $f \in E(M)$ and C is the generalized conjugacy class of an idempotent $e \in E(M)$, then

$$X(M)_{\chi_J,C} = \chi_J(e) = \begin{cases} 1, & \text{if } MfM \subseteq MeM \\ 0, & \text{else.} \end{cases}$$

We have thus proved the following result, where the reader should consult Appendix C for the definition of the Möbius function of a poset.

Proposition 7.20. *The character table $X(M)$ of an $\mathscr{R}$-trivial monoid is the zeta function of the lattice $\Lambda(M) = \{MeM \mid e \in E(M)\}$. Hence $X(M)^{-1}$ is the Möbius function μ of $\Lambda(M)$.*

From Proposition 7.20, we obtain the following specialization of Proposition 7.11.

Corollary 7.21. *Let M be an $\mathscr{R}$-trivial monoid and V a finite dimensional $\mathbb{C}M$-module. Then, for a regular $\mathscr{J}$-class J, one has that*

$$[V : S_J] = \sum_{\substack{MeM\in\Lambda(M),\\ MeM\subseteq MJM}} \chi_V(e) \cdot \mu(MeM, MJM)$$

holds.

Proof. Since $\chi_V = \sum_{MJM\in\Lambda(M)}[V : S_J] \cdot \chi_{S_J}$, we have

$$[V : S_J] = \sum_{MeM\in\Lambda(M)} \chi_V(e)X(M)^{-1}_{MeM,MJM}$$

$$= \sum_{\substack{MeM\in\Lambda(M),\\ MeM\subseteq MJM}} \chi_V(e) \cdot \mu(MeM, MJM)$$

by Proposition 7.11. ☐

7.3 The character table of the full transformation monoid

In this section, we use the results of Section 5.3 to compute the character table of the full transformation monoid, a result first achieved by Putcha [Put96]. Fix $n \geq 1$. For $1 \leq r \leq n$, we let $\mathcal{P}_r$ denote the set of all partitions of r and we put $\mathcal{P}(n) = \bigcup_{r=1}^{n} \mathcal{P}_r$. Both the simple $\mathbb{C}T_n$-modules and the generalized conjugacy classes of T_n are indexed by $\mathcal{P}(n)$. We order $\mathcal{P}(n)$ as follows. Let $\lambda = (\lambda_1, \ldots, \lambda_\ell)$ and $\mu = (\mu_1, \ldots, \mu_m)$ be elements of $\mathcal{P}(n)$. Then we declare $\mu \prec \lambda$ if $|\mu| < |\lambda|$ or $|\mu| = |\lambda|$ and μ precedes λ in lexicographic order, that is, there exists j such that $\mu_i = \lambda_i$ for $i < j$ and $\mu_j < \lambda_j$.

We use the idempotents $e_1, \ldots, e_n$ defined in Section 5.3 as the representatives of the $\mathscr{J}$-classes of T_n. Recall that e_i fixes $[i]$ and sends $[n] \setminus [i]$ to 1. We identify G_{e_r} with S_r when convenient (via restriction). If λ is a partition of r with $1 \leq r \leq n$, then C_λ will denote the conjugacy class of G_{e_r} consisting of elements whose restriction to $[r]$ has cycle type λ and $\overline{C_\lambda}$ will denote the corresponding generalized conjugacy class of T_n. We order the simple $\mathbb{C}T_n$-modules by putting $S_\mu^\sharp < S_\lambda^\sharp$ if $\mu \prec \lambda$ and, similarly, we order the generalized conjugacy classes if T_n by setting $\overline{C_\mu} < \overline{C_\lambda}$ if $\mu \prec \lambda$.

Theorem 7.22. *With respect to the above ordering of simple modules and generalized conjugacy classes, $X(T_n) = UX$ where*

$$X = \begin{bmatrix} X(S_1) & 0 & \cdots & 0 \\ 0 & X(S_2) & 0 & \vdots \\ \vdots & 0 & \ddots & 0 \\ 0 & \cdots & 0 & X(S_n) \end{bmatrix}$$

and U is the unipotent upper triangular matrix $\mathcal{P}(n) \times \mathcal{P}(n)$ matrix given by

$$U_{\mu,\lambda} = \begin{cases} 1, & \text{if } \mu = (1^r) \text{ and } \lambda = (m - r + 1, 1^{r-1}) \text{ with } m \geq r \\ 1, & \text{if } \mu \neq (1^r), \ \mu \subseteq \lambda \text{ and } \lambda \setminus \mu \text{ is a horizontal strip} \\ 0, & \text{else} \end{cases}$$

for $\mu, \lambda \in \mathcal{P}(n)$ with $|\mu| = r$ and $|\lambda| = m$.

Proof. This follows from Corollary 7.17 and Proposition 6.6. □

Example 7.23. We compute the character table for T_3. We have $\mathcal{P}(3) = \{(1), (1^2), (2), (1^3), (2, 1), (3)\}$. The matrix U is the transpose of the matrix L computed in Example 6.7 and so we have

$$X(T_3) = \begin{bmatrix} 1 & 0 & 1 & 0 & 0 & 1 \\ 0 & 1 & 0 & 0 & 1 & 0 \\ 0 & 0 & 1 & 0 & 1 & 1 \\ 0 & 0 & 0 & 1 & 0 & 0 \\ 0 & 0 & 0 & 0 & 1 & 0 \\ 0 & 0 & 0 & 0 & 0 & 1 \end{bmatrix} \cdot \begin{bmatrix} 1 & 0 & 0 & 0 & 0 & 0 \\ 0 & 1 & -1 & 0 & 0 & 0 \\ 0 & 1 & 1 & 0 & 0 & 0 \\ 0 & 0 & 0 & 1 & -1 & 1 \\ 0 & 0 & 0 & 2 & 0 & -1 \\ 0 & 0 & 0 & 1 & 1 & 1 \end{bmatrix} = \begin{bmatrix} 1 & 1 & 1 & 1 & 1 & 1 \\ 0 & 1 & -1 & 2 & 0 & -1 \\ 0 & 1 & 1 & 3 & 1 & 0 \\ 0 & 0 & 0 & 1 & -1 & 1 \\ 0 & 0 & 0 & 2 & 0 & -1 \\ 0 & 0 & 0 & 1 & 1 & 1 \end{bmatrix}.$$

7.4 The Burnside-Brauer theorem

A classical result of Burnside [Bur55] says that if a finite dimensional $\mathbb{C}G$-module V affords a faithful representation of a finite group G, then every simple $\mathbb{C}G$-module is a composition factor of some tensor power of V. Brauer [Bra64] refined this result by bounding the number of tensor powers needed in terms of the number of distinct values taken on by χ_V. Here we prove the analogues for finite monoids; this is original to the text. In Chapter 10, we will prove a result of R. Steinberg [Ste62], which says that the direct sum of the tensor powers of a faithful representation of a monoid provides a faithful module for the monoid algebra. This also implies the result of Burnside, but without a good bound on the number of tensor powers needed. These kinds of results for representations of finite monoids over finite fields can be found in [Kuh94a, KK94].

We begin with a well-known lemma about idempotents of group algebras.

Lemma 7.24. *Let G be a finite group. Suppose that $e = \sum_{g \in G} c_g g$ in $\mathbb{C}G$ is a nonzero idempotent. Then $c_1 \neq 0$.*

Proof. Because $e \neq 0$, we have $\dim e\mathbb{C}G > 0$. Let χ be the character of the regular representation of G over $\mathbb{C}$, which we extend linearly to $\mathbb{C}G$. Then

$$\dim e\mathbb{C}G = \chi(e) = \sum_{g \in G} c_g \chi(g) = c_1 \cdot |G|$$

since by (B.2)

$$\chi(g) = \begin{cases} |G|, & \text{if } g = 1 \\ 0, & \text{else.} \end{cases}$$

Therefore, $c_1 = (\dim e\mathbb{C}G)/|G| \neq 0$.   □

We now prove the monoid analogue of the Burnside-Brauer theorem. We shall, in fact, weaken the faithfulness hypothesis. Let us say that a homomorphism $\varphi \colon M \longrightarrow N$ of monoids is an **LI**-*morphism*[1] if φ separates e from $eMe \setminus \{e\}$ for all idempotents $e \in E(M)$, that is, $\varphi^{-1}(\varphi(e)) \cap eMe = \{e\}$. Obviously an injective homomorphism is an **LI**-morphism and the converse holds for a group homomorphism. The notion of an **LI**-morphism will become fundamental in Chapter 11.

Example 7.25. Let $M = \{1, x, x^2\}$ where $x^2 = x^3$. Then $\varphi \colon M \longrightarrow \mathbb{C}$ given by $\varphi(1) = 1$ and $\varphi(x) = 0 = \varphi(x^2)$ is an **LI**-morphism but is not injective.

Theorem 7.26. *Let M be a finite monoid and let V be a finite dimensional $\mathbb{C}M$-module affording a representation $\rho \colon M \longrightarrow \mathrm{End}_{\mathbb{C}}(V)$ which is an **LI**-morphism. Suppose that the character χ of V takes on r distinct values. Then every simple $\mathbb{C}M$-module is a composition factor of $V^{\otimes i}$ for some $0 \leq i \leq r - 1$. In particular, this applies if ρ is faithful.*

Proof. Let S be a simple $\mathbb{C}M$-module with apex $e \in E(M)$. Put $A = \mathbb{C}M$ and let $R = \mathrm{rad}(A)$. Observe that $eAe = \mathbb{C}[eMe]$. As $eS \neq 0$, there is a primitive idempotent f of eAe such that f is primitive in A and $S \cong Af/Rf$ by Lemma 4.24. Write

$$f = \sum_{m \in eMe} c_m m.$$

By definition of an apex $I_e S = 0$. On the other hand, $fS \neq 0$. Thus $f \notin \mathbb{C}I_e$. Define a homomorphism $\varphi \colon eAe \longrightarrow \mathbb{C}G_e$ by

$$\varphi(m) = \begin{cases} m, & \text{if } m \in G_e \\ 0, & \text{if } m \in I_e \end{cases}$$

for $m \in eMe$ and note that $\ker \varphi = \mathbb{C}I_e$. Therefore,

$$\varphi(f) = \sum_{g \in G_e} c_g g$$

is a nonzero idempotent of $\mathbb{C}G_e$ and hence $c_e \neq 0$ by Lemma 7.24.

---

[1] The notation **LI** in semigroup theory denotes the class of locally trivial semigroups, that is, semigroups S such that $eSe = \{e\}$ for any idempotent $e \in S$; it can be shown that an **LI**-morphism is precisely a homomorphism such that the inverse image of any idempotent is a locally trivial semigroup.

Let $\chi_1, \ldots, \chi_s$ be the distinct values taken on by $\chi|_{eMe}$ and let

$$M_j = \{m \in eMe \mid \chi(m) = \chi_j\}.$$

Without loss of generality assume that $\chi_1 = \chi(e) = \dim eV$. Put

$$b_j = \sum_{m \in M_j} c_m.$$

Suppose now that $[V^{\otimes i} : S] = 0$ for all $0 \le i \le s - 1$. We follow here the convention that $\chi_j^0 = 1$ even if $\chi_j = 0$. As $\chi_{V^{\otimes i}}(m) = \chi_V(m)^i$ for $m \in M$, an application of Proposition 7.19 yields

$$0 = [V^{\otimes i} : S] = \sum_{m \in eMe} c_m \chi^i(m) = \sum_{j=1}^s \chi_j^i \sum_{m \in M_j} c_m = \sum_{j=1}^s \chi_j^i b_j$$

for all $0 \le i \le s - 1$. By nonsingularity of the Vandermonde matrix, we conclude that $b_j = 0$ for all $1 \le j \le s$.

Applying Proposition 7.14, we have that

$$M_1 = \{m \in eMe \mid \chi_{eV}(m) = \dim eV\}.$$

Because V affords a representation of M which is an **LI**-morphism, it follows that eV affords a representation of $\psi \colon eMe \longrightarrow \mathrm{End}_{\mathbb{C}}(eV)$ such that $\psi^{-1}(1_{eV}) = \{e\}$. Lemma 7.8(iv) then implies that $M_1 = \{e\}$. Thus $0 = b_1 = c_e \ne 0$. As $s \le r$, this contradiction concludes the proof. $\square$

We remark that it is necessary to include the trivial representation $V^{\otimes 0}$ because if M is a monoid with a zero element z and if $zV = 0$, then $zV^{\otimes i} = 0$ for all $i > 0$ and so the trivial representation is not a composition factor of any positive tensor power of V.

If G is a finite group and V is a finite dimensional $\mathbb{C}G$-module affording a faithful representation of G whose character takes on r distinct values, then $\bigoplus_{i=0}^{r-1} V^{\otimes i}$ contains every simple $\mathbb{C}G$-module as a composition factor by the Burnside-Brauer theorem and hence is a faithful $\mathbb{C}G$-module because $\mathbb{C}G$ is semisimple. We observe that the analogous result fails in a very strong sense for monoids.

Let $N_t = \{0, 1, \ldots, t\}$ where 1 is the identity and $xy = 0$ for $x, y \in N_t \setminus \{1\}$. Define a faithful two-dimensional representation $\rho \colon N_t \longrightarrow M_2(\mathbb{C})$ by

$$\rho(0) = \begin{bmatrix} 0 & 0 \\ 0 & 0 \end{bmatrix}, \quad \rho(1) = \begin{bmatrix} 1 & 0 \\ 0 & 1 \end{bmatrix}, \quad \rho(j) = \begin{bmatrix} 0 & j \\ 0 & 0 \end{bmatrix}, \quad \text{for } 2 \le j \le t.$$

Let V be the corresponding $\mathbb{C}N_t$-module. The character of ρ takes on two values: 0 and 2. However, $V^{\otimes 0} \oplus V^{\otimes 1}$ is 3-dimensional and so cannot be a faithful $\mathbb{C}N_t$-module for $t \ge 9$ by dimension considerations. In fact, given any integer $k \ge 0$, we can choose t sufficiently large so that $\bigoplus_{i=0}^k V^{\otimes i}$ is not a faithful $\mathbb{C}N_t$-module (again by dimension considerations). Thus, the minimum k such that $\bigoplus_{i=0}^{k} V^{\otimes i}$ is a faithful $\mathbb{C}N_t$-module cannot be bounded as a function of only the number of values assumed by the character χ_V (independently of the monoid in question).

7.5 The Cartan matrix

In this section we provide, following an idea of Thiéry [Thi12], a method to compute the Cartan matrix of $\mathbb{C}M$ for a finite monoid M.

If A, B are $\mathbb{k}$-algebras, then the tensor product $A \otimes B$ has a unique $\mathbb{k}$-algebra structure such that the product of basic tensors is given by

$$(a \otimes b)(a' \otimes b') = aa' \otimes bb'.$$

Moreover, if V is an A-module and W is a B-module, then the tensor product $V \otimes W$ has a unique $A \otimes B$-module structure such that $(a \otimes b)(v \otimes w) = av \otimes bw$ on basic tensors.

If A is a finite dimensional $\mathbb{k}$-algebra and M is a finite dimensional left A-module, then $D(M) = \operatorname{Hom}_{\mathbb{k}}(M, \mathbb{k})$ is a right A-module, i.e., an A^{op}-module. Moreover, it is known that D takes simples to simples, projectives to injectives, injectives to projectives, and preserves indecomposability. In fact, D provides a contravariant equivalence $A\text{-mod} \longrightarrow A^{op}\text{-mod}$ (cf. Theorem A.25). The *enveloping algebra* of A, defined by $A^e = A \otimes A^{op}$, has the property that its left modules correspond to A-A-bimodules [CE99]. We shall need the following consequence of [SY11, Proposition 11.3] and its proof, which is beyond the scope of this text.

Theorem 7.27. *Let $\mathbb{k}$ be an algebraically closed field and let A be a finite dimensional algebra over $\mathbb{k}$. Suppose that $e_1, \ldots, e_n$ form a complete set of orthogonal primitive idempotents for A. Then the $e_i \otimes e_j$ with $1 \leq i, j \leq n$ form a complete set of orthogonal primitive idempotents of $A^e = A \otimes A^{op}$. Moreover, if $S_i = Ae_i/\operatorname{rad}(A)e_i$ is the simple module corresponding to e_i, then $A^e(e_i \otimes e_j)/\operatorname{rad}(A^e)(e_i \otimes e_j) \cong S_i \otimes D(S_j)$.*

As a corollary, we can reinterpret the Cartan matrix.

Corollary 7.28. *Let $\mathbb{k}$ be an algebraically closed field and let A be a finite dimensional algebra over $\mathbb{k}$. Retaining the notation of Theorem 7.27, if C is the Cartan matrix of A, then $C_{ij} = [A : S_i \otimes D(S_j)]$ (viewing A as an A-A-bimodule).*

Proof. By definition $C_{ij} = [Ae_j : S_i] = \dim \operatorname{Hom}_A(Ae_i, Ae_j) = \dim e_i Ae_j$. But $e_i Ae_j = (e_i \otimes e_j)A$ if we view A as an A^e-module. Therefore,

$$C_{ij} = \dim \operatorname{Hom}_{A^e}(A^e(e_i \otimes e_j), A) = [A : S_i \otimes D(S_j)]$$

by Theorem 7.27 and Proposition A.24. This completes the proof.   □

Next we observe that the enveloping algebra of $\mathbb{C}M$ is nothing more than $\mathbb{C}[M \times M^{op}]$.

Proposition 7.29. *Let* $\Bbbk$ *be a field and let* M, N *be finite monoids. Then* $\Bbbk[M \times N] \cong \Bbbk M \otimes \Bbbk N$. *In particular,* $\Bbbk M^e \cong \Bbbk[M \times M^{op}]$.

Proof. The isomorphism takes (m, n) to $m \otimes n$. The details are left to the reader.   □

Let $e_1, \ldots, e_s \in E(M)$ be a complete set of idempotent representatives of the regular $\mathscr{J}$-classes of M. Then the (e_i, e_j) with $1 \leq i, j \leq s$ form a complete set of idempotent representatives of the regular $\mathscr{J}$-classes of $M \times M^{op}$ and $G_{(e_i, e_j)} = G_{e_i} \times G_{e_j}^{op}$. The simple $\mathbb{C}[M \times M^{op}]$-modules are then the modules

$$V^{\sharp} \otimes D(W^{\sharp}) = (V \otimes D(W))^{\sharp}$$

with $V \in \operatorname{Irr}_{\mathbb{C}}(G_{e_i})$, $W \in \operatorname{Irr}_{\mathbb{C}}(G_{e_j})$ because (e_i, e_j) is an apex for $V^{\sharp} \otimes D(W^{\sharp})$ and $e_i V^{\sharp} \otimes D(W^{\sharp}) e_j = e_i V^{\sharp} \otimes D(e_j W^{\sharp}) = V \otimes D(W)$.

Theorem 7.30. *Let* M *be a finite monoid and* $e_1, \ldots, e_s$ *form a complete set of idempotent representatives of the regular* $\mathscr{J}$-*classes of* M. *Denote by* L *the decomposition matrix of* M *over* $\mathbb{C}$. *Let* B *be a* $\mathbb{C}M$-$\mathbb{C}M$-*bimodule. Then* $e_i B e_j$ *is a* $\mathbb{C}G_{e_i}$-$\mathbb{C}G_{e_j}$-*module for* $1 \leq i, j \leq s$. *Define matrices* $Y(B)$ *and* $Z(B)$ *by*

$$Y(B)_{V,W} = [B : V \otimes D(W)]$$

for $V, W \in \operatorname{Irr}_{\mathbb{C}}(M)$ *and*

$$Z(B)_{S,S'} = [e_i B e_j : S \otimes D(S')]$$

for $S \in \operatorname{Irr}_{\mathbb{C}}(G_{e_i})$ *and* $S' \in \operatorname{Irr}_{\mathbb{C}}(G_{e_j})$. *Then* $Z(B) = LY(B)L^T$ *holds.*

Proof. Since Y and Z are additive as functions of $[B]$ on $G_0(\mathbb{C}[M \times M^{op}])$, it is enough to handle the case where B is simple, i.e., $B = V \otimes D(W)$ with V, W simple. Note that $Y(V \otimes D(W))$ has 1 in position V, W and 0 in all other positions. Then we have

$$
\begin{aligned}
[e_i(V \otimes D(W))e_j] &= [e_i V \otimes D(e_j W)] \\
&= \sum_{S \in \operatorname{Irr}_{\mathbb{C}}(G_{e_i})} \sum_{S' \in \operatorname{Irr}_{\mathbb{C}}(G_{e_j})} [e_i V : S][e_j W : S'] \cdot [S \otimes D(S')] \\
&= \sum_{S \in \operatorname{Irr}_{\mathbb{C}}(G_{e_i})} \sum_{S' \in \operatorname{Irr}_{\mathbb{C}}(G_{e_j})} L_{S,V} L_{S',W} \cdot [S \otimes D(S')] \\
&= \sum_{\substack{S \in \operatorname{Irr}_{\mathbb{C}}(G_{e_i}), \\ S' \in \operatorname{Irr}_{\mathbb{C}}(G_{e_j})}} (LY(V \otimes D(W))L^T)_{S,S'} \cdot [S \otimes D(S')].
\end{aligned}
$$

We conclude that $Z(V \otimes D(W)) = LY(V \otimes D(W))L^T$.   □

It follows from Corollary 7.28 that $Y(\mathbb{C}M)$ is the Cartan matrix of $\mathbb{C}M$. There is an isomorphism $G_{e_i} \times G_{e_j} \longrightarrow G_{e_i} \times G_{e_j}^{op}$ given by $(g, h) \mapsto (g, h^{-1})$. The character χ of $e_i\mathbb{C}Me_j$ as a $\mathbb{C}[G_{e_i} \times G_{e_j}]$-module is given by

$$\chi(g, h) = |\{m \in e_iMe_j \mid gmh^{-1} = m\}| = |\{m \in e_iMe_j \mid gm = mh\}|.$$

The character θ of $S \otimes D(S')$ as a $\mathbb{C}[G_{e_i} \times G_{e_j}]$-module is given by $\theta(g, h) = \chi_S(g)\overline{\chi_{S'}(h)}$ using Proposition B.3. Therefore, we have that

$$Z(\mathbb{C}M)_{S,S'} = \frac{1}{|G_{e_i}||G_{e_j}|} \sum_{g \in G_{e_i}} \sum_{h \in G_{e_j}} |\{m \in e_iMe_j \mid gm = mh\}| \cdot \overline{\chi_S(g)}\chi_{S'}(h).$$

$$(7.5)$$

As a consequence of Theorem 7.30 and the above discussion, we obtain the following result.

Theorem 7.31. *Let M be a finite monoid and $e_1, \ldots, e_s$ form a complete set of idempotent representatives of the regular $\mathscr{J}$-classes of M. Then the Cartan matrix C of $\mathbb{C}M$ is given by $L^{-1}Z(\mathbb{C}M)(L^{-1})^T$ where L is the decomposition matrix of M over $\mathbb{C}$ and $Z(\mathbb{C}M)$ is as in (7.5).*

For example, suppose that each maximal subgroup of M is trivial. Let $e_1, \ldots, e_s$ form a complete set of idempotent representatives of the regular $\mathscr{J}$-classes of M and let S_i be the unique simple module for $\Bbbk G_{e_i} \cong \Bbbk$. Then $Z(\mathbb{C}M)_{S_i,S_j} = |e_iMe_j|$ and $L_{S_i,S_j^\sharp} = \dim e_iS_j^\sharp$.

7.6 Exercises

7.1. Give an alternative proof of Theorem 7.7 using Proposition 7.19 (which does not depend on it).

7.2. Let M be a monoid. Show that the one-dimensional irreducible characters of M form a commutative inverse monoid under pointwise product. (Hint: show that $\chi^*(m) = \overline{\chi(m)}$.)

7.3. Provide an alternative proof of Proposition 7.9 using the Fitting decomposition (Theorem 5.38).

7.4. Prove that generalized conjugacy is the smallest equivalence relation $\sim$ on a finite monoid M such that $mn \sim nm$ and $m \sim m^{\omega+1}$ for all $m, n \in M$.

7.5. Let M be a finite monoid. Prove that $m, n \in M$ are generalized conjugates if and only if $\chi(m) = \chi(n)$ for every irreducible character χ of M.

7.6. Let M be a finite monoid and $m, n \in M$. Prove that m, n are generalized conjugates if and only if there exist $x, x' \in M$ such that $xx'x = x$, $x'xx' = x'$, $xm^{\omega+1}x' = n^{\omega+1}$ and $x'n^{\omega+1}x = m^{\omega+1}$.

7.7. Let M be a finite monoid and $e, f \in E(M)$. Prove that e, f are generalized conjugates if and only if $MeM = MfM$. (Hint: apply Theorem 1.11.)

7.8. Prove that $f, g \in I_n$ belong to the same generalized conjugacy class if and only if $|\operatorname{rec}(f)| = |\operatorname{rec}(g)|$ and the permutations $f|_{\operatorname{rec}(f)}$ and $g|_{\operatorname{rec}(g)}$ have the same cycle type.

7.9. Compute the character table of I_3.

7.10. Let $f \in T_n$. Define $i \in \{1, \dots, n\}$ to be *recurrent* if $f^m(i) = i$ for some $m > 0$. Denote by $\operatorname{rec}(f)$ the set of recurrent points of f.

(a) Prove that $\operatorname{rec}(f)$ is the image of f^{ω}.
(b) Prove that f leaves $\operatorname{rec}(f)$ invariant and that $f|_{\operatorname{rec}(f)}$ is a permutation.
(c) Prove that two mappings $f, g \in T_n$ are in the same generalized conjugacy class if and only if $|\operatorname{rec}(f)| = |\operatorname{rec}(g)|$ and the permutations $f|_{\operatorname{rec}(f)}$ and $g|_{\operatorname{rec}(g)}$ have the same cycle type.

7.11. Compute the character table of T_4.

7.12. Complete the details of the proof of Proposition 7.29.

7.13. Compute the Cartan matrix of $\mathbb{C}T_3$.

7.14 (Kudryavtseva-Mazorchuk [KM09]). Let M be a finite regular monoid. Prove that $f \colon M \longrightarrow \mathbb{C}$ is a class function if and only if $f(mn) = f(nm)$ for all $m, n \in M$.

7.15. Let V be a finite dimensional $\mathbb{C}M$-module for a finite monoid M and let S be a simple $\mathbb{C}M$-module. Prove that the generating function

$$f(t) = \sum_{n=0}^{\infty} [V^{\otimes n} : S] t^n$$

is a rational function.

7.16. Let M, N be finite monoids. Describe the character table $X(M \times N)$ in terms of $X(M)$ and $X(N)$.

7.17. Let A be a finite dimensional $\Bbbk$-algebra and let V, W be finite dimensional left A-modules. Prove that $V \otimes D(W) \cong \operatorname{Hom}_{\Bbbk}(W, V)$ as an A^e-module where $(af)(w) = af(w)$ and $(fa)(w) = f(aw)$ for $a \in A$, $w \in W$ and $f \in \operatorname{Hom}_{\Bbbk}(W, V)$.

The Representation Theory of Inverse Monoids

8

Categories and Groupoids

In this chapter, we consider a generalization of monoid algebras that will be used in the next chapter to study inverse monoid algebras, namely the algebra of a small category. Further examples include incidence algebras of posets (cf. Appendix C) and path algebras of quivers. We show that the Clifford-Munn-Ponizovskiĭ theory applies equally well to categories. The parametrization of the simple modules for the algebra of a finite category given here could also be obtained from a result of Webb [Web07], reducing to the monoid case, and the Clifford-Munn-Ponizovskiĭ theory, but we give a direct proof. Since category algebras are contracted semigroup algebras, these results also follow from the original results of Munn and Ponizovskiĭ (cf. [CP61, Chapter 5]). A basic reference on category theory is Mac Lane [Mac98]. Category algebras were considered at least as far back as Mitchell [Mit72].

8.1 Categories

A *small category* C is a category whose objects and arrows each form a set. We write C_0 for the object set of C and C_1 for the set of arrows. The domain and range mappings will be denoted by $\boldsymbol{d}$ and $\boldsymbol{r}$, respectively. The identity at c will be denoted 1_c and the hom set of arrows from c to d will be denoted $C(c, d)$. Sometimes we write C_c for the endomorphism monoid $C(c, c)$.

Let us consider some key examples.

Example 8.1 (Monoids). A monoid M can be viewed as a category with a single object $*$ and with arrow set M. Then $\boldsymbol{d}(m) = * = \boldsymbol{r}(m)$ for all $m \in M$, $1_* = 1$ and the composition is the product in M. Functors between monoids, viewed as categories, are exactly homomorphisms.

Example 8.2 (Posets). A poset P can be considered a category by taking the object set to be P and the arrow set to be $\{(p, q) \mid p \le q\}$. Here $\boldsymbol{d}(p, q) = p$

and $r(p, q) = q$. Composition is given by $(q, r)(p, q) = (p, r)$ and $1_p = (p, p)$. The category is finite if and only if the poset is finite. Functors between posets, viewed as categories, are precisely order-preserving maps.

Example 8.3 (Path categories). Let Q be a directed graph with vertex set Q_0 and edge set Q_1, called a *quiver* in this context. Then the *path category* of Q, or *free category* on Q, is the category Q^* with object set Q_0 and with hom set $Q^*(v, w)$ consisting of all (directed) paths from v to w, including an empty path 1_v if $v = w$. Composition is given by concatenation of paths, where we follow the convention that paths are concatenated from right to left, i.e., if $e_1, \ldots, e_n$ are edges with $d(e_{i+1}) = r(e_i)$, then their concatenation is

$$e_n e_{n-1} \cdots e_1 \colon d(e_1) \longrightarrow r(e_n).$$

For example, if Q has a single vertex, then Q^* is the free monoid on Q_1, viewed as a one-object category. The path category of Q is finite if and only if Q is finite and *acyclic*, that is, has no directed cycles.

The category Q^* has the following universal property. Given a category C and a pair of maps $f_0 \colon Q_0 \longrightarrow C_0$ and $f_1 \colon Q_1 \longrightarrow C_1$ such that, for each edge $e \colon v \longrightarrow w$ of Q, one has $f_1(e) \colon f_0(v) \longrightarrow f_0(w)$, there is a unique functor $F \colon Q^* \longrightarrow C$ such that $F|_{Q_0} = f_0$ and $F|_{Q_1} = f_1$, that is, the diagram

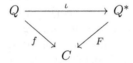

commutes, where ι is the inclusion and edges are viewed as paths of length 1.

Example 8.4 (Injective maps). Let FI_n be the category whose objects are subsets of $\{1, \ldots, n\}$ and whose morphisms are injective mappings. The study of the representation theory of the direct limit of the categories FI_n has become quite popular in recent years, cf. [CEFN14].

A number of the above examples are *EI-categories*, that is, categories in which each endomorphism is an isomorphism. A monoid is an EI-category if and only if it is a group. The category associated with a poset and the path category of an acyclic quiver are EI-categories, as is FI_n. There is a significant amount of literature on EI-categories and their representation theory, cf. [tD87, Lüc89, Web07, Web08, Li11, MS12a, Li14].

Fix now a category C and a field $\Bbbk$. The *category algebra* of C, denoted $\Bbbk C$, has underlying $\Bbbk$-vector space $\Bbbk C_1$. The product on basis elements is given by

$$fg = \begin{cases} f \circ g, & \text{if } d(f) = r(g) \\ 0, & \text{else} \end{cases}$$

and is then extended linearly.

Example 8.5. If M is a monoid, viewed as a one-object category, then the category algebra of M is the usual monoid algebra $\Bbbk M$. If P is a finite poset, viewed as a category, then the reader should check that $\Bbbk P \cong I(P, \Bbbk)^{op}$, the opposite of the incidence algebra of P over $\Bbbk$ (see Appendix C for the definition of the incidence algebra of a poset). If Q is a quiver, then $\Bbbk Q^*$ is usually denoted by $\Bbbk Q$ and is called the *path algebra* of Q. The path algebra $\Bbbk Q$ is finite dimensional if and only if Q is finite and acyclic (that is, it has no directed cycles).

Proposition 8.6. *Let C be a category with finitely many objects and $\Bbbk$ a field. Then $\Bbbk C$ is a unital $\Bbbk$-algebra with identity element $\sum_{c \in C_0} 1_c$. Moreover, the set of identity elements $\{1_c \mid c \in C_0\}$ is a complete set of orthogonal idempotents.*

Proof. We just verify that $\Bbbk C$ is unital. The remaining assertions are straightforward. Indeed, if f is an arrow of C, then

$$\sum_{c \in C_0} 1_c \cdot f \cdot \sum_{c \in C_0} 1_c = 1_{r(f)} f 1_{d(f)} = f$$

and the result follows. $\qquad\qquad\qquad\qquad\qquad\qquad\qquad\qquad\qquad\qquad\qquad$ $\square$

Next we provide a concrete interpretation of what are $\Bbbk C$-modules. Let $\mathrm{rep}_\Bbbk(C)$ be the category of all functors $F \colon C \longrightarrow \Bbbk\text{-mod}$. The morphisms are natural transformations. We remind the reader that if $F, G \colon C \longrightarrow D$ are two functors, then a *natural transformation* $\eta \colon F \longrightarrow G$ is a collection of morphisms $\eta_c \colon F(c) \longrightarrow G(c)$, for each object c of C, such that the diagram

$$
\begin{array}{ccc}
F(c) & \xrightarrow{\ F(f)\ } & F(c') \\
\eta_c \downarrow & & \downarrow \eta_{c'} \\
G(c) & \xrightarrow[\ G(f)\ ]{} & G(c')
\end{array}
$$

commutes for each arrow $f \colon c \longrightarrow c'$ of C.

For example, if M is a monoid, viewed as a one-object category, then a functor $F \colon M \longrightarrow \Bbbk\text{-mod}$ is the same thing as a homomorphism from M into the endomorphism monoid of a finite dimensional $\Bbbk$-vector space. If Q is a quiver, then a functor $F \colon Q^* \longrightarrow \Bbbk\text{-mod}$ is the same thing as an assignment of a finite dimensional $\Bbbk$-vector space $F(v)$ to each vertex v and of a linear transformation $F(e) \colon F(v) \longrightarrow F(w)$ to each edge $e \colon v \longrightarrow w$ of Q by the universal property in Example 8.3.

Theorem 8.7. *Let C be a category with finitely many objects and $\Bbbk$ a field. Then the categories $\mathrm{rep}_\Bbbk(C)$ and $\Bbbk C$-mod are equivalent.*

Proof. First let V be a $\Bbbk C$-module. Define a functor $F \colon C \longrightarrow \Bbbk\text{-mod}$ by putting $F(c) = 1_c V$ for an object $c \in C_0$ and if $g \colon c \longrightarrow d$, then

$$F(g)\colon 1_cV \longrightarrow 1_dV$$

is given by $F(g)v = gv$. Note that $1_dg = g$ implies that $gv \in 1_dV$. It is straightforward to verify that F is a functor.

Conversely, suppose that $F\colon C \longrightarrow \Bbbk\text{-mod}$ is a functor and put

$$V = \bigoplus_{c \in C_0} F(c).$$

If $g\colon c \longrightarrow d$ is an arrow of C and $v \in F(c)$, define $gv = F(g)(v)$. If $v \in F(c')$ with $c' \neq c$, put $gv = 0$. This yields a mapping $\varphi\colon C_1 \longrightarrow \mathrm{End}_\Bbbk(V)$ whose natural extension $\Phi\colon \Bbbk C \longrightarrow \mathrm{End}_\Bbbk(V)$ is a $\Bbbk$-algebra homomorphism, thereby making V into a $\Bbbk C$-module. Moreover, the reader can check that the above two constructions are mutual inverses up to natural isomorphism using that the set $\{1_c \mid c \in C_0\}$ is a complete set of orthogonal idempotents. $\square$

As a consequence, it follows that equivalent categories have Morita equivalent algebras.

Corollary 8.8. *Let C and D be equivalent categories with finitely many objects and let $\Bbbk$ be a field. Then $\Bbbk C$-mod is equivalent to $\Bbbk D$-mod.*

Proof. Clearly, if C and D are equivalent, then $\mathrm{rep}_\Bbbk(C)$ and $\mathrm{rep}_\Bbbk(D)$ are equivalent. The corollary then follows from Theorem 8.7. $\square$

The set of idempotents in a category C will be denoted $E(C)$. We can define Green's relations on a category C exactly as we do for a monoid. Put, for $f, g \in C_1$,

(i) $f \mathcal{J} g$ if and only if $C_1 f C_1 = C_1 g C_1$
(ii) $f \mathcal{L} g$ if and only if $C_1 f = C_1 g$
(iii) $f \mathcal{R} g$ if and only if $f C_1 = g C_1$.

If C is a category, then we can define a monoid $M(C)$ by

$$M(C) = C_1 \cup \{1\} \cup \{z\}$$

where 1 is an identity element, z is a zero element, and

$$fg = \begin{cases} f \circ g, & \text{if } d(f) = r(g) \\ z, & \text{else} \end{cases}$$

defines the product of elements of $f, g \in C_1$. Note that if C is a monoid, viewed as a 1-element category, then $C \neq M(C)$.

Proposition 8.9. *Let C be a category and $f, g \in C_1$.*

(i) $C_1 f C_1 \subseteq C_1 g C_1$ if and only if $M(C)fM(C) \subseteq M(C)gM(C)$
(ii) $C_1 f \subseteq C_1 g$ if and only if $M(C)f \subseteq M(C)g$
(iii) $f C_1 \subseteq g C_1$ if and only if $fM(C) \subseteq gM(C)$.

In particular, Green's relations on C coincide with the restriction of Green's relations on $M(C)$ to C_1.

Proof. We prove only (i), as the other cases are similar. Trivially, if $C_1 f C_1 \subseteq C_1 g C_1$, then $M(C) f M(C) \subseteq M(C) g M(C)$. Conversely, if $f = ugv$ with $u, v \in M(C)$, then $u, v \neq z$. If $u, v \in C_1$, there is nothing to prove. If $u = 1$ or $v = 1$, then we can replace u by $1_{r(g)}$, respectively, v by $1_{d(g)}$ and maintain the equality. Thus $f \in C_1 g C_1$. $\qquad\square$

It follows that properties of Green's relations in finite monoids, like Theorem 1.11 and Theorem 1.13, have analogues in finite categories. If $e \in E(C)$ is an idempotent, then $eC_1 e$ is a monoid with identity e and we denote by G_e the group of units of this monoid. Again we call G_e the *maximal subgroup* of C at e. Note that if $e \colon c \longrightarrow c$, then $eC_1 e = eC_c e$ (where $C_c = C(c, c)$) and hence G_e is the maximal subgroup at e of the monoid C_c. We put $I_e = eC_1 e \setminus G_e = eC_c e \setminus G_e$ for an idempotent $e \colon c \longrightarrow c$.

The following lemma shows that $\Bbbk C$ and $\Bbbk M(C)$ are very closely related.

Lemma 8.10. *Let C be a finite category. Then the elements z,*

$$\varepsilon = \sum_{c \in C_0} (1_c - z)$$

and $1 - \varepsilon - z$ form a complete set of orthogonal central idempotents of $\Bbbk M(C)$. Moreover, one has the following isomorphisms.

(i) $\Bbbk M(C)\varepsilon = \varepsilon \Bbbk M(C)\varepsilon \cong \Bbbk C$.
(ii) $\Bbbk M(c)z = z\Bbbk M(C)z \cong \Bbbk \cong (1-\varepsilon-z)\Bbbk M(C)(1-\varepsilon-z) = \Bbbk M(C)(1-\varepsilon-z)$
(iii) $\Bbbk M(C) \cong \Bbbk^2 \times \Bbbk C$.

Proof. It is immediate that $\{z\} \cup \{1_c - z \mid c \in C_0\}$ is a set of orthogonal idempotents and hence ε, z are orthogonal idempotents. It then follows that $\varepsilon, z, 1 - \varepsilon - z$ form a complete set of orthogonal idempotents. Trivially, we have that z is a central idempotent and that $\Bbbk M(C)z = \Bbbk z$. If $g \in C_1$, then

$$(1_c - z)g = \begin{cases} g - z, & \text{if } c = r(g) \\ 0, & \text{else} \end{cases}$$

and

$$g(1_c - z) = \begin{cases} g - z, & \text{if } c = d(g) \\ 0, & \text{else.} \end{cases}$$

It follows that $\varepsilon g = g - z = g\varepsilon$ and hence $g(1 - \varepsilon - z) = g - (g - z) - z = 0$. Since $1, z$ are central and ε commutes with elements of C_1, we conclude that ε is central. Therefore, ε, z and $1 - \varepsilon - z$ form a complete set of orthogonal central idempotents and $\Bbbk M(C)(1 - \varepsilon - z) = \Bbbk(1 - \varepsilon - z)$.

Finally, it is immediate from the definitions that there is a surjective homomorphism of $\Bbbk$-algebras $\varphi \colon \Bbbk M(C) \longrightarrow \Bbbk C$ given $\varphi(1) = \sum_{c \in C_0} 1_c$, $\varphi(z) = 0$

and $\varphi(g) = g$ for $g \in C_1$. It is clear that $z, 1 - \varepsilon - z \in \ker \varphi$ and hence form a basis for $\ker \varphi$ by dimension considerations (or direct computation). We conclude that $\varphi|_{kM(C)\varepsilon} \colon kM(C)\varepsilon \longrightarrow kC$ is an isomorphism.

Item (iii) follows from (i) and (ii). $\square$

Let C be a finite category. If S is a simple kC-module, we say that an idempotent $e \colon c \longrightarrow c$ is an *apex* for S if $eS \neq 0$ and $I_eS = 0$. Note that in this case, $1_cS \neq 0$ is a simple kC_c-module with apex e.

If $e \in E(C)$, then we can define the functors $\mathrm{Ind}_{G_e}(V) = kL_e \otimes_{kG_e} V$ and $\mathrm{Coind}_{G_e}(V) = \mathrm{Hom}_{kG_e}(kR_e, V)$ exactly as for monoids, using Green's relations in C. Also, we define

$$\mathrm{Res}_{G_e}(V) = eV, \quad T_e(V) = kCeV, \quad N_e(V) = \{v \mid ekCV = 0\},$$

as was the case for monoids.

The following is a generalization of Theorem 5.5 to finite categories. It can also be deduced from a result of [Web07] and Clifford-Munn-Ponizovskiĭ theory. In fact, it follows directly from the original work of Clifford, Munn and Ponizovskiĭ (cf. [CP61, Chapter 5]) because the algebra of a finite category is a contracted semigroup algebra (the contracted semigroup algebra of the semigroup $M(C) \setminus \{1\}$); since we have developed here the theory for finite monoids, rather than finite semigroups, we are not at liberty to proceed in this fashion. See also [MS12a,LS12,BD12,Lin14,DE15,BD15] for applications.

Theorem 8.11. *Let C be a finite category and k a field.*

(i) There is a bijection between isomorphism classes of simple kC-modules with apex $e \in E(C)$ and isomorphism classes of simple kG_e-modules induced by

$$S \longmapsto \mathrm{Res}_{G_e}(S) = eS$$
$$V \longmapsto V^\sharp = \mathrm{Ind}_{G_e}(V)/N_e(\mathrm{Ind}_{G_e}(V)) = \mathrm{Ind}_{G_e}(V)/\mathrm{rad}(\mathrm{Ind}_{G_e}(V))$$
$$\cong \mathrm{soc}(\mathrm{Coind}_{G_e}(V)) = T_e(\mathrm{Coind}_{G_e}(V))$$

for S a simple kC-module with apex e and V a simple kG_e-module.
(ii) Every simple kC-module has an apex (unique up to $\mathscr{J}$-equivalence).
(iii) If V is a simple kG_e-module, then every composition factor of $\mathrm{Ind}_{G_e}(V)$ and $\mathrm{Coind}_{G_e}(V)$ has apex f with $C_1eC_1 \subseteq C_1fC_1$. Moreover, $V^\sharp$ is the unique composition factor of either of these two modules with apex e and it appears in both with multiplicity one.

Proof. By Lemma 8.10, we can identify simple kC-modules with simple $kM(C)$-modules that are not annihilated by the central idempotent ε. Since z and ε are orthogonal central idempotents, any simple $kM(C)$-module with apex z is annihilated by ε. Also, simple $kM(C)$-modules with apex 1 are annihilated by ε because $\varepsilon \in kI(1)$. If $e \in E(M(C)) \setminus \{1, z\}$, then $\varepsilon e = e - z$. Therefore, if S is a simple $kM(C)$-module with apex e, then $\varepsilon S \supseteq \varepsilon eS =$

$(e - z)S = eS \neq 0$ (as $z \in I(e)$) and so ε does not annihilate any simple $\Bbbk M(C)$-module with apex in C_1. Thus, applying Theorem 5.5, simple $\Bbbk C$-modules are in bijection with simple $\Bbbk M(C)$-modules whose apex is different from 1 and z. In light of Proposition 8.9, which lets us identify Green's relations on $M(C) \cap C_1$ with Green's relations on C, the result follows. □

This theorem specializes to the following result for EI-categories; see [tD87, Section 11 of Chapter 1] and [Lüc89, Web07].

Corollary 8.12. *Let C be a finite EI-category and $\Bbbk$ a field. Let $c_1, \ldots, c_n$ represent the distinct isomorphism classes of objects of C and let G_{c_i} be the automorphism group at the object c_i. Then there is a bijection between simple $\Bbbk C$-modules and simple $\Bbbk G_{c_i}$-modules for $i = 1, \ldots, n$.*

The bijection is as follows. If V is a simple $\Bbbk C$-module, then there exists a unique i such that $1_{c_i} V \neq 0$ and this latter module is a simple $\Bbbk G_{c_i}$-module. Conversely, if V is a simple $\Bbbk G_{c_i}$-module, then

$$\left(\Bbbk L_{1_{c_i}} \otimes_{\Bbbk G_{c_i}} V\right) / \mathrm{rad}(\Bbbk L_{1_{c_i}} \otimes_{\Bbbk G_{c_i}} V)$$

is the corresponding simple $\Bbbk C$-module.

Proof. Since C is an EI-category, the identities are its only idempotents. The result then follows from Theorem 8.11 once one observes that two identities $1_c, 1_d$ of C are $\mathscr{J}$-equivalent if and only if c, d are isomorphic. Trivially, if c, d are isomorphic, then $1_c, 1_d$ are $\mathscr{J}$-equivalent. Conversely, if $1_c, 1_d$ are $\mathscr{J}$-equivalent, then by the category version of Theorem 1.11, there exist $a, b \in C_1$ with $ab = 1_c$ and $ba = 1_d$. Thus c, d are isomorphic. □

8.2 Groupoids

A *groupoid* G is a category in which each arrow is an isomorphism. That means, for each arrow $f: c \longrightarrow d$, there is an arrow $f^{-1}: d \longrightarrow c$ such that $f^{-1}f = 1_c$ and $ff^{-1} = 1_d$. The inverse arrow is unique. For each object $c \in G_0$, the endomorphism monoid G_c is, in fact, a group. A group is, of course, the same thing as a one-object groupoid.

Assume from now on that G is a finite groupoid. We shall write $\bar{c}$ for the isomorphism class of an object $c \in G_0$. Fix representatives $c_1, \ldots, c_s$ of the isomorphism classes of objects on G. For each object $c \in G_0$ with $\bar{c} = \bar{c}_i$, fix an arrow $p_c: c_i \longrightarrow c$. We may take $p_{c_i} = 1_{c_i}$.

Proposition 8.13. *Let G be a groupoid and let $c_1, \ldots, c_s$ be representatives of the isomorphism classes of objects on G. Then G is equivalent to the disjoint union groupoid $\coprod_{i=1}^{s} G_{c_i}$. Hence if $\Bbbk$ is a field, $\Bbbk G$-mod is equivalent to $\prod_{i=1}^{s} \Bbbk G_{c_i}$-mod.*

Proof. The second statement is a consequence of the first, Theorem 8.7 and Corollary 8.8. To prove the first statement, define $F\colon G \longrightarrow \coprod_{i=1}^{s} G_{c_i}$ by $F(c) = c_i$ if $\bar{c} = \bar{c}_i$ on objects and on arrows, for $g\colon c \longrightarrow d$, define $F(g) = p_d^{-1} g p_c$. It is straightforward to verify that F and the inclusion functor are mutual quasi-inverses. $\square$

Let us make the second equivalence in Proposition 8.13 more explicit. Let $A = \Bbbk G$ and $B = \Bbbk \left[\coprod_{i=1}^{s} G_{c_i} \right]$. Set $e = \sum_{i=1}^{s} 1_{c_i}$. Then e is an idempotent with $eAe = B$ and $AeA = A$. Thus Res_e gives an equivalence between A-mod and B-mod with quasi-inverse Ind_e by Theorem 4.13. But in B, the idempotents $1_{c_1}, \ldots, 1_{c_s}$ form a complete set of orthogonal central idempotents and also $1_{c_i} \cdot B \cdot 1_{c_i} = \Bbbk G_{c_i}$. Thus $B \cong \prod_{i=1}^{s} \Bbbk G_{c_i}$ and, therefore, we have the following corollary.

Corollary 8.14. *Let G be a groupoid and let $c_1, \ldots, c_s$ be representatives of the isomorphism classes of objects of G. Then there is an equivalence of categories*

$$\Bbbk G\text{-mod} \longrightarrow \prod_{i=1}^{s} \Bbbk G_{c_i}\text{-mod}$$

sending a $\Bbbk G$-module V to $(1_{c_1} V, \ldots, 1_{c_s} V)$. The quasi-inverse equivalence is given on objects by

$$(V_1, \ldots, V_s) \longmapsto \bigoplus_{i=1}^{s} \Bbbk L_i \otimes_{\Bbbk G_{c_i}} V_i$$

where $L_i = \boldsymbol{d}^{-1}(c_i)$ (which is a free right G_{c_i}-set). In particular, the simple $\Bbbk G$-modules are the modules of the form $\Bbbk L_i \otimes_{\Bbbk G_{c_i}} S$ with S a simple $\Bbbk G_{c_i}$-module.

Proof. Note that $\Bbbk G 1_{c_i} = \Bbbk L_i$ and so $\mathrm{Ind}_{1_{c_i}}(V) = \Bbbk L_i \otimes_{\Bbbk G_{c_i}} V$ for V a $\Bbbk G_{c_i}$-module. Also L_i is a free right G_{c_i}-set because of the cancellation law in the groupoid G. The result now follows from the discussion preceding the corollary. $\square$

We now prove the analogue of Proposition 8.13 at the level of algebras.

Theorem 8.15. *Let G be a finite groupoid and $\Bbbk$ a field. Let $c_1, \ldots, c_s$ be representatives of the isomorphism classes of objects of G. Let n_i be the cardinality of the isomorphism class of c_i. Then we have*

$$\Bbbk G \cong \prod_{i=1}^{s} M_{n_i}(\Bbbk G_{c_i})$$

and hence $\Bbbk G$ is semisimple if and only if $|G_{c_i}|$ is not divisible by the characteristic of $\Bbbk$ for $i = 1, \ldots, s$.

Proof. Let $e_i = \sum_{c \in \overline{c}_i} 1_c$. Then $e_1, \ldots, e_s$ are orthogonal central idempotents of $\Bbbk G$ and $1 = \sum_{i=1}^s e_i$. Moreover, if H_i denotes the full subgroupoid of G on the objects of $\overline{c}_i$ (with all morphisms of G between these objects), then $e_i \Bbbk G e_i = \Bbbk H_i$. It follows that $\Bbbk G \cong \prod_{i=1}^s \Bbbk H_i$ and hence it suffices to show that $\Bbbk H_i \cong M_{n_i}(\Bbbk G_{c_i})$. Thus, without loss of generality, we may assume that all objects of G are isomorphic.

So suppose from now on that $G_0 = \{x_1, \ldots, x_n\}$ and that $p_i \colon x_1 \longrightarrow x_i$ is an arrow for each $i = 1, \ldots, n$. We may choose $p_1 = 1_{x_1}$. Let E_{ij} be the standard matrix unit of $M_n(\Bbbk G_{x_1})$. Define $\alpha \colon \Bbbk G \longrightarrow M_n(\Bbbk G_{x_1})$ on a basis element $g \colon x_i \longrightarrow x_j$ by

$$\alpha(g) = p_j^{-1} g p_i E_{ji}.$$

Observe that if $g \colon x_i \longrightarrow x_j$ and $h \colon x_k \longrightarrow x_\ell$, then

$$\alpha(g)\alpha(h) = p_j^{-1} g p_i p_\ell^{-1} h p_k E_{ji} E_{\ell k} = \begin{cases} p_j^{-1} g h p_k E_{jk}, & \text{if } i = \ell \\ 0, & \text{else.} \end{cases}$$

Therefore, $\alpha(g)\alpha(h) = \alpha(gh)$ and so α is a homomorphism, as required.

The inverse of α is given by

$$\beta((a_{ij})) = \sum_{i,j} p_i a_{ij} p_j^{-1}$$

where we identify $\Bbbk G_{x_1}$ with $1_{x_1} \Bbbk G 1_{x_1}$. Indeed, if $g \colon x_i \longrightarrow x_j$, then $\beta(\alpha(g)) = \beta(p_j^{-1} g p_i E_{ji}) = p_j p_j^{-1} g p_i p_i^{-1} = g$. Conversely, $\alpha(\beta(g E_{ij})) = \alpha(p_i g p_j^{-1}) = p_i^{-1} p_i g p_j^{-1} p_j E_{ij} = g E_{ij}$. This completes the proof. $\quad\square$

In summary the representation theory of finite groupoids is essentially the same as the representation theory of finite groups.

8.3 Exercises

8.1. Show that if P, Q are posets, then order-preserving maps from P to Q are in bijection with functors from the category associated with P to the category associated with Q.

8.2. Let P be a poset and $\Bbbk$ a field. Prove that $\Bbbk P \cong I(P, \Bbbk)^{op}$. (See Appendix C for the definition of the incidence algebra of a poset.)

8.3. Prove the universal property of Q^* in Example 8.3.

8.4. Prove that if C and D are equivalent categories and A is any category, then the categories A^C and A^D of functors from C to A and D to A, respectively, are equivalent.

8.5. Let Q be a finite quiver and $\Bbbk$ a field. Prove that $\Bbbk Q$ is finite dimensional if and only if Q is acyclic.

8.6. Let $\Bbbk$ be a field and $Q = (Q_0, Q_1)$ a finite acyclic quiver.

(a) Prove that $\{1_v \mid v \in Q_0\}$ is a complete set of orthogonal primitive idempotents of $\Bbbk Q$.
(b) Prove that $\mathrm{rad}(\Bbbk Q)$ is the ideal spanned by paths of length at least 1 and that $\Bbbk Q / \mathrm{rad}(\Bbbk Q) \cong \Bbbk^{|Q_0|}$.
(c) Describe explicitly the simple $\Bbbk Q$-modules.
(d) Describe explicitly the projective indecomposable $\Bbbk Q$-modules.

8.7. Let P be a finite poset and $\Bbbk$ a field.

(a) Prove that $\{(p, p) \mid p \in P\}$ is a complete set of orthogonal primitive idempotents of $\Bbbk P$.
(b) Prove that $\mathrm{rad}(\Bbbk P)$ is spanned by all arrows (p, q) with $p < q$ and $\Bbbk P / \mathrm{rad}(\Bbbk P) \cong \Bbbk^{|P|}$.
(c) Describe explicitly the simple $\Bbbk P$-modules.
(d) Describe explicitly the projective indecomposable $\Bbbk P$-modules.

8.8. Let P be a finite poset and $\Bbbk$ a field. Let Q be the Hasse diagram of P. Express $\Bbbk P$ as a quotient of $\Bbbk Q$.

8.9. Let $\Bbbk$ be a field of characteristic 0. Prove that

$$\Bbbk FI_n / \mathrm{rad}(\Bbbk FI_n) \cong \prod_{r=0}^{n} M_{\binom{n}{r}}(\Bbbk S_r).$$

(Hint: if G is the groupoid of bijections between subsets of $[n]$, prove that $\Bbbk FI_n / \mathrm{rad}(\Bbbk FI_n) \cong \Bbbk G$.)

8.10. Fill in the missing details of the proof of Theorem 8.7.

8.11. Let C be a finite EI-category. Prove that L_{1_c} is the set of isomorphisms with domain c for $c \in C_0$.

8.12. Let G be a finite group acting on a finite set X. Let H be the groupoid with $H_0 = X$, $H_1 = G \times X$, $d(g, x) = x$, $r(g, x) = gx$, product

$$(g', gx)(g, x) = (g'g, x),$$

identities $(1, x)$ and inverses $(g, x)^{-1} = (g^{-1}, gx)$ where $x \in X$ and $g, g' \in G$. Let $\Bbbk$ be a field.

(a) Express $\Bbbk H$ as a direct product of matrix algebras over group algebras of subgroups of G.
(b) Prove that $\varphi \colon \Bbbk G \longrightarrow \Bbbk H$ given by

$$\varphi(g) = \sum_{x \in X} (g, x)$$

for $g \in G$ is a $\Bbbk$-algebra homomorphism.

8.13. Let G be a finite groupoid. A representation $\rho\colon \mathbb{C}G \longrightarrow M_n(\mathbb{C})$ is a *-*representation* if $\rho(g^{-1}) = \rho(g)^*$ for each arrow g of G where A^* is the Hermitian adjoint of a matrix A.

(a) Give a direct proof that each *-representation of a finite groupoid is completely reducible.
(b) Prove that each complex representation of a finite groupoid is equivalent to a *-representation. (Hint: imitate the proof that each complex representation of a finite group is equivalent to a unitary representation.)
(c) Give another proof that $\mathbb{C}G$ is a semisimple algebra.

8.14. Let $G = (G_0, G_1)$ be a finite groupoid. A mapping $f\colon G_1 \longrightarrow \mathbb{C}$ is called a *class function* on G if

  (i) $f(g) = 0$ for all $g \in G_1$ with $\boldsymbol{d}(g) \neq \boldsymbol{r}(g)$.
  (ii) $f(hgh^{-1}) = f(g)$ for all $g, h \in G_1$ with $\boldsymbol{d}(h) = \boldsymbol{d}(g) = \boldsymbol{r}(g)$.

(a) Prove that if $\rho\colon \mathbb{C}G \longrightarrow M_n(\mathbb{C})$ is a representation, then the *character* $\chi_\rho\colon G_1 \longrightarrow \mathbb{C}$ defined by $\chi_\rho(g) = \mathrm{Tr}(\rho(g))$ is a class function.
(b) Prove that if ρ, ψ are inequivalent irreducible representations of $\mathbb{C}G$ (that is, the corresponding simple $\mathbb{C}G$-modules are not isomorphic), then χ_ρ and χ_ψ are orthogonal with respect to the inner product on $\mathbb{C}^{G_1}$ defined by

$$\langle f, g \rangle = \sum_{h \in G_1} f(h)\overline{g(h)}.$$

(c) Let $\rho_1, \ldots, \rho_s$ form a complete set of representatives of the equivalence classes of irreducible representations of $\mathbb{C}G$. Let χ_i be the character of ρ_i, for $i = 1, \ldots, s$. Prove that $\chi_1, \ldots, \chi_s$ form an orthogonal basis for the space of class functions on G.
(d) Prove that the center $Z(\mathbb{C}G)$ of $\mathbb{C}G$ consists of all elements of the form $\sum_{g \in G_1} c_g g$ with $f(g) = c_g$ a class function on G.

8.15. Let M be a finite monoid. Define a category $K(M)$ (called the *Karoubi envelope* of M) by $K(M)_0 = E(M)$ and

$$K(M)_1 = \{(e, m, f) \in E(M) \times M \times E(M) \mid fme = m\}.$$

Here $\boldsymbol{d}(e, m, f) = e$ and $\boldsymbol{r}(e, m, f) = f$. The product is given by

$$(f, n, g)(e, m, f) = (e, nm, g).$$

  (i) Verify that $K(M)$ is a category.
  (ii) Prove that (e, m, f) is an isomorphism if and only if $Me = Mm$ and $fM = mM$.
  (iii) Prove that two idempotents $e, f \in E(M)$ are isomorphic in $K(M)$ if and only if $MeM = MfM$.
  (iv) Prove that $K(M)(e, e) \cong eMe$.
  (v) Prove that $\Bbbk K(M)$ is Morita equivalent to $\Bbbk M$ for any field $\Bbbk$.

9

The Representation Theory of Inverse Monoids

In this chapter we develop the representation theory of inverse monoids following the approach of the author [Ste06, Ste08] using groupoid algebras and Möbius inversion. In fact, this work has a precursor in the work of Rukolaĭne [Ruk78, Ruk80], who used alternating sums of idempotents to achieve the same effect as Möbius inversion and used Brandt inverse semigroups instead of groupoids. The author only became aware of the work of Rukolaĭne after [Ste06, Ste08] were published. However, our more explicit approach lets one take advantage of the detailed knowledge of the Möbius function for a number of naturally occurring lattices. We will, for instance, exploit this for our simple character theoretic proof of Solomon's computation [Sol02] of the tensor powers of the natural module for the symmetric inverse monoid.

In this chapter, we work with inverse monoids because we provided an exposition in Chapter 3 of the apparatus of inverse semigroup theory in the context of monoids. But, in fact, finite inverse semigroups have unital semigroup algebras and everything in this chapter works *mutatis mutandis* for inverse semigroups.

There are generalizations of these results to infinite inverse semigroups [Ste10d] and compact inverse monoids [HS15]. Recently, I. Stein has generalized these results to a larger class of semigroups [Ste15].

9.1 The groupoid of an inverse monoid

Canonically associated with each inverse monoid is a groupoid. The author learned of this construction from Lawson's book [Law98]. The *groupoid* $G(M)$ of an inverse monoid M has object set $E(M)$ and arrow set $\{[m] \mid m \in M\}$, a set in bijection with M via $m \mapsto [m]$. One defines $\boldsymbol{d}([m]) = m^*m$ and $\boldsymbol{r}([m]) = mm^*$. The product is given by $[m][n] = [mn]$ if $\boldsymbol{d}([m]) = \boldsymbol{r}([n])$. The

© Springer International Publishing Switzerland 2016   137
B. Steinberg, *Representation Theory of Finite Monoids*,
Universitext, DOI 10.1007/978-3-319-43932-7_9

identity at an object $e \in E(S)$ is $1_e = [e]$ and the inverse operation is given by $[m]^{-1} = [m^*]$.

Proposition 9.1. *If M is a finite inverse monoid, then $G(M)$ is a finite groupoid. Moreover, the automorphism group $G(M)_e$ at an object $e \in E(M)$ can be identified with the maximal subgroup G_e and $\mathbf{d}^{-1}(e)$ can be identified with the $\mathscr{L}$-class L_e as a right G_e-set.*

Proof. First observe that if $\mathbf{d}([m]) = \mathbf{r}([n])$, then $\mathbf{d}([mn]) = \mathbf{d}([n])$ and $\mathbf{r}([mn]) = \mathbf{r}([m])$ by Proposition 3.15. Associativity follows from the associativity of multiplication in M. If $[m] \colon e \longrightarrow f$, then $e = m^*m$ and $f = mm^*$. Thus $[m][e] = [me] = [m]$ and $[f][m] = [fm] = [m]$, whence $1_e = [e]$ and $1_f = [f]$. We deduce that $G(M)$ is a category. Finally, if $m \in M$, then $[m] \colon m^*m \longrightarrow mm^*$, $[m^*] \colon mm^* \longrightarrow m^*m$ and

$$[m^*][m] = [m^*m] = 1_{m^*m} \quad \text{and} \quad [m][m^*] = [mm^*] = 1_{mm^*}.$$

It follows that $[m^*] = [m]^{-1}$ and so $G(M)$ is a groupoid, as required.

Note that $[m] \in G(M)_e$ if and only if $\mathbf{d}([m]) = e = \mathbf{r}([m])$, if and only if $m^*m = e = mm^*$, if and only if $m \in G_e$ by Corollary 3.6. It is then immediate that $g \mapsto [g]$ gives an isomorphism between G_e and $G(M)_e$. Also it follows that $m \mapsto [m]$ is a bijection between L_e and $\mathbf{d}^{-1}(e)$ by Proposition 3.10 and clearly this bijection identifies these two sets as right G_e-sets. $\qquad \square$

For example, the groupoid $G(I_n)$ of the symmetric inverse monoid I_n is isomorphic to the groupoid whose objects are subsets of $\{1, \dots, n\}$ and whose arrows are bijections. If $M = E(M)$ is a lattice, then each element of $G(M)$ is an identity.

Proposition 9.2. *Let M be a finite inverse monoid. Then idempotents $e, f \in E(M)$ are isomorphic in $G(M)$ if and only if $MeM = MfM$.*

Proof. If e is isomorphic to f in $G(M)$, then there exists $m \in M$ with $m^*m = e$ and $mm^* = f$. But then $MeM = MmM = MfM$. Conversely, if $MeM = MfM$, then by Corollary 1.14, there exists $m \in M$ with $Me = Mm$ and $mM = fM$. By Proposition 3.7 and Proposition 3.10, we then have $m^*m = e$ and $mm^* = f$. Thus $[m] \colon e \longrightarrow f$ is an isomorphism. $\qquad \square$

9.2 The isomorphism of algebras

The main result of this section is that the algebras of an inverse monoid M and of its groupoid $G(M)$ are isomorphic. In the next section, we shall exploit this isomorphism to obtain a better understanding of the representation theory of the inverse monoid M. An analogue of this theorem for infinite inverse semigroups appears in [Ste10d]. The reader is referred to Appendix C for the definitions of the zeta function ζ and Möbius function μ of a poset.

Theorem 9.3. *Let M be a finite inverse monoid and $\Bbbk$ a field. Then there is an isomorphism $\alpha\colon \Bbbk M \longrightarrow \Bbbk G(M)$ given by*

$$\alpha(m) = \sum_{n \leq m} [n]$$

for $m \in M$. The inverse isomorphism is given by

$$\beta([m]) = \sum_{n \leq m} \mu(n, m)n,$$

for $m \in M$, where μ is the Möbius function of the poset $(M, \leq)$.

Proof. We first verify that α is a homomorphism; that is, we must verify that if $m_1, m_2 \in M$, then $\alpha(m_1)\alpha(m_2) = \alpha(m_1 m_2)$. We begin by computing

$$\alpha(m_1)\alpha(m_2) = \sum_{n_1 \leq m_1} [n_1] \sum_{n_2 \leq m_2} [n_2] = \sum_{\substack{n_1 \leq m_1, n_2 \leq m_2, \\ n_1^* n_1 = n_2 n_2^*}} [n_1 n_2]. \qquad (9.1)$$

But Lemma 3.16 implies that each $n \leq m_1 m_2$ can be uniquely factored as $n = n_1 n_2$ with $n_1 \leq m_1$, $n_2 \leq m_2$ and $n_1^* n_1 = n_2 n_2^*$. Therefore, the right-hand side of (9.1) is precisely

$$\sum_{n \leq m_1 m_2} [n] = \alpha(m_1 m_2).$$

This proves that α is a homomorphism.

If $m \in M$, we compute

$$\beta(\alpha(m)) = \sum_{n \leq m} \sum_{n' \leq n} \mu(n', n)n' = \sum_{n' \leq m} \left(\sum_{n' \leq n \leq m} \mu(n', n)\zeta(n, m) \right) n'$$

$$= \sum_{n' \leq m} \delta(n', m)n' = m$$

$$\alpha(\beta([m])) = \sum_{n \leq m} \sum_{n' \leq n} \mu(n, m)[n'] = \sum_{n' \leq m} \left(\sum_{n' \leq n \leq m} \zeta(n', n)\mu(n, m) \right) [n']$$

$$= \sum_{n' \leq m} \delta(n', m)[n'] = [m].$$

This completes the proof that α, β are inverse isomorphisms. $\square$

As a corollary, we obtain that $\Bbbk M$ is isomorphic to a product of matrix algebras over the group algebras of its maximal subgroups.

Corollary 9.4. *Let M be a finite inverse monoid and $\Bbbk$ a field. Let $e_1, \ldots, e_s$ be idempotent representatives of the $\mathscr{J}$-classes of M and let $n_i = |E(J_{e_i})|$. Then there is an isomorphism*

$$\Bbbk M \cong \prod_{i=1}^{s} M_{n_i}(\Bbbk G_{e_i}).$$

Consequently, $\Bbbk M$ is semisimple if and only if the characteristic of $\Bbbk$ does not divide the order of any of the maximal subgroups G_{e_i}.

Proof. By Proposition 9.1, Proposition 9.2, and Theorem 8.15, we have an isomorphism

$$\Bbbk G(M) \cong \prod_{i=1}^{s} M_{n_i}(\Bbbk G_{e_i}).$$

The required isomorphism now follows from Theorem 9.3. The final statement follows from Maschke's theorem and the fact that a finite dimensional algebra A is semisimple if and only if $M_n(A)$ is semisimple (for any $n \geq 1$) as A and $M_n(A)$ are Morita equivalent. □

Let us specialize to the case of a lattice, which was considered early on by Solomon [Sol67].

Corollary 9.5. *Let L be a finite lattice and $\Bbbk$ a field. Then $\Bbbk L \cong \Bbbk^L$ is a commutative semisimple $\Bbbk$-algebra. If, for $x \in L$, we put*

$$e_x = \sum_{y \leq x} \mu(y, x) y$$

with μ the Möbius function of L, then we have that $\{e_x \mid x \in L\}$ is a complete set of orthogonal primitive idempotents of $\Bbbk L$.

Proof. The groupoid $G(L)$ consists of just the identities $[x]$ and these identities form a complete set of orthogonal primitive idempotents, which are, in addition, central. Moreover, $[x]\Bbbk G(L)[x] \cong \Bbbk G_x \cong \Bbbk$ and so $\Bbbk G(L) \cong \Bbbk^L$. As $e_x = \beta([x])$ (using the notation of Theorem 9.3), the result follows from Theorem 9.3. □

Next we consider the case of a more general commutative inverse monoid.

Corollary 9.6. *Let M be a finite commutative inverse monoid and let $\Bbbk$ be a field. Then the isomorphism*

$$\Bbbk M \cong \prod_{e \in E(M)} \Bbbk G_e$$

holds.

Proof. Proposition 3.11 says that the $\mathscr{J}$-classes of a commutative inverse monoid are its maximal subgroups, and hence no two distinct idempotents of M are $\mathscr{J}$-equivalent. The corollary now follows from Corollary 9.4.   □

Define the *rank* of a partial injective map to be the cardinality of its image. Then it is straightforward to check that two elements of the symmetric inverse monoid I_n are $\mathscr{J}$-equivalent if and only if they have the same rank. As usual, for $k \geq 0$, put $[k] = \{1, \ldots, k\}$ (and so $[0] = \emptyset$). Then $1_{[0]}, \ldots, 1_{[n]}$ are idempotent representatives of the $\mathscr{J}$-classes of I_n. There are $\binom{n}{k}$ idempotents of rank k and the maximal subgroup at $1_{[k]}$ can be identified with the symmetric group S_k of degree k; indeed, it consists of all bijections from $[k]$ to $[k]$. We then deduce the following corollary of Theorem 9.3, which we believe was first written in an explicit fashion by Solomon [Sol02].

Corollary 9.7. *Let $\Bbbk$ be a field. Then there is an isomorphism*

$$\Bbbk I_n \cong \prod_{k=0}^{n} M_{\binom{n}{k}}(\Bbbk S_k)$$

where I_n is the symmetric inverse monoid of degree n and S_k is the symmetric group of degree k.

9.3 Decomposing representations of inverse monoids

Let M be a finite inverse monoid. Then $\mathbb{C}M$ is semisimple by Corollary 9.4. Proposition 7.19 provides a method to compute the composition factors of a finite dimensional $\mathbb{C}M$-module from its character by inverting the character table. However, in practice, computing the character table of a monoid is not an easy task. We present here, for inverse monoids, an alternative means for computing composition factors from the character that exploits the isomorphism $\mathbb{C}M \cong \mathbb{C}G(M)$. This method requires only knowing the character tables of the maximal subgroups and the Möbius function of the lattice $E(M)$ of idempotents. As an application, we provide simpler character theoretic proofs of results of Solomon [Sol02] decomposing the exterior and tensor powers of the standard module for the symmetric inverse monoid.

We first need a lemma relating the Möbius function μ_M of M and the Möbius function $\mu_{E(M)}$ of $E(M)$.

Lemma 9.8. *Let $m \in M$. Then the equality*

$$\sum_{n \leq m} \mu_M(n, m)n = \sum_{f \leq m^*m} \mu_{E(M)}(f, m^*m)mf$$

holds.

Proof. If P is a partially ordered set and $p \leq q$, then it is well known and easy to show that $\mu_P(p, q)$ only depends on the isomorphism type of the poset $[p, q] = \{r \in P \mid p \leq r \leq q\}$ (see Exercise C.5). Let $P = \{n \in M \mid n \leq m\}$ and let $Q = \{f \in E(M) \mid f \leq m^*m\}$. Define $\psi \colon P \longrightarrow Q$ and $\gamma \colon Q \longrightarrow P$ by $\psi(n) = n^*n$ and $\gamma(f) = mf$. Then $n_1 \leq n_2 \leq m$ implies $n_1^*n_1 \leq n_2^*n_2 \leq m^*m$ by Proposition 3.9, and so ψ is a well-defined order-preserving map. Similarly, $f \leq e \leq m^*m$ implies that $mf \leq me \leq m$, again by Proposition 3.9, and so γ is order-preserving. We claim that ψ and γ are mutual inverses. Indeed, if $n \leq m$ and $f \leq m^*m$, then $\gamma(\psi(n)) = mn^*n = n$ and $\psi(\gamma(f)) = (mf)^*mf = fm^*mf = f$.

It then follows that, for $f \leq m^*m$, we have

$$\mu_M(\gamma(f), m) = \mu_M(\gamma(f), \gamma(m^*m)) = \mu_{E(M)}(f, m^*m).$$

We may now compute

$$\sum_{n \leq m} \mu_M(n, m)n = \sum_{n \leq m} \mu_M(\gamma(\psi(n)), m)\gamma(\psi(n))$$

$$= \sum_{f \leq m^*m} \mu_M(\gamma(f), m)\gamma(f)$$

$$= \sum_{f \leq m^*m} \mu_{E(M)}(f, m^*m)mf$$

as required. $\qquad \square$

We are now ready to give a formula for decomposing a $\mathbb{C}M$-module from its character. We retain the notation from Theorem 5.5. Note that if S is a simple $\mathbb{C}G_e$-module, for $e \in E(M)$, then $S^\sharp = \mathrm{Ind}_{G_e}(S)$ because $\mathbb{C}M$ is semisimple (see the discussion following Theorem 5.19).

Theorem 9.9. *Let M be a finite inverse monoid and let V be a finite dimensional $\mathbb{C}M$-module. Let $e_1, \ldots, e_s$ form a complete set of idempotent representatives of the $\mathscr{J}$-classes of M. Then one has that*

$$V = \bigoplus_{i=1}^{s} \bigoplus_{[S] \in \mathrm{Irr}_{\mathbb{C}}(G_{e_i})} m_S \cdot \mathrm{Ind}_{G_{e_i}}(S) = \bigoplus_{i=1}^{s} \bigoplus_{[S] \in \mathrm{Irr}_{\mathbb{C}}(G_{e_i})} m_S \cdot S^\sharp$$

is the decomposition of V into a direct sum of simple modules where

$$m_S = \frac{1}{|G_{e_i}|} \sum_{g \in G_{e_i}} \overline{\chi_S(g)} \sum_{f \leq e_i} \chi_V(gf) \cdot \mu(f, e_i)$$

with μ the Möbius function of $E(M)$, for $S \in \mathrm{Irr}_{\mathbb{C}}(G_{e_i})$. In particular, V is determined up to isomorphism by χ_V.

Proof. Let $\beta\colon \mathbb{C}G(M) \longrightarrow \mathbb{C}M$ be the isomorphism from Theorem 9.3. Then, applying Lemma 9.8, we have that

$$\beta([m]) = \sum_{n \leq m} \mu_M(n, m)n = \sum_{f \leq m^*m} \mu(f, m^*m)mf.$$

We can view every $\mathbb{C}M$-module as a $\mathbb{C}G(M)$-module via the isomorphism β. Observe that if S is a simple $\mathbb{C}G_{e_i}$-module, then since e_i is an apex of $S^\sharp$, $h < g$ implies $MhM \subsetneq MgM$ (by Proposition 3.13) and $\mu(g, g) = 1$ for $g \in G_{e_i}$, we have that

$$[g]v = \sum_{h \leq g} \mu(h, g)hv = \mu(g, g)gv = gv$$

for all $g \in G_{e_i}$. Hence $[e_i]S^\sharp \cong e_i S^\sharp \cong S$ as a $\Bbbk G_{e_i}$-module. Therefore, $S^\sharp$ is the simple $\mathbb{C}G(M)$-module corresponding to S in Corollary 8.14 and hence $[V : S^\sharp] = [[e_i]V : S]$ (again by Corollary 8.14). An argument similar to that in the proof of Proposition 7.14 shows that the character of $[e_i]V$ as a $\mathbb{C}G_{e_i}$-module is $\chi_V \circ \beta|_{G_{e_i}}$. It follows that if $[S] \in \mathrm{Irr}_\mathbb{C}(G_{e_i})$, then (extending χ_V linearly to $\mathbb{C}M$) we have that

$$\begin{aligned}
[V : S^\sharp] &= [[e_i]V : S] \\
&= \langle \chi_V \circ \beta|_{G_{e_i}}, \chi_S \rangle_{G_{e_i}} \\
&= \frac{1}{|G_{e_i}|} \sum_{g \in G_{e_i}} \chi_V(\beta([g]))\overline{\chi_S(g)} \\
&= \frac{1}{|G_{e_i}|} \sum_{g \in G_{e_i}} \sum_{f \leq e_i} \chi_V(gf)\mu(f, e_i)\overline{\chi_S(g)} \\
&= \frac{1}{|G_{e_i}|} \sum_{g \in G_{e_i}} \overline{\chi_S(g)} \sum_{f \leq e_i} \chi_V(gf) \cdot \mu(f, e_i)
\end{aligned}$$

as required. $\qquad\square$

We now prove two results of Solomon concerning the tensor and exterior powers of the natural module for the symmetric inverse monoid [Sol02]. The original proofs of Solomon involved inverting the character table and applying some symmetric function theory. The proof here is elementary and based on Theorem 9.9. A more general version of these results appears in [Ste08].

Let I_n be the symmetric inverse monoid of degree n. The *natural module* for $\mathbb{C}I_n$ is $V = \mathbb{C}^n$ where if $\{e_1, \ldots, e_n\}$ is the standard basis and $m \in I_n$, then

$$me_i = \begin{cases} e_{m(i)}, & \text{if } m(i) \text{ is defined} \\ 0, & \text{else.} \end{cases}$$

We continue to identify the maximal subgroup at $1_{[r]}$ with the symmetric group S_r in the natural way for $0 \leq r \leq n$.

Let $S(p, r)$ denote the *Stirling number of the second kind* with parameters $p, r \geq 0$. By definition, $S(p, r)$ is the number of partitions of a p-element set into r nonempty subsets. A well-known formula for it is

$$S(p, r) = \frac{1}{r!} \sum_{k=0}^{r} (-1)^{r-k} \binom{r}{k} k^p.$$

See [Sta97, Page 34] for details.

Lemma 9.10. *If X is a set, then the equality*

$$\sum_{Y \subseteq X} (-1)^{|X|-|Y|} = \begin{cases} 1, & \text{if } X = \emptyset \\ 0, & \text{else} \end{cases}$$

holds.

Proof. If $X = \emptyset$, the result is clear. If $|X| = n \geq 1$, then

$$\sum_{Y \subseteq X} (-1)^{|X|-|Y|} = \sum_{k=0}^{n} (-1)^{n-k} \binom{n}{k} = (1 - 1)^n = 0$$

as required.   □

The following theorem, decomposing the tensor powers of the natural module, is due to Solomon [Sol02].

Theorem 9.11. *Let V be the natural module for $\mathbb{C}I_n$ and let $[S] \in \mathrm{Irr}_{\mathbb{C}}(S_r)$ for $0 \leq r \leq n$. Then one has that $[V^{\otimes p} : \mathrm{Ind}_{S_r}(S)] = \dim S \cdot S(p, r)$.*

Proof. Throughout this proof, we shall use that $E(I_n)$ is isomorphic to the power set $\mathscr{P}([n])$ under the inclusion ordering via $X \mapsto 1_X$ and that

$$\mu(X, Y) = (-1)^{|Y|-|X|}$$

if $X \subseteq Y$ by Corollary C.5. Let θ be the character afforded by V. If $m \in I_n$, we put $\mathrm{Fix}(m) = \{i \in [n] \mid m(i) = i\}$. It is easy to see that $\theta(m) = |\mathrm{Fix}(m)|$ and hence $\chi_{V^{\otimes p}}(m) = \theta^p(m) = |\mathrm{Fix}(m)|^p$.

Using Theorem 9.9 we conclude that

$$[V^{\otimes p} : \mathrm{Ind}_{S_r}(S)] = \frac{1}{r!} \sum_{g \in S_r} \overline{\chi_S(g)} \sum_{Y \subseteq [r]} (-1)^{r-|Y|} \theta^p(g|_Y). \qquad (9.2)$$

Let $g \in S_r$. Observe that $\theta^p(g|_Y) = |\mathrm{Fix}(g|_Y)|^p = |\mathrm{Fix}(g) \cap Y|^p$. Let $X = \mathrm{Fix}(g) \subseteq [r]$ and $Z = [r] \setminus X$. Performing the change of variables $Y \mapsto (Y \cap X, Y \cap Z)$ (with inverse $(U, V) \mapsto U \cup V$), we obtain

$$\sum_{Y \subseteq [r]} (-1)^{r-|Y|} \theta^p(g|_Y) = \sum_{U \subseteq X, V \subseteq Z} (-1)^{r-|U|-|V|} |U|^p$$

$$= \sum_{U \subseteq X} (-1)^{|X|-|U|} |U|^p \sum_{V \subseteq Z} (-1)^{|Z|-|V|} \qquad (9.3)$$

$$= \begin{cases} \sum_{U \subseteq [r]} (-1)^{r-|U|} |U|^p, & \text{if } Z = \emptyset \\ 0, & \text{else} \end{cases}$$

where the last equality follows from Lemma 9.10. But $Z = \emptyset$ if and only if $[r] = X = \mathrm{Fix}(g)$, if and only if $g = 1_{[r]}$. Substituting the rightmost term of (9.3) into (9.2) and using that $\chi_S(1_{[r]}) = \dim S$ yields

$$[V^{\otimes p} : \mathrm{Ind}_{S_r}(S)] = \frac{\dim S}{r!} \sum_{U \subseteq [r]} (-1)^{r-|U|} |U|^p$$

$$= \dim S \cdot \frac{1}{r!} \sum_{k=0}^{r} (-1)^{r-k} \binom{r}{k} k^p$$

$$= \dim S \cdot S(p, r)$$

as required. $\qquad\qquad\qquad\qquad\qquad\qquad\qquad\qquad\qquad\qquad\square$

If $\Bbbk$ is any field, we can define the *natural module* $V = \Bbbk^n$ for $\Bbbk I_n$ exactly as we did for the field of complex numbers. We now present a direct sum decomposition of the tensor powers of V at this level of generality which immediately implies Theorem 9.11 for $\mathbb{C}$. Essentially, we are finding the decomposition of $V^{\otimes p}$ into $M_{\binom{n}{r}}(\Bbbk S_r)$-modules coming from the direct product decomposition in Corollary 9.7. This result seems to be new.

Theorem 9.12. *Let $\Bbbk$ be a field and let V be the natural module for $\Bbbk I_n$. Then we have the isomorphism*

$$V^{\otimes p} \cong \bigoplus_{r=0}^{p} S(p, r) \cdot \Bbbk L_{1_{[r]}}.$$

Consequently, if the characteristic of $\Bbbk$ is 0 or greater than n, then

$$[V^{\otimes p} : \mathrm{Ind}_{S_r}(S)] = \dim S \cdot S(p, r)$$

for $S \in \mathrm{Irr}_{\Bbbk}(S_r)$.

Proof. First we observe that the basis for $V^{\otimes p}$ can be indexed by mappings $f \colon [p] \longrightarrow [n]$ by putting $e_f = e_{f(1)} \otimes \cdots \otimes e_{f(p)}$. If $g \in I_n$, then one checks that

$$g e_f = \begin{cases} e_{gf}, & \text{if } f([p]) \subseteq \mathrm{dom}(g) \\ 0, & \text{else} \end{cases} \qquad (9.4)$$

where $\mathrm{dom}(g)$ denotes the domain of g. Notice that because g is injective, if $f([p]) \subseteq \mathrm{dom}(g)$, then gf and f have the same associated partition into fibers over elements of its range. Let $\Pi_{p,r}$ be the set of all set partitions of $[p]$ into r nonempty subsets; so $\Pi_{p,r}$ has $S(p,r)$ elements. Then, for $\pi \in \Pi_{p,r}$, set V_π to be the $\Bbbk$-span of all e_f such that π is the partition associated with f. We have that V_π is a $\Bbbk I_n$-submodule of $V^{\otimes p}$ by (9.4) and the preceding discussion. Moreover, one has that

$$V^{\otimes p} = \bigoplus_{r=0}^{p} \bigoplus_{\pi \in \Pi_{p,r}} V_\pi \tag{9.5}$$

where we note that if $p > 0$, then $\Pi_{p,0} = \emptyset$. If $p = 0$, one should make the obvious conventions about functions from the empty set and partitions (equal equivalence relations) on the empty set.

From (9.5) it is clear that to prove the first statement of the theorem it suffices to show that if $\pi \in \Pi_{p,r}$, then $V_\pi \cong \Bbbk L_{1_{[r]}}$. Let X_π be the set of all mappings $f \colon [p] \longrightarrow [n]$ with associated partition π. Then elements of X_π are in bijection with injective mappings $f' \colon [p]/\pi \longrightarrow [n]$. On the other hand, the $\mathscr{L}$-class $L_{1_{[r]}}$ consists of all injective mappings $f \colon [r] \longrightarrow [n]$ (viewed as partial injective mappings on $[n]$). So if we fix a bijection of $h \colon [p]/\pi \longrightarrow [r]$ and let $k \colon [p] \longrightarrow [p]/\pi$ be the canonical projection, then we have a bijection $\alpha \colon L_{1_{[r]}} \longrightarrow X_\pi$ given by $\alpha(f) = fhk$. Moreover, if $g \in I_n$, then $gf \in L_{1_{[r]}}$ if and only if the range of f is contained in the domain of g. If the latter occurs, then $\alpha(gf) = gfhk = g\alpha(f)$. It now follows easily from (9.4) that the mapping $\varphi \colon \Bbbk L_{1_{[r]}} \longrightarrow V_\pi$ defined on the basis $L_{1_{[r]}}$ by $\varphi(f) = e_{\alpha(f)}$ is a $\Bbbk I_n$-module isomorphism.

To deduce the second statement, we have that $\Bbbk S_r$ is semisimple, for all $0 \leq r \leq n$, under the hypothesis on $\Bbbk$. Therefore, $\Bbbk I_n$ is also semisimple. Moreover, (since $\Bbbk$ is a splitting field for S_r) one has that

$$\Bbbk L_{1_{[r]}} \cong \Bbbk L_{1_{[r]}} \otimes_{\Bbbk S_r} \Bbbk S_r \cong \Bbbk L_{1_{[r]}} \otimes_{\Bbbk S_r} \left(\bigoplus_{S \in \mathrm{Irr}_\Bbbk(S_r)} \dim S \cdot S \right)$$

$$\cong \bigoplus_{S \in \mathrm{Irr}_\Bbbk(S_r)} \dim S \cdot \left(\Bbbk L_{1_{[r]}} \otimes_{\Bbbk S_r} S \right) = \bigoplus_{S \in \mathrm{Irr}_\Bbbk(S_r)} \dim S \cdot \mathrm{Ind}_{S_r}(S)$$

from which the second statement follows. $\qquad \square$

Next we consider the exterior powers of V. Again, the result is due to Solomon [Sol02].

Theorem 9.13. *Let V be the natural module for $\mathbb{C}I_n$. Then the exterior power $V^{\wedge p}$ is simple, for any $0 \leq p \leq n$. Moreover, $V^{\wedge p} = \mathrm{Ind}_{S_p}(S_{(1^p)})$ where $S_{(1^p)}$ affords the sign representation of S_p.*

Proof. Let $\{e_1, \ldots, e_n\}$ be the standard basis for V. For each p-element subset $X \subseteq [n]$ with $X = \{i_1 < \cdots < i_p\}$, put $e_X = e_{i_1} \wedge \cdots \wedge e_{i_p}$. Then the e_X with $|X| = p$ form a basis for $V^{\wedge p}$. If $m \in I_n$, one verifies directly that

$$
m e_X = \begin{cases} \mathrm{sgn}(m|_X)e_X, & \text{if } m|_X \in S_X \\ \pm e_{m(X)}, & \text{if } |m(X)| = p, \ X \neq m(X) \\ 0, & \text{else.} \end{cases} \tag{9.6}
$$

If θ is the character afforded by $V^{\wedge p}$, then we deduce that

$$
\theta(m) = \sum_{\substack{1_Y \leq m^* m, \\ |Y|=p, m|_Y \in S_Y}} \mathrm{sgn}(m|_Y).
$$

Let S be a simple $\mathbb{C}S_r$-module with $0 \leq r \leq n$. Then by Theorem 9.9, we have that

$$
\begin{aligned}
[V^{\wedge p} : \mathrm{Ind}_{S_r}(S)] &= \frac{1}{r!} \sum_{g \in S_r} \overline{\chi_S(g)} \sum_{X \subseteq [r]} (-1)^{r-|X|} \sum_{\substack{Y \subseteq X, \\ |Y|=p, g|_Y \in S_Y}} \mathrm{sgn}(g|_Y) \\
&= \frac{1}{r!} \sum_{g \in S_r} \overline{\chi_S(g)} \sum_{\substack{Y \subseteq [r], \\ |Y|=p, g|_Y \in S_Y}} \mathrm{sgn}(g|_Y) \sum_{Y \subseteq X \subseteq [r]} (-1)^{r-|X|} \\
&= \frac{1}{r!} \sum_{g \in S_r} \overline{\chi_S(g)} \sum_{\substack{Y \subseteq [r], \\ |Y|=p, g|_Y \in S_Y}} \mathrm{sgn}(g|_Y) \sum_{U \subseteq [r] \setminus Y} (-1)^{r-|Y|-|U|}.
\end{aligned}
$$

Using Lemma 9.10, we see that the rightmost term of the above equation vanishes unless $p = r$ and in this case it is

$$
\frac{1}{p!} \sum_{g \in S_p} \mathrm{sgn}(g)\overline{\chi_S(g)} = \langle \chi_{S_{(1^p)}}, \chi_S \rangle_{S_p}.
$$

We conclude that $V^{\wedge p} \cong \mathrm{Ind}_{S_p}(S_{(1^p)})$, completing the proof. □

One could also prove Theorem 9.13 by giving an explicit isomorphism between $V^{\wedge p}$ and $\mathrm{Ind}_{S_p}(S_{(1^p)})$. Analogues of the previous results for wreath products $G \wr I_n$ with G a finite group are given in [Ste08].

9.4 The character table of the symmetric inverse monoid

The character table of the symmetric inverse monoid was first computed by Munn [Mun57a]. We follow here the approach of Solomon [Sol02]. A generalization of this technique to arbitrary inverse monoids was given by the author [Ste08]. Let us fix $n \geq 0$. We shall present the character table $X(I_n)$

of the symmetric inverse monoid I_n (over $\mathbb{C}$) as a product of a block diagonal matrix consisting of the character tables of symmetric groups of degree at most n and an explicit unipotent upper triangular matrix on both the left and the right.

The reader is referred to Section B.4 for the representation theory of the symmetric group and basics about partitions. We simply recall that $\lambda = (\lambda_1, \ldots, \lambda_s)$ is a *partition* of $k \geq 1$ if $\lambda_1 \geq \cdots \geq \lambda_s \geq 1$ and $\lambda_1 + \cdots + \lambda_s = k$. We also consider there to be an empty partition of 0. We write $|\lambda| = k$ to indicate that λ is a partition of k. We denote by $\mathcal{P}_k$ be the set of partitions of k and we put $\mathcal{Q} = \bigcup_{k=0}^n \mathcal{P}_k$. We totally order $\mathcal{Q}$ by $\alpha \prec \beta$ if $|\alpha| < |\beta|$ or $|\alpha| = |\beta|$ and α precedes β in lexicographic order meaning that if $\alpha = (\alpha_1, \ldots, \alpha_m)$ and $\lambda = (\lambda_1, \ldots, \lambda_s)$, then α lexicographically precedes λ if there exists i such that $\alpha_j = \lambda_j$ for all $j < i$ and $\alpha_i < \lambda_i$.

It is well known from elementary group theory that the conjugacy classes of the symmetric group S_k are indexed by the elements of $\mathcal{P}_k$. The conjugacy class C_λ corresponding to $\lambda = (\lambda_1, \ldots, \lambda_s)$ is the set of permutations in S_k of cycle type λ. Let us take the elements $1_{[k]}$ with $0 \leq k \leq n$ as idempotent representatives of the $\mathscr{J}$-classes of I_n and identify $G_{1_{[k]}}$ with S_k as we have done before. Then it follows that the generalized conjugacy classes of I_n can be indexed by $\mathcal{Q}$ where $\overline{C}_\lambda$, for $|\lambda| = k$, is the generalized conjugacy class of I_n that intersects S_k in C_λ.

The simple $\mathbb{C}S_k$-modules can be naturally parameterized by $\mathcal{P}_k$, as well; see Section B.4. We shall not explicitly use the nature of the parametrization here except in Theorem 9.16. Let S_λ be the simple module corresponding to $\lambda \in \mathcal{P}_k$ and let χ_λ be its character. We set ζ_λ to be the character of $\mathrm{Ind}_{S_k}(S_\lambda)$, the corresponding simple module for $\mathbb{C}I_n$. Let $X(S_k)$ be the character table of S_k, which we index by $\mathcal{P}_k$, so that

$$X(S_k)_{\alpha,\beta} = \chi_\alpha(C_\beta).$$

We index the character table $X(I_n)$ by $\mathcal{Q}$, so that

$$X(I_n)_{\alpha,\beta} = \zeta_\alpha(\overline{C}_\beta).$$

If $\alpha, \beta \in \mathcal{Q}$, we put

$$\binom{\alpha}{\beta} = \prod_{i \geq 0} \binom{a_i}{b_i}$$

where a_i is the number of parts of α equal to i and b_i is the number of parts of β equal to i with the convention $\binom{0}{0} = 1$ and $\binom{n}{m} = 0$ if $n < m$.

Proposition 9.14. *If $\binom{\alpha}{\beta} > 0$, then $\beta \preceq \alpha$. Moreover, $\binom{\alpha}{\alpha} = 1$.*

Proof. If $\binom{\alpha}{\beta} > 0$, then $a_i \geq b_i$ for all $i \geq 1$ (retaining the above notation). As $|\alpha| = \sum_{i \geq 1} i a_i$ and $|\beta| = \sum_{i \geq 1} i b_i$, we conclude that $|\beta| \leq |\alpha|$ and that if $|\alpha| = |\beta|$, then $\alpha = \beta$. Thus $\beta \preceq \alpha$. The second statement is immediate from the definition. $\square$

All the preparations are finished to compute the character table $X(I_n)$ of the symmetric inverse monoid of degree n.

Theorem 9.15. *Let $n \geq 0$. Then $X(I_n) = XT$ where*

$$
X = \begin{bmatrix} X(S_0) & 0 & \cdots & 0 \\ 0 & X(S_1) & 0 & \vdots \\ \vdots & 0 & \ddots & 0 \\ 0 & \cdots & 0 & X(S_n) \end{bmatrix}
$$

and T, given by

$$
T_{\beta,\alpha} = \binom{\alpha}{\beta},
$$

is unipotent upper triangular.

Proof. The fact that T is unipotent upper triangular is immediate from Proposition 9.14.

For $K \subseteq [n]$ with $|K| = k$, let $p_K \colon [k] \longrightarrow K$ be the unique order-preserving bijection. Note that $p_{[k]} = 1_{[k]}$. Then since $L_{1_{[k]}}$ consists of the partial injective mappings with domain $[k]$ and two mappings are in the same right S_k-orbit if and only if they have the same range, it follows that the mappings p_K with $|K| = k$ form a transversal for $L_{1_{[k]}}/S_k$. Therefore, if S_λ with $\lambda \in \mathcal{P}_k$ is a simple $\mathbb{C}S_k$-module, then as a $\mathbb{C}$-vector space

$$
\mathrm{Ind}_{S_k}(S_\lambda) = \mathbb{C}L_{1_{[k]}} \otimes_{\mathbb{C}S_k} S_\lambda = \bigoplus_{K \subseteq [n], |K|=k} p_K \otimes S_\lambda.
$$

Hence, if $\{v_1, \ldots, v_{f_\lambda}\}$ is a basis for S_λ, then the $p_K \otimes v_j$ with $K \subseteq [n]$, $|K| = k$ and $1 \leq j \leq f_\lambda$ form a basis for $\mathrm{Ind}_{S_k}(S_\lambda)$.

Let $\alpha \in \mathcal{Q}$ with $|\alpha| = r$ and let $g \in C_\alpha \subseteq S_r$. We know that g annihilates $\mathrm{Ind}_{S_k}(S_\lambda)$ unless $r \geq k$, so let us assume this. Then g preserves the summand $p_K \otimes S_\lambda$ if and only if $K \subseteq [r]$ and $g(K) = K$, in which case

$$
g(p_K \otimes v_j) = p_K(p_K^* g|_K p_K) \otimes v_j = p_K \otimes (p_K^* g|_K p_K)v_j.
$$

We conclude from this that

$$
\zeta_\lambda(\overline{C}_\alpha) = \sum_{\substack{K \subseteq [r], \\ |K|=k, g(K)=K}} \chi_\lambda(p_K^* g|_K p_K). \tag{9.7}
$$

Note that if $g(K) = K$, then the cycle type of $p_K^* g|_K p_K$ in S_k is the same as cycle type of $g|_K$ in S_K. Now $K \subseteq [r]$ is invariant under g if and only if K is a union of orbits of g on $[r]$. If $g|_K$ is to have cycle type $\beta \in \mathcal{P}_k$, then (retaining the above notation) $g|_K$ must have b_i orbits on K of size i. As g has a_i orbits of size i on $[r]$, there are $\binom{a_i}{b_i}$ ways to choose these b_i orbits of size i and hence

$\binom{\alpha}{\beta}$ ways to choose $K \subseteq [r]$ such that $g(K) = K$ and $g|_K$ has cycle type β. As $\chi_\lambda(p_K^* g|_K p_K) = \chi_\lambda(C_\beta)$ for all such K, we conclude from (9.7) that

$$\zeta_\lambda(\overline{C}_\alpha) = \sum_{\beta \in \mathcal{P}_k} \binom{\alpha}{\beta} \chi_\lambda(C_\beta) = \sum_{\beta \in \mathcal{P}_k} X(S_k)_{\lambda,\beta} T_{\beta,\alpha} = \sum_{\beta \in \mathcal{Q}} X_{\lambda,\beta} T_{\beta,\alpha}$$

from which it follows that $X(I_n) = XT$.   □

Solomon also computed the unipotent upper triangular matrix U from Corollary 7.17 for I_n in [Sol02, Proposition 3.11]. Equivalently, he computed the decomposition matrix, i.e., the matrix of the isomorphism

$$\text{Res}: G_0(\mathbb{C}I_n) \longrightarrow \prod_{r=0}^{n} G_0(\mathbb{C}S_r).$$

An equivalent result can be found in Putcha [Put96, Theorem 2.5].

Theorem 9.16. *The character table of I_n is given by $X(I_n) = UX$ where X is as in Theorem 9.15 and U is the unipotent upper triangular matrix given by*

$$U_{\alpha,\beta} = \begin{cases} 1, & \text{if } \alpha \subseteq \beta \text{ and } \beta \setminus \alpha \text{ is a horizontal strip} \\ 0, & \text{else} \end{cases}$$

for $\alpha, \beta \in \mathcal{Q}$.

Proof. By Corollary 7.17, we have that $X(I_n) = UX$ where U is the transpose of the decomposition matrix. Let $r = |\alpha|$ and $m = |\beta|$ with $r \leq m$. We retain the notation of Section 5.3. Note that $1_{[m]} L 1_{[r]}$ can be identified with the set $I_{m,r}$ of injective mappings from $[r]$ to $[m]$. Thus $1_{[m]}(\mathbb{C}L1_{[r]} \otimes_{\mathbb{C}S_r} S_\alpha) \cong \mathbb{C}I_{m,r} \otimes_{\mathbb{C}S_r} S_\alpha$ as a $\mathbb{C}S_m$-module by Proposition 4.4. Corollary 5.13 shows that this module is isomorphic to $S_\alpha \boxtimes S_{m-r}$ as a $\mathbb{C}S_m$-module. The result now follows from Pieri's rule (Theorem B.13).   □

As an example, we compute the character table of I_3.

Example 9.17. For $n = 3$, we have that

$$\mathcal{Q} = \{(), (1), (1^2), (2), (1^3), (2,1), (3)\}.$$

It is a routine exercise in group representation theory to compute the character tables of S_0, S_1, S_2, S_3. One obtains

$$X = \begin{bmatrix} 1 & 0 & 0 & 0 & 0 & 0 & 0 \\ 0 & 1 & 0 & 0 & 0 & 0 & 0 \\ 0 & 0 & 1 & -1 & 0 & 0 & 0 \\ 0 & 0 & 1 & 1 & 0 & 0 & 0 \\ 0 & 0 & 0 & 0 & 1 & -1 & 1 \\ 0 & 0 & 0 & 0 & 2 & 0 & -1 \\ 0 & 0 & 0 & 0 & 1 & 1 & 1 \end{bmatrix}.$$

Direct computation yields

$$T = \begin{bmatrix} 1 & 1 & 1 & 1 & 1 & 1 & 1 \\ 0 & 1 & 2 & 0 & 3 & 1 & 0 \\ 0 & 0 & 1 & 0 & 3 & 0 & 0 \\ 0 & 0 & 0 & 1 & 0 & 1 & 0 \\ 0 & 0 & 0 & 0 & 1 & 0 & 0 \\ 0 & 0 & 0 & 0 & 0 & 1 & 0 \\ 0 & 0 & 0 & 0 & 0 & 0 & 1 \end{bmatrix}, \quad U = \begin{bmatrix} 1 & 1 & 0 & 1 & 0 & 0 & 1 \\ 0 & 1 & 1 & 1 & 0 & 1 & 1 \\ 0 & 0 & 1 & 0 & 1 & 1 & 0 \\ 0 & 0 & 0 & 1 & 0 & 1 & 1 \\ 0 & 0 & 0 & 0 & 1 & 0 & 0 \\ 0 & 0 & 0 & 0 & 0 & 1 & 0 \\ 0 & 0 & 0 & 0 & 0 & 0 & 1 \end{bmatrix}, \quad X(I_3) = \begin{bmatrix} 1 & 1 & 1 & 1 & 1 & 1 & 1 \\ 0 & 1 & 2 & 0 & 3 & 1 & 0 \\ 0 & 0 & 1 & -1 & 3 & -1 & 0 \\ 0 & 0 & 1 & 1 & 3 & 1 & 0 \\ 0 & 0 & 0 & 0 & 1 & -1 & 1 \\ 0 & 0 & 0 & 0 & 2 & 0 & -1 \\ 0 & 0 & 0 & 0 & 1 & 1 & 1 \end{bmatrix}$$

In the exercises, we shall ask the reader to compute the character table of I_4.

Remark 9.18. Grood gave a construction of a $\mathbb{C}I_n$-module isomorphic to the simple module $\mathbb{C}L_{1_{[r]}} \otimes_{\mathbb{C}S_r} S_\lambda$ via n-polytabloids of shape λ in [Gro02] and provided a direct proof of their simplicity. This case is easier than that of $\mathbb{C}T_n$.

9.5 Exercises

9.1. Let M be a finite commutative inverse monoid and $\mathbb{k}$ an algebraically closed field of characteristic 0. Prove that $\mathbb{k}M \cong \mathbb{k}^M$. For $\mathbb{k} = \mathbb{C}$, describe a complete set of orthogonal primitive idempotents of $\mathbb{k}M$ in terms of the characters of maximal subgroups and the Möbius function of the lattice of idempotents of M.

9.2. Let M be a finite inverse monoid with central idempotents and $\mathbb{k}$ a field. Prove that $\mathbb{k}M \cong \prod_{e \in E(M)} \mathbb{k}G_e$. (Hint: prove that $mm^* = m^*m$ for all $m \in M$.)

9.3. Verify directly that if V is the natural module for $\mathbb{C}I_n$, then $V^{\wedge p} \cong \operatorname{Ind}_{S_p}(S_{(1^p)})$.

9.4. Compute the character table of I_4.

9.5. Let M be a finite inverse monoid and $\mathbb{k}$ a field. Let J be a $\mathscr{J}$-class of M. Prove that $\mathbb{k}E(J)$ is a simple $\mathbb{k}M$-module with respect to the module structure defined by

$$m \odot e = \begin{cases} mem^*, & \text{if } e \leq m^*m \\ 0, & \text{else} \end{cases}$$

for $m \in M$ and $e \in E(J)$. (Hint: prove that $\mathbb{k}E(J) \cong \operatorname{Ind}_{G_e}(\mathbb{k})$ where $e \in E(J)$ and $\mathbb{k}$ is the trivial $\mathbb{k}G_e$-module; can you give a direct proof?)

9.6. Let M be a finite inverse monoid and let $d_1, \ldots, d_s$ be the degrees of a complete set of inequivalent irreducible representations of M over $\mathbb{C}$. Prove that $|M| = d_1^2 + \cdots + d_s^2$.

9.7. Let M be a finite inverse monoid. A representation $\rho\colon M \longrightarrow M_n(\mathbb{C})$ is called a *-*representation* if $\rho(m^*) = \rho(m)^*$, for all $m \in M$, where A^* denotes the Hermitian adjoint of a matrix A.

(a) Give a direct proof that a *-representation is completely reducible.
(b) Let $\ell_2 M$ be the Hermitian inner product space with orthonormal basis $\{v_m \mid m \in M\}$.
　　(i) Prove that $\ell_2 M$ is a $\mathbb{C}M$-module with respect to the Preston-Wagner representation

$$mv_n = \begin{cases} v_{mn}, & \text{if } n \in m^*mM \\ 0, & \text{else} \end{cases}$$

　　for $m, n \in M$.
　　(ii) Prove that $\langle mv_{m_1}, v_{m_2} \rangle = \langle v_{m_1}, m^*v_{m_2} \rangle$ for all $m, m_1, m_2 \in M$.
　　(iii) Prove that the representation afforded by $\ell_2 M$ with respect to the basis $\{v_m \mid m \in M\}$ is a *-representation.
　　(iv) Prove that $\ell_2 M$ is isomorphic to the regular $\mathbb{C}M$-module.
　　(v) Give another proof that $\mathbb{C}M$ is a semisimple algebra.
(c) Prove that each complex representation of a finite inverse monoid is equivalent to a *-representation, a result of Munn [Mun78]. (Hint: use Exercise 8.13 and the isomorphism of complex algebras $\alpha\colon \mathbb{C}M \longrightarrow \mathbb{C}G(M)$; this approach is due to Hajji, Handelman, and the author [HSH13].)

9.8. Let M be a finite inverse monoid. Use Exercise 8.14 to describe the center of $\mathbb{C}M$ and generalize to describe the center of $\Bbbk M$ for an arbitrary field $\Bbbk$.

9.9. Let PT_n be the monoid of all partial mappings of $[n]$ and $\Bbbk$ a field of characteristic 0. Prove that $\Bbbk PT_n / \operatorname{rad}(\Bbbk PT_n) \cong \Bbbk I_n$.

9.10 (I. Stein). Let PT_n be the monoid of all partial mappings of $[n]$ and $\Bbbk$ a field. Prove that $\Bbbk PT_n \cong \Bbbk FE_n$ where FE_n is the EI-category with objects the subsets of $[n]$ and with arrows surjective mappings. (Hint: imitate the proof of Theorem 9.3.)

Part V

The Rhodes Radical

10

Bi-ideals and R. Steinberg's Theorem

In this chapter, we prove R. Steinberg's Theorem [Ste62] that the direct sum of the tensor powers of a faithful representation of a monoid yields a faithful representation of the monoid algebra. We also commence the study of a special family of ideals, called bi-ideals, which will be at the heart of the next chapter.

The results of this chapter should more properly be viewed as about bialgebras, but we have chosen not to work at that level of generality in order to keep things more concrete. The approach we follow here is influenced by Passman [Pas14]. The bialgebraic approach was pioneered by Rieffel [Rie67].

10.1 Annihilators of tensor products

Fix a field $\Bbbk$ for the chapter. If A, B are $\Bbbk$-algebras, recall that their tensor product $A \otimes B$ (over $\Bbbk$) is a $\Bbbk$-algebra with respect to the product defined by

$$(a \otimes b)(a' \otimes b') = aa' \otimes bb'$$

on basic tensors. Moreover, if V is an A-module and W is a B-module, then tensor product $V \otimes W$ is an $A \otimes B$-module via the action defined on basic tensors by $(a \otimes b)(v \otimes w) = av \otimes bw$.

Lemma 10.1. *Let A and B be finite dimensional $\Bbbk$-algebras and let I and J be ideals of A and B, respectively. Then the kernel of the natural projection $\pi \colon A \otimes B \longrightarrow A/I \otimes B/J$, given on basic tensors by $\pi(a \otimes b) = (a+I) \otimes (b+J)$, is $A \otimes J + I \otimes B$.*

Proof. It is clear that π is a homomorphism and $A \otimes J + I \otimes B \subseteq \ker \pi$. Choose a basis $a_1, \dots, a_n$ for A such that $a_1, \dots, a_i$ is a basis for I and $a_{i+1}+I, \dots, a_n+I$ is a basis for A/I and, similarly, choose a basis $b_1, \dots, b_m$ for B such that $b_1, \dots, b_j$ is a basis for J and $b_{j+1}+J, \dots, b_m+J$ is a basis for B/J. Suppose that

B. Steinberg, *Representation Theory of Finite Monoids*,
Universitext, DOI 10.1007/978-3-319-43932-7_10

$$u = \sum_{r=1}^{n} \sum_{s=1}^{m} c_{rs}(a_r \otimes b_s)$$

belongs to $\ker \pi$, where $c_{rs} \in \Bbbk$ for all r, s. Then we have

$$0 = \sum_{r=i+1}^{n} \sum_{s=j+1}^{m} c_{rs}(a_r + I) \otimes (b_s + J)$$

and so $c_{rs} = 0$ if $(r, s) \in \{i+1, \ldots, n\} \times \{j+1, \ldots, m\}$ because the elements $(a_r + I) \otimes (b_s + J)$ with $i+1 \le r \le n$ and $j+1 \le s \le m$ form a basis for $A/I \otimes B/J$. Therefore, we have $u \in A \otimes J + I \otimes B$, as required. □

The next lemma connects the annihilator ideal of a tensor product of modules with the annihilators of the factors.

Lemma 10.2. *Let A, B be finite dimensional $\Bbbk$-algebras, V an A-module with annihilator ideal I, and W a B-module with annihilator ideal J. Then the annihilator of $V \otimes W$ in $A \otimes B$ is $A \otimes J + I \otimes B$.*

Proof. By Lemma 10.1, the ideal $A \otimes J + I \otimes B$ of $A \otimes B$ is the kernel of the natural surjective homomorphism $A \otimes B \longrightarrow A/I \otimes B/J$. Note that V is a faithful A/I-module, W is a faithful B/J-module, and what we are trying to prove is that $V \otimes W$ is a faithful $A/I \otimes B/J$-module. Thus without loss of generality, we may assume that $I = 0 = J$ and we aim to prove that $V \otimes W$ is a faithful $A \otimes B$-module.

Suppose that $c \in A \otimes B$ annihilates $V \otimes W$ and write $c = \sum_{i=1}^{k} a_i \otimes b_i$ where we may assume without loss of generality that $b_1, \ldots, b_k$ are linearly independent over $\Bbbk$. Suppose that $\gamma : V \longrightarrow \Bbbk$ is a functional. Then we can define a linear mapping $\Gamma : V \otimes W \longrightarrow W$ by $\Gamma(v \otimes w) = \gamma(v)w$.

We have, for all $v \in V$ and $w \in W$, that

$$0 = c(v \otimes w) = \sum_{i=1}^{k} a_i v \otimes b_i w.$$

Applying Γ then yields that $\sum_{i=1}^{k} \gamma(a_i v) b_i w = 0$. Fixing v, we see that $\sum_{i=1}^{k} \gamma(a_i v) b_i$ annihilates W. Since W is a faithful B-module and the b_i are assumed linearly independent, we must have that $\gamma(a_i v) = 0$ for all i. As this holds for all functionals γ, we conclude that $a_i v = 0$ for all $v \in V$ and thus a_i annihilates V. As V is faithful, we have $a_i = 0$ for all i. This shows that $c = 0$. □

Note that we did not assume that the modules V and W were finite dimensional in Lemma 10.2.

10.2 Bi-ideals and R. Steinberg's theorem

We continue to hold fixed the field $\Bbbk$ and now fix a finite monoid M for the remainder of the chapter. Put $A = \Bbbk M$ and note that we have a homomorphism

$$\Delta \colon A \longrightarrow A \otimes A,$$

called the *comultiplication*, given on elements $m \in M$ by $\Delta(m) = m \otimes m$. Notice that if V and W are A-modules, then the A-module structure on $V \otimes W$ that we have been considering previously is induced by its $A \otimes A$-module structure via inflation along the homomorphism Δ.

The trivial representation of M induces a homomorphism $\varepsilon \colon \Bbbk M \longrightarrow \Bbbk$ of $\Bbbk$-algebras given by $\varepsilon(m) = 1$ for all $m \in M$, which is sometimes called the *augmentation map*. The *augmentation ideal* of $\Bbbk M$ is then

$$\mathrm{Aug}(\Bbbk M) = \ker \varepsilon = \Bbbk \{m - m' \mid m, m' \in M\}$$

where the second equality is easily checked (cf. the proof that (iii) implies (ii) of Proposition 10.3 below).

Motivated by Lemma 10.2, we make the following definition. An ideal I of $A = \Bbbk M$ is a *bi-ideal*[1] if $\Delta(I) \subseteq A \otimes I + I \otimes A$ and $I \subseteq \mathrm{Aug}(\Bbbk M)$. The next proposition characterizes bi-ideals in monoid theoretic terms.

Proposition 10.3. *Let I be an ideal of $\Bbbk M$. Then the following are equivalent.*

(i) I is a bi-ideal.
(ii) $I = \Bbbk \{m - m' \mid m, m' \in M, m - m' \in I\}$.
(iii) There is a monoid homomorphism $\varphi \colon M \longrightarrow N$ such that the kernel of the induced homomorphism $\Phi \colon \Bbbk M \longrightarrow \Bbbk N$ is I.

Proof. We continue to let $A = \Bbbk M$. To see that (iii) implies (ii), we may assume without loss of generality that φ is onto. For each $n \in N$, fix $\tilde{n} \in M$ with $\varphi(\tilde{n}) = n$. Suppose that $a \in I = \ker \Phi$. Then

$$a = \sum_{m \in M} c_m m = \sum_{n \in N} \sum_{m \in \varphi^{-1}(n)} c_m m.$$

Therefore, if we put

$$b_n = \sum_{m \in \varphi^{-1}(n)} c_m,$$

---

[1] These are precisely the ideals for which A/I is a quotient bialgebra [Pas14].

then $0 = \varphi(a) = \sum_{n \in N} b_n n = 0$. Consequently, each $b_n = 0$ and hence we have that

$$a = \sum_{n \in N} \sum_{m \in \varphi^{-1}(n)} c_m m = \sum_{n \in N} \sum_{m \in \varphi^{-1}(n)} c_m m - \sum_{n \in N} b_n \tilde{n}$$

$$= \sum_{n \in N} \sum_{m \in \varphi^{-1}(n)} c_m (m - \tilde{n}).$$

Thus I is spanned by differences of elements of M.

To see that (ii) implies (i), suppose that $m - m' \in I$ with $m, m' \in M$. Then we calculate

$$\Delta(m - m') = \Delta(m) - \Delta(m') = m \otimes m - m' \otimes m'$$
$$= m \otimes m - m \otimes m' + m \otimes m' - m' \otimes m'$$
$$= m \otimes (m - m') + (m - m') \otimes m'$$

which is in $A \otimes I + I \otimes A$. Thus $\Delta(I) \subseteq A \otimes I + I \otimes A$. Trivially, $I \subseteq \mathrm{Aug}(\Bbbk M)$. This shows that I is a bi-ideal.

Suppose now that I is a bi-ideal. Define a congruence $\equiv$ on M by $m \equiv m'$ if and only if $m - m' \in I$. Let $N = M/\equiv$ and let $\varphi \colon M \longrightarrow N$ be the canonical homomorphism. Let $\Phi \colon \Bbbk M \longrightarrow \Bbbk N$ be the extension. Then we have a commutative diagram

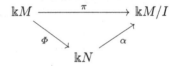

of surjective homomorphisms where $\alpha([m]_\equiv) = \pi(m) = m + I$ for $m \in M$. Our goal is to show that $I = \ker \Phi$ or, equivalently, that α is an isomorphism. Because $I \subseteq \mathrm{Aug}(\Bbbk M)$, we have that $N \cap \ker \alpha = \emptyset$, i.e., $0 \notin \alpha(N)$. Since $\alpha|_N \colon N \longrightarrow \Bbbk M/I$ is an injective homomorphism of monoids by construction, it suffices to show that $\alpha(N)$ is linearly independent. Assume that this is not the case. Then by considering a minimum size dependence relation in $\alpha(N)$ (and using that $0 \notin \alpha(N)$), we can find $n, n_1, \dots, n_k \in N$ such that $\alpha(n_1), \dots, \alpha(n_k)$ are linearly independent, $n \neq n_i$ for all i and

$$\alpha(n) = \sum_{i=1}^{k} c_i \alpha(n_i) \tag{10.1}$$

with the $c_i \in \Bbbk \setminus \{0\}$. Choose $m, m_1, \dots, m_k \in M$ with $\varphi(m) = n$ and $\varphi(m_i) = n_i$ for $i = 1, \dots, k$ and observe that $m - \sum_{i=1}^{k} c_i m_i \in I$ by (10.1).

Consider the A-module $V = A/I \otimes A/I$. By Lemma 10.2 and the assumption that I is a bi-ideal, we have that I annihilates V. Thus $m - \sum_{i=1}^{k} c_i m_i$ annihilates V. Let $e = 1 + I \in A/I$. Then we compute

$$m(e \otimes e) = me \otimes me = \alpha(n) \otimes \sum_{i=1}^{k} c_i \alpha(n_i) = \sum_{i=1}^{k} c_i(\alpha(n) \otimes \alpha(n_i))$$

$$m(e \otimes e) = \sum_{i=1}^{k} c_i m_i(e \otimes e) = \sum_{i=1}^{k} c_i(\alpha(n_i) \otimes \alpha(n_i)).$$

Subtracting these equations yields

$$0 = \sum_{i=1}^{k} c_i(\alpha(n) - \alpha(n_i)) \otimes \alpha(n_i).$$

As $\alpha(n_1), \ldots, \alpha(n_k)$ are linearly independent, it follows that

$$\sum_{i=1}^{k} A/I \otimes \alpha(n_i) = \bigoplus_{i=1}^{k} A/I \otimes \alpha(n_i)$$

and so $c_i(\alpha(n) - \alpha(n_i)) = 0$ for all i. But then since $c_i \neq 0$, we conclude $\alpha(n) = \alpha(n_i)$, contradicting the injectivity of $\alpha|_N$. Thus α is an isomorphism and $I = \ker \Phi$. This completes the proof. □

Proposition 10.3 sets up a bijection between bi-ideals of $\Bbbk M$ and congruences on M. More explicitly, if $\equiv$ is a congruence on M, then

$$I_\equiv = \Bbbk\{m - m' \mid m \equiv m'\}$$

is the corresponding bi-ideal. Conversely, if I is a bi-ideal, then $\equiv_I$ defined by $m \equiv_I m'$ if and only if $m - m' \in I$ is the corresponding congruence on M.

Next we consider a module theoretic way of obtaining bi-ideals.

Proposition 10.4. *Let $\mathcal{F}$ be a family of $\Bbbk M$-modules closed under tensor product and containing the trivial module. Let I be the intersection of the annihilators of the modules in $\mathcal{F}$. Then I is a bi-ideal.*

Proof. Let $U = \bigoplus_{V \in \mathcal{F}} V$. Then I is the annihilator of U. Also, by hypothesis $I \subseteq \mathrm{Aug}(\Bbbk M)$. Closure of $\mathcal{F}$ under tensor products guarantees that I annihilates $U \otimes U = \bigoplus_{V,W \in \mathcal{F}} V \otimes W$ and hence $\Delta(I) \subseteq A \otimes I + I \otimes A$ by Lemma 10.2. Therefore, I is a bi-ideal. □

All the preparation has now been completed in order to prove R. Steinberg's theorem [Ste62].

Theorem 10.5. *Let $\Bbbk$ be a field and M a finite monoid. Suppose that V is a $\Bbbk M$-module affording a faithful representation $\varphi \colon M \longrightarrow \mathrm{End}_{\Bbbk}(V)$. Then*

$$\mathcal{T}(V) = \bigoplus_{n=0}^{\infty} V^{\otimes n}$$

is a faithful $\Bbbk M$-module.

Proof. The family $\mathcal{F}$ of modules $V^{\otimes n}$ with $n \geq 0$ is closed under tensor product and contains the trivial module. Thus the intersection I of the annihilator ideals of the tensor powers of V is a bi-ideal by Proposition 10.4. But I is precisely the annihilator ideal of $\mathcal{T}(V)$. Trivially, I is contained in the annihilator of V. By Proposition 10.3 and faithfulness of φ, we conclude that

$$I = \Bbbk\{m - n \in I \mid m, n \in M\} \subseteq \Bbbk\{m - n \mid \varphi(m) = \varphi(n)\} = 0,$$

as required. □

We can, in fact, replace the infinite direct sum with a finite one.

Corollary 10.6. *Let M be a finite monoid and $\Bbbk$ a field. Let V be a $\Bbbk M$-module affording a faithful representation $\varphi \colon M \longrightarrow \mathrm{End}_{\Bbbk}(V)$. Then there exists $r \geq 0$ such that $\bigoplus_{n=0}^{r} V^{\otimes n}$ is a faithful $\Bbbk M$-module.*

Proof. We have that $\mathcal{T}(V)$ is faithful $\Bbbk M$-module by Theorem 10.5. Let I_k be the annihilator of $\bigoplus_{n=0}^{k} V^{\otimes n}$. Then $I_0 \supseteq I_1 \supseteq \cdots$ and $\bigcap_{k=0}^{\infty} I_k = 0$. As $\Bbbk M$ is finite dimensional, there exists r such that $I_r = 0$. This completes the proof. □

In fact, it is not difficult to refine the proof of Corollary 10.6 to show that one can take $r = |M|$. As a further corollary, we obtain the following result, which should be contrasted with Theorem 7.26.

Corollary 10.7. *Let M be a finite monoid and $\Bbbk$ a field. Let V be a $\Bbbk M$-module affording a faithful representation $\varphi \colon M \longrightarrow \mathrm{End}_{\Bbbk}(V)$. Then every simple $\Bbbk M$-module is a composition factor of a tensor power $V^{\otimes n}$ for some $n \geq 0$.*

Proof. Let S be a simple A-module and write $S = Ae/\mathrm{rad}(A)e$ where e is a primitive idempotent of A. By Corollary 10.6, there exists $r \geq 0$ such that $\bigoplus_{n=0}^{r} V^{\otimes n}$ is faithful. Therefore, $0 \neq \bigoplus_{n=0}^{r} eV^{\otimes n}$ and hence $eV^{\otimes n} \neq 0$ for some $n \geq 0$. But this means that S is a composition factor of $V^{\otimes n}$ by Proposition A.24. □

We end this chapter with an application of R. Steinberg's theorem that will be explored further in the next chapter. Namely, we characterize those monoids admitting a faithful representation over $\Bbbk$ by upper triangular matrices as precisely those monoids whose irreducible representations over $\Bbbk$ are all one-dimensional.

Proposition 10.8. *Let M be a finite monoid and $\Bbbk$ a field. The M admits a faithful representation by upper triangular matrices over $\Bbbk$ if and only if each simple $\Bbbk M$-module is one-dimensional.*

Proof. If each simple $\Bbbk M$-module is one-dimensional, then the regular module $\Bbbk M$ has only one-dimensional composition factors and hence affords a representation by upper triangular matrices according to Lemma 6.8. Since the regular module affords a faithful representation, this completes one direction of the proof.

For the converse, suppose that M admits a faithful representation by upper triangular matrices over $\Bbbk$, afforded by a module V. Then by Corollary 10.7 each simple $\Bbbk M$-module is a composition factor of a tensor power $V^{\otimes n}$ with $n \geq 0$. But these tensor powers are triangularizable modules by Corollary 6.9, and hence have only one-dimensional composition factors by Lemma 6.8.   □

It follows from Proposition 10.8 and Lemma 6.8 that if M has a faithful representation over $\Bbbk$ by upper triangular matrices, then each representation of M over $\Bbbk$ is equivalent to one by upper triangular matrices. We say that a monoid M is *triangularizable* over $\Bbbk$ if the equivalent conditions of Proposition 10.8 hold.

Corollary 10.9. *If M is a finite $\mathscr{R}$-trivial monoid and $\Bbbk$ is a field, then M admits a faithful representation over $\Bbbk$ by upper triangular matrices. Moreover, every representation of M is equivalent to one by upper triangular matrices.*

Proof. This follows from Proposition 10.8 because every irreducible representation of M over $\Bbbk$ is one-dimensional by Corollary 5.7.   □

10.3 Exercises

10.1. Let P, Q be finite posets. Prove that $I(P \times Q, \Bbbk) \cong I(P, \Bbbk) \otimes I(Q, \Bbbk)$ where $I(X, \Bbbk)$ denotes the incidence algebra of the poset X over the field $\Bbbk$ (see Appendix C).

10.2. Let M be a finite monoid and $\varphi \colon M \longrightarrow N$ a surjective monoid homomorphism. Let $\Phi \colon \Bbbk M \longrightarrow \Bbbk N$ be the induced homomorphism where $\Bbbk$ is a field. Fix, for each $n \in N$, an element $\overline{n} \in M$ with $\varphi(\overline{n}) = n$. Prove that the set of nonzero elements of the form $m - \overline{\varphi(m)}$ is a basis for $\ker \Phi$.

10.3. Let M be a finite monoid and $\Bbbk$ a field. Let I be an ideal of $\Bbbk M$ and put $J = \Bbbk\{m - m' \in I \mid m, m' \in M\}$.

(a) Prove that J is a bi-ideal.
(b) Prove that J is the largest bi-ideal contained in I.
(c) Put $V = \Bbbk M/I$. Prove that J is the intersection of the annihilators of the tensor powers $V^{\otimes n}$ with $n \geq 0$.

10.4. Let M be a finite monoid and $\Bbbk$ a field. Let V be a $\Bbbk M$-module affording a representation $\rho\colon M \longrightarrow \mathrm{End}_{\Bbbk}(V)$. Prove that the annihilator ideal of $\mathcal{T}(V) = \bigoplus_{n=0}^{\infty} V^{\otimes n}$ is spanned by the elements $m - m'$ with $m, m' \in M$ and $\rho(m) = \rho(m')$.

10.5. Prove that Theorem 10.5 is valid for infinite monoids.

10.6. Suppose that $\rho\colon M \longrightarrow \mathrm{End}_{\Bbbk}(V)$ is a faithful representation of a finite monoid M on a finite dimensional vector space V. Prove that $\bigoplus_{i=0}^{n} V^{\otimes i}$ is a faithful $\Bbbk M$-module for $n = |M|$.

10.7. Let M be a finite monoid and $\Bbbk$ a field of characteristic 0. Prove that there is a faithful representation $\rho\colon M \longrightarrow M_n(\Bbbk)$ where $n = |M| - 1$.

10.8 (G. Bergman). Let M be a monoid and $\Bbbk$ a field. Suppose that L is a left ideal of $\Bbbk M$ whose socle S both is simple and contains a nonzero element of the form $m - m'$ with $m, m' \in M$. Prove that if V is a $\Bbbk M$-module affording a faithful representation of M, then V contains a submodule isomorphic to L.

10.9. Let V be a $\mathbb{C}T_n$-module affording a faithful representation of T_n. Prove that V contains a submodule isomorphic to the natural module $\mathbb{C}^n$. (Hint: apply Exercise 10.8 to the left ideal spanned by the constant mappings.)

11

The Rhodes Radical and Triangularizability

In this chapter we provide a correspondence between nilpotent bi-ideals and a certain class of congruences on a finite monoid. We characterize the largest nilpotent bi-ideal, which is called the *Rhodes radical* because it was first described by Rhodes [Rho69b] in the case of an algebraically closed field of characteristic zero. For simplicity, we only give complete details in characteristic zero. As an application, we characterize those monoids with a faithful representation by upper triangular matrices over an algebraically closed field $\Bbbk$ (the general case will be left to the reader in the exercises). These are precisely the monoids with a basic algebra over $\Bbbk$ and so it also characterizes these monoids. Our treatment of these topics is based, for the most part, on that of Almeida, Margolis, the author, and Volkov [AMSV09].

11.1 The Rhodes radical and nilpotent bi-ideals

Let $\Bbbk$ be a field and M a finite monoid. We saw in Chapter 10 that bi-ideals in $\Bbbk M$ are precisely kernels of homomorphisms $\Phi\colon \Bbbk M \longrightarrow \Bbbk N$ induced by monoid homomorphisms $\varphi\colon M \longrightarrow N$. It follows from Proposition 10.3 that each ideal I of $\Bbbk M$ contains a largest bi-ideal, namely the $\Bbbk$-span of all differences $m - n$ with $m, n \in M$ and $m - n \in I$; see Exercise 10.3. In particular, there is a largest nilpotent bi-ideal, namely

$$\mathrm{rad}_{\Bbbk}(M) = \Bbbk\{m - n \mid m, n \in M, \ m - n \in \mathrm{rad}(\Bbbk M)\}.$$

We call $\mathrm{rad}_{\Bbbk}(M)$ the *Rhodes radical* of M with respect to $\Bbbk$. The corresponding congruence $\equiv_{\Bbbk}$ on M is defined by

$$m \equiv_{\Bbbk} n \iff m - n \in \mathrm{rad}(\Bbbk M)$$

© Springer International Publishing Switzerland 2016
B. Steinberg, *Representation Theory of Finite Monoids*,
Universitext, DOI 10.1007/978-3-319-43932-7_11

and is called the *Rhodes radical congruence* on M. Equivalently, $m \equiv_k n$ if and only if $\rho(m) = \rho(n)$ for every irreducible representation ρ of M over k. In particular, every semisimple kM-module is, in fact, an inflation of a semisimple $k[M/\equiv_k]$-module.

Let us first investigate when the Rhodes radical coincides with the Jacobson radical.

Proposition 11.1. *Let M be a finite monoid and k a field. Then the following are equivalent.*

(i) $\mathrm{rad}_k(M) = \mathrm{rad}(kM)$.
(ii) $kM/\mathrm{rad}(kM)$ *is the monoid algebra of* $M/\equiv_k$.
(iii) $k[M/\equiv_k]$ *is semisimple.*
(iv) *The semisimple kM-modules are closed under tensor product.*

Proof. Let $N = M/\equiv_k$. The implication (i) implies (ii) is clear from Proposition 10.3. Trivially, (ii) implies (iii). Assume that (iii) holds. As every semisimple kM-module is an inflation of a kN-module and the kN-modules are closed under tensor product, (iv) follows because each kN-module is semisimple by assumption. Finally, to see that (iv) implies (i), suppose that the family $\mathcal{F}$ of semisimple kM-modules is closed under tensor product. Then since the trivial module is simple and $\mathrm{rad}(kM)$ is the intersection of the annihilators of the elements of $\mathcal{F}$, it follows from Proposition 10.4 that $\mathrm{rad}(kM)$ is a bi-ideal and hence coincides with $\mathrm{rad}_k(M)$.   □

Since every simple module for a triangularizable monoid is one-dimensional by Proposition 10.8, and one-dimensional modules are closed under tensor product, it follows that Proposition 11.1 applies to triangularizable monoids.

Corollary 11.2. *If M is a triangularizable monoid over k, then $\mathrm{rad}(kM) = \mathrm{rad}_k(M)$ and hence $M/\equiv_k$ has a semisimple algebra over k.*

Recall from Section 7.4 that a monoid homomorphism $\varphi\colon M \longrightarrow N$ is an **LI**-morphism if φ separates e from $eMe \setminus \{e\}$ for all $e \in E(M)$, that is, $\varphi^{-1}(\varphi(e)) \cap eMe = \{e\}$ for all $e \in E(M)$. Let us say that a congruence $\equiv$ on M is an **LI**-*congruence* if the canonical projection $\pi\colon M \longrightarrow M/\equiv$ is an **LI**-morphism. Our aim is to show that if k is a field of characteristic zero, then nilpotent bi-ideals of kM are in bijection with **LI**-congruences. In particular, $\equiv_k$ will be the largest **LI**-congruence on M. We first need some algebraic properties of **LI**-morphisms.

Lemma 11.3. *Let $\varphi\colon M \longrightarrow N$ be an **LI**-morphism and suppose that $e, m, f \in M$ satisfy $\varphi(e) = \varphi(m) = \varphi(f)$ and $e, f \in E(M)$. Then $emf = ef$.*

Proof. First note that since $fef \in fMf$ and $\varphi(fef) = \varphi(f)$, we have $fef = f$. Similarly, since $emfe \in eMe$ and $\varphi(emfe) = \varphi(e)$, we have $emfe = e$. Thus $emf = emfef = ef$.   □

Now we can prove the main result of this section.

Theorem 11.4. *Let* $\Bbbk$ *be a field of characteristic* 0 *and* $\varphi\colon M \longrightarrow N$ *a homomorphism of finite monoids. Let* $\Phi\colon \Bbbk M \longrightarrow \Bbbk N$ *be the extension of* φ. *Then the following are equivalent.*

(i) φ *is an* **LI***-morphism.*
(ii) $I_\varphi = \ker \Phi$ *is nilpotent.*

Moreover, the implication (i) implies (ii) remains valid over any field $\Bbbk$.

Proof. First suppose that I_φ is nilpotent (and that the characteristic of $\Bbbk$ is 0). Let $e \in E(M)$ and $m \in M$ be such that $\varphi(e) = \varphi(eme)$. Then $eme - e \in I_\varphi$ and hence $(eme - e)^k = 0$ for some $k > 0$. Thus the minimal polynomial $f(x)$ of eme, as an element of the unital $\Bbbk$-algebra $\Bbbk[eMe]$ with identity e, divides the polynomial $p(x) = (x - 1)^k$. But if eme has index c and period d, then $(eme)^c = (eme)^{c+d}$ and so if $q(x) = x^c(x^d - 1)$, then $q(eme) = 0$. Therefore, $f(x)$ divides $(x - 1)^k$ and $x^c(x^d - 1)$, and consequently $f(x) = x - 1$, as the polynomial $x^d - 1$ has no repeated roots over a field $\Bbbk$ of characteristic 0. Thus $eme = e$ (as $f(eme) = 0$), completing the proof that φ is an **LI**-morphism.

Suppose now that φ is an **LI**-morphism. We drop the assumption that $\Bbbk$ is of characteristic 0. By a theorem of Wedderburn (Theorem A.11), to prove that I_φ is nilpotent, it suffices to show that each element of its spanning set $\{m_1 - m_2 \mid \varphi(m_1) = \varphi(m_2)\}$ is nilpotent. We shall need the following subresult. Let $e \in E(N)$ and let $S = \varphi^{-1}(e)$. Note that S is a subsemigroup of M. Suppose that $e_1, e_2 \in E(S)$ and $a \in \mathrm{Aug}(\Bbbk M) \cap \Bbbk S$. We claim that $e_1 a e_2 = 0$. Indeed, we can write $a = \sum_{s \in S} c_s s$ with $\sum_{s \in S} c_s = 0$. Therefore, by Lemma 11.3, we compute

$$e_1 a e_2 = \sum_{s \in S} c_s e_1 s e_2 = \sum_{s \in S} c_s e_1 e_2 = 0.$$

Now let $\varphi(m_1) = n = \varphi(m_2)$ and put $e = n^\omega$. Fix $k > 0$ such that $n^k = e$ and put $S = \varphi^{-1}(e)$. Let $r = |S|$. We claim that $(m_1 - m_2)^{(2r+1)k} = 0$. First note that $(m_1 - m_2)^k \in \Bbbk S$ and hence $(m_1 - m_2)^{kr} \in \Bbbk S^r$. Applying Lemma 1.5, it follows that

$$(m_1 - m_2)^{kr} = \sum_{i=1}^{d} c_i a_i e_i b_i$$

with the $a_i, b_i \in S$, $e_i \in E(S)$ and $c_i \in \Bbbk$. As $(m_1 - m_2)^k \in \mathrm{Aug}(\Bbbk M) \cap \Bbbk S$, it follows from the claim in the previous paragraph that $e_i b_i (m_1 - m_2)^k a_j e_j = 0$ for all i, j and hence

$$(m_1 - m_2)^{(2r+1)k} = (m_1 - m_2)^{kr}(m_1 - m_2)^k(m_1 - m_2)^{kr}$$

$$= \sum_{i,j=1}^{d} c_i c_j a_i e_i b_i (m_1 - m_2)^k a_j e_j b_j$$

$$= 0.$$

This completes the proof. $\square$

Corollary 11.5. *Let M be a finite monoid and $\Bbbk$ a field. Then there is an inclusion-preserving bijection between **LI**-congruences and nilpotent bi-ideals. In particular, $\equiv_\Bbbk$ is the largest **LI**-congruence on M.*

The reader is referred to [RS09, Chapter 6] for an explicit description of the largest **LI**-congruence on a finite monoid. Let us give an application to the case of $\mathscr{R}$-trivial monoids. Recall from Corollary 2.7 that if M is an $\mathscr{R}$-trivial monoid, then the set

$$\Lambda(M) = \{MeM \mid e \in E(M)\}$$

is a lattice ordered by inclusion, which we view as a monoid via its meet.

Corollary 11.6. *Let M be a finite $\mathscr{R}$-trivial monoid and $\Bbbk$ a field. Then the surjective homomorphism $\overline{\sigma}\colon \Bbbk M \longrightarrow \Bbbk\Lambda(M)$ induced by the natural homomorphism $\sigma\colon M \longrightarrow \Lambda(M)$ given by $\sigma(m) = Mm^\omega M$ is the semisimple quotient, i.e., $\ker\overline{\sigma} = \mathrm{rad}(\Bbbk M)$.*

Proof. By Corollary 2.7 the map σ is a surjective homomorphism. If $e \in E(M)$, then $M(eme)^\omega M = MeM$ implies $eme \in G_e = \{e\}$ by Corollary 1.16 and so $eme = e$. It follows that σ is an **LI**-morphism and hence $\ker\overline{\sigma}$ is nilpotent by Theorem 11.4. Thus it suffices to observe that $\Bbbk\Lambda(M)$ is semisimple by Corollary 9.5 and apply Corollary A.10. $\qquad\square$

We end this section by stating without proof the analogue of Theorem 11.4 in positive characteristic. Details can be found in [AMSV09] and the exercises. Let p be a prime. Then a homomorphism $\varphi\colon M \longrightarrow N$ of finite monoids is called an **LG**$_p$-morphism if, for each $e \in E(M)$, one has that $\varphi^{-1}(\varphi(e)) \cap eMe$ is a p-group. Notice that each **LI**-morphism is an **LG**$_p$-morphism. By an **LG**$_p$-congruence, we mean a congruence $\equiv$ on M such that the natural projection $\pi\colon M \longrightarrow M/\equiv$ is an **LG**$_p$-morphism.

Theorem 11.7. *Let $\Bbbk$ be a field of characteristic $p > 0$ and $\varphi\colon M \longrightarrow N$ a homomorphism of finite monoids. Let $\Phi\colon \Bbbk M \longrightarrow \Bbbk N$ be the extension of φ. Then the following are equivalent.*

*(i) φ is an **LG**$_p$-morphism.*
(ii) $I_\varphi = \ker\Phi$ is nilpotent.

Corollary 11.8. *Let M be a finite monoid and let $\Bbbk$ be a field of characteristic $p > 0$. Then there is an inclusion-preserving bijection between **LG**$_p$-congruences and nilpotent bi-ideals. In particular, $\equiv_\Bbbk$ is the largest **LG**$_p$-congruence on M.*

The largest **LG**$_p$-congruence on a finite monoid was determined by Rhodes and Tilson [Rho69a, Til69] (see also [AMSV09]).

11.2 Triangularizable monoids and basic algebras

Let $\Bbbk$ be an algebraically closed field. A finite dimensional $\Bbbk$-algebra A is said to be *basic* if $A/\operatorname{rad}(A) \cong \Bbbk^n$ for some $n > 0$. Equivalently, by Wedderburn's theorem, A is basic if every simple A-module is one-dimensional. Basic algebras play an important role in representation theory because each finite dimensional $\Bbbk$-algebra is Morita equivalent to a unique (up to isomorphism) basic algebra, cf. [DK94, ARS97, Ben98, ASS06]. Our aim is to characterize monoids with basic algebras. We provide all the details in characteristic zero; for the proofs in characteristic $p > 0$ the reader is referred to [AMSV09] or the exercises.

First we reformulate Proposition 10.8 in this language.

Proposition 11.9. *Let $\Bbbk$ be an algebraically closed field and M a finite monoid. Then the following are equivalent.*

(i) $\Bbbk M$ is basic.
(ii) M admits a faithful representation by upper triangular matrices over $\Bbbk$.
(iii) Each representation of M over $\Bbbk$ is equivalent to one by upper triangular matrices.

If $\Bbbk M$ is basic, then the simple (and hence semisimple) $\Bbbk M$-modules are closed under tensor product and so $\operatorname{rad}(\Bbbk M) = \operatorname{rad}_{\Bbbk}(M)$ by Proposition 11.1. It follows, again by Proposition 11.1, that $M/\equiv_{\Bbbk}$ has a semisimple algebra over $\Bbbk$. This leads to our first characterization of triangularizable monoids over $\Bbbk$.

Proposition 11.10. *Let $\Bbbk$ be an algebraically closed field of characteristic 0 and M a finite monoid. Then the following are equivalent.*

(i) $\Bbbk M$ is basic.
(ii) M admits a faithful representation by upper triangular matrices over $\Bbbk$.
(iii) There is an **LI**-*morphism $\varphi \colon M \longrightarrow N$ with N a commutative inverse monoid.*
(iv) $M/\equiv_{\Bbbk}$ is a commutative inverse monoid.

Proof. If (iv) holds, then since $\equiv_{\Bbbk}$ is an **LI**-congruence on M, it follows that (iii) holds. For the converse, assume that (iii) holds. Then $\varphi(M)$ is a commutative inverse monoid by Corollary 3.12 and hence, without loss of generality, we may assume that φ is surjective. But then $\ker\varphi \subseteq \equiv_{\Bbbk}$, because $\equiv_{\Bbbk}$ is the largest **LI**-congruence on M by Corollary 11.5, and so $M/\equiv_{\Bbbk}$ is a quotient of N. As quotients of commutative inverse monoids are again commutative inverse monoids by Corollary 3.5, we conclude that (iv) holds. This yields the equivalence of (iii) and (iv).

We already have the equivalence of (i) and (ii). Suppose that (iv) holds. We shall prove (i). Let us put $N = M/\equiv_{\Bbbk}$. Then $\Bbbk N \cong \prod_{e \in E(N)} \Bbbk G_e$ by Corollary 9.6 and hence $\Bbbk N$ is semisimple. Therefore, we have $\Bbbk M/\operatorname{rad}(\Bbbk M) \cong \Bbbk N$

by Proposition 11.1. But if G is a finite abelian group, then $\Bbbk G \cong \Bbbk^{|G|}$ by Proposition B.2 and hence $\Bbbk N \cong \Bbbk^{|N|}$. We conclude that $\Bbbk M$ is basic.

Now suppose that (i) holds. Then, as discussed above, the semisimple $\Bbbk M$-modules are closed under tensor product and hence $\Bbbk M/\operatorname{rad}(\Bbbk M) \cong \Bbbk N$, where $N = M/\equiv_\Bbbk$, by Proposition 11.1. In particular, $\Bbbk N$ is semisimple and commutative because $\Bbbk M$ is basic. We conclude that N is commutative and regular, where the latter conclusion uses Theorem 5.19. But a commutative regular monoid must be an inverse monoid by Theorem 3.2. Thus (iv) holds, completing the proof.   □

We also state here the corresponding result in characteristic p. A proof can be found in [AMSV09].

Proposition 11.11. *Let $\Bbbk$ be an algebraically closed field of characteristic $p > 0$ and M a finite monoid. Then the following are equivalent.*

(i) $\Bbbk M$ is basic.
(ii) M admits a faithful representation by upper triangular matrices over $\Bbbk$.
(iii) There is a surjective $\mathbf{LG}_p$-morphism $\varphi\colon M \longrightarrow N$ with N a commutative inverse monoid whose maximal subgroups have order prime to p.
(iv) $M/\equiv_\Bbbk$ is a commutative inverse monoid whose maximal subgroups have order prime to p.

We now wish to describe in purely algebraic terms when a monoid admits a surjective $\mathbf{LI}$-morphism to a commutative inverse monoid. We say that a finite monoid M is *rectangular*[1] if, for all $e, f \in E(M)$ with $MeM = MfM$, one has $efe = e$.

Remark 11.12. The reason for the name "rectangular monoid" is that in the semigroup theory literature, a semigroup satisfying the identities $x^2 = x$ and $xyx = x$ is called a *rectangular band*. Every rectangular band is isomorphic to one whose underlying set is a direct product $A \times B$ with multiplication $(a,b)(a',b') = (a,b')$, whence the term "rectangular." One can check that a monoid M is a rectangular monoid if and only if the idempotents of each regular $\mathscr{J}$-class of M form a rectangular band.

Example 11.13. Let us show that bands are rectangular monoids. Let M be a band (i.e., $M = E(M)$). If $MeM = MfM$, then there exists $a \in M$ such that $Ma = Me$ and $aM = fM$ by Proposition 1.22. Then $ea = e$ and $fa = a$, whence $efa = ea = e$. Therefore, we have $MeM = MefM$. But $efef = ef$ and so $MeM = MefM = MefeM$. We conclude, using Corollary 1.16, that $efe \in J_e \cap MeM = G_e$ and so $efe = e$ as $G_e = \{e\}$.

Let us explore the structure of rectangular monoids.

---

[1] The class of all rectangular monoids is often denoted by **DO** in the literature.

Proposition 11.14. *Let M be a rectangular monoid.*

(i) *If $e, f \in E(M)$ with $MeM = MfM$, then ef, fe are idempotents belonging to $J_e = J_f$.*

(ii) *Each regular $\mathscr{J}$-class of M is a subsemigroup.*

(iii) *If $e \in E(M)$ and $e \in MmM \cap MnM$, then $eme, ene, emne \in G_e$ and $emene = emne$.*

Proof. By definition of a rectangular monoid, $efe = e$ and so $efef = ef$ and $fefe = fe$. Clearly $ef, fe \in J_e$. This proves (i). To prove (ii), let J be a regular $\mathscr{J}$-class and $m, n \in J$. Then there exist idempotents $e, f \in E(M)$ such that $Mm = Me$ and $nM = fM$ by Proposition 1.22. Therefore, $e, f \in J$ and so $ef \in J$ by (i). Choose $u, v \in M$ such that $um = e$ and $nv = f$. Note that $umnv = ef \in J$ and hence $mn \in J$. This establishes (ii).

Turning to (iii), we have that $eme, ene, emne \in J_e$ by (ii) and Proposition 2.3. Hence these elements belong to G_e by Corollary 1.16. Since $em, ne \in J_e$, we can find, by Proposition 1.22, idempotents $f, f' \in E(J_e)$ such that $Mf = Mem$ and $f'M = neM$. Notice that this implies $emf = em$ and $f'ne = ne$. Then $fef = f$ and $f'ef' = f'$ by definition of a rectangular monoid. Also, as $ff' \in E(J_e)$ by (i), we have that $eff'e = e$. Consequently,

$$emne = emff'ne = emfeff'ef'ne = emeff'ene = emene,$$

as required.   □

If G is a group, we put $G^0 = G \cup \{0\} \subseteq \mathbb{C}G$. Then G^0 is an inverse monoid that is commutative if G is commutative.

Proposition 11.15. *Let M be a rectangular monoid. Let $e_1, \ldots, e_s$ form a complete set of idempotent representatives of the regular $\mathscr{J}$-classes of M. Then the mapping*

$$\varphi \colon M \longrightarrow \prod_{i=1}^{s} G_{e_i}^0$$

defined by

$$\varphi(m)_i = \begin{cases} e_i m e_i, & \text{if } e_i \in MmM \\ 0, & \text{if } m \in I(e_i) \end{cases}$$

is an **LI**-*morphism.*

Proof. By Proposition 11.14, if $e_i \in MmM$, then $e_i m e_i \in G_{e_i}$ and so φ makes sense as a mapping. Since $I(e_i)$ is an ideal, the only thing required to check that φ is a homomorphism is that if $e_i \in MmM \cap MnM$, then $e_i m e_i n e_i = e_i m n e_i \in G_{e_i}$. But this is the content of Proposition 11.14(iii). Let us proceed to verify that φ is an **LI**-morphism.

Suppose that $e \in E(M)$. Then $MeM = Me_iM$ for some i. Let $m \in eMe$ and assume that $\varphi(m) = \varphi(e)$. Then, since $\varphi(e)_i \neq 0$, we conclude that

$\varphi(m)_i \neq 0$ and so $e_i m e_i = \varphi(m)_i = \varphi(e)_i = e_i e e_i = e_i$ (using that M is a rectangular monoid). As M is a rectangular monoid and $eme = m$, we compute that

$$e = e e_i e = e e_i m e_i e = e e_i e m e e_i e = e m e = m.$$

This concludes the proof that φ is an **LI**-morphism.   □

We are now prepared to state and prove the main theorem of this chapter. It may be helpful for the reader to recall that a direct product of commutative inverse monoids is again a commutative inverse monoid.

Theorem 11.16. *Let* $\Bbbk$ *be an algebraically closed field of characteristic* 0 *and* M *a finite monoid. Then the following are equivalent.*

(i) $\Bbbk M$ *is basic.*
(ii) M *admits a faithful representation by upper triangular matrices over* $\Bbbk$.
(iii) There is an **LI***-morphism* $\varphi \colon M \longrightarrow N$ *with* N *a commutative inverse monoid.*
(iv) $M/\!\equiv_{\Bbbk}$ *is a commutative inverse monoid.*
(v) M *is a rectangular monoid with abelian maximal subgroups.*

Proof. The equivalence of (i)–(iv) is the content of Proposition 11.10. We check the equivalence of (iii) and (v). If M is rectangular with abelian maximal subgroups, then Proposition 11.15 furnishes an **LI**-morphism to a commutative inverse monoid. Suppose that $\varphi \colon M \longrightarrow N$ is an **LI**-morphism with N a commutative inverse monoid. Let $e \in E(M)$. Then $\ker \varphi|_{G_e} = \{e\}$ by definition of an **LI**-morphism. Therefore, $\varphi|_{G_e}$ is injective and we may deduce that G_e is abelian. Let $e, f \in E(M)$ with $MeM = MfM$. Then $N\varphi(e)N = N\varphi(f)N$. But the $\mathscr{J}$-classes of a commutative inverse monoid are maximal subgroups, by Proposition 3.11, and hence contain a unique idempotent. Therefore, we must have $\varphi(e) = \varphi(f)$. But then $\varphi(efe) = \varphi(e)$ and so, by definition of a **LI**-morphism, we conclude that $efe = e$. This shows that M is a rectangular monoid and completes the proof of the theorem.   □

It is straightforward to describe the irreducible representations of a rectangular monoid and hence of a triangularizable monoid.

Proposition 11.17. *Let* M *be a rectangular monoid and* $\Bbbk$ *a field. Suppose that* $e \in E(M)$ *and let* $V \in \mathrm{Irr}_{\Bbbk}(G_e)$. *Suppose that* V *affords the representation* $\rho \colon G_e \longrightarrow M_n(\Bbbk)$. *Then the representation* $\rho^{\sharp} \colon M \longrightarrow M_n(\Bbbk)$ *given by*

$$\rho^{\sharp}(m) = \begin{cases} \rho(eme), & \text{if } e \in MmM \\ 0, & \text{if } m \in I(e) \end{cases}$$

is afforded by the simple module $V^{\sharp} \in \mathrm{Irr}_{\Bbbk}(M)$ *corresponding to* V.

Proof. One can verify that $\rho^\sharp$ is a representation of M as a straightforward consequence of Proposition 11.14. It is obvious that $\rho^\sharp(e) = I$ and $\rho^\sharp|_{G_e} = \rho$. Hence $\rho^\sharp$ is irreducible by irreducibility of ρ. Also, by construction, $\rho^\sharp(I_e) = 0$. We conclude that e is an apex for $\rho^\sharp$. It follows that the module corresponding to $\rho^\sharp$ is $V^\sharp$, retaining the notation of Theorem 5.5. $\square$

Remark 11.18. Proposition 11.17 can be interpreted as follows. Let M be a rectangular monoid and $\Bbbk$ a field. If $e \in E(M)$, then there is a surjective homomorphism $\varphi_e \colon \Bbbk M \longrightarrow \Bbbk G_e$ of $\Bbbk$-algebras defined by

$$\varphi_e(m) = \begin{cases} eme, & \text{if } e \in MmM \\ 0, & \text{if } m \in I(e). \end{cases}$$

Proposition 11.17 then states that the simple $\Bbbk M$-modules with apex e are precisely the inflations of the simple $\Bbbk G_e$-modules under φ_e.

11.3 Exercises

11.1. A finite semigroup S is said to be *locally trivial* if $eSe = \{e\}$ for all $e \in E(S)$. Prove that a homomorphism $\varphi \colon M \longrightarrow N$ of finite monoids is an **LI**-morphism if and only if $\varphi^{-1}(f)$ is a locally trivial semigroup for each $f \in E(N)$.

11.2. Let $\Bbbk$ be a field. Prove that any finite submonoid of the multiplicative monoid of $\Bbbk^n$ is a commutative inverse monoid.

11.3. Prove that a finite monoid M is rectangular if and only if $E(J)$ is a rectangular band for each regular $\mathscr{J}$-class J.

11.4. Prove that a finite monoid M is rectangular if and only if there is an **LI**-morphism $\varphi \colon M \longrightarrow N$ where N is a finite inverse monoid with central idempotents. (Hint: you will find Exercise 3.14 helpful.)

11.5. Let M be a finite rectangular monoid and $e \in E(M)$.

(a) Prove that $I(e)$ is a prime ideal.
(b) Prove that $\varphi_e \colon M \setminus I(e) \longrightarrow G_e$ given by $\varphi_e(m) = eme$ for $m \in M \setminus I(e)$ is a monoid homomorphism.
(c) Prove that if $f \in E(M)$ and $MeM \subseteq MfM$, then $efmfe = eme$ for all $m \in M \setminus I(e)$ and, in particular, $efe = e$.

11.6. Prove that a finite monoid M is triangularizable over an algebraically closed field of characteristic 0 if and only if:

(i) the regular $\mathscr{J}$-classes of M are subsemigroups;
(ii) if J is a regular $\mathscr{J}$-class, then $\langle E(J) \rangle = E(J)$;
(iii) each maximal subgroup of M is abelian.

11.7. Let M be a finite monoid. Define $m \equiv m'$ if, for all regular $\mathscr{J}$-classes J of M and all $x, y \in J$, one has that

$$xmy \in J \iff xm'y \in J$$

and, moreover, if $xmy, xm'y \in J$, then $xmy = xm'y$. Prove that $\equiv$ is the largest **LI**-congruence on M.

11.8. Let M be a rectangular monoid and $\Bbbk$ a field. Fix a complete set $e_1, \ldots, e_s$ of idempotent representatives of the regular $\mathscr{J}$-classes of M. Prove that $\Bbbk M / \operatorname{rad}(\Bbbk M) \cong \prod_{i=1}^{s} \Bbbk G_{e_i} / \operatorname{rad}(\Bbbk G_{e_i})$.

11.9. Prove that a finite monoid M has a faithful completely reducible representation over $\mathbb{C}$ if and only if the largest **LI**-congruence on M is the equality relation.

11.10 (J. Rhodes). Prove that a finite monoid M has a faithful irreducible representation over $\mathbb{C}$ if and only if it has an ideal I such that M acts faithfully on both the left and right of I and there is an idempotent $e \in I$ with $I = IeI$, $|I(e)| \leq 1$ and such that the maximal subgroup G_e has a faithful irreducible representation over $\mathbb{C}$. (Hint: you may find it helpful to use Exercises 11.7 and 5.8.)

11.11. Prove Theorem 11.7.

11.12. Let M be a finite monoid and $\Bbbk$ a field of characteristic 0. Prove that M admits a faithful representation over $\Bbbk$ by upper triangular matrices if and only if M is a rectangular monoid with abelian maximal subgroups and $x^n - 1$ splits over $\Bbbk$ into linear factors where n is the least common multiple of the orders of the elements of the maximal subgroups of M.

11.13. Let M be a finite monoid and $\Bbbk$ a field of characteristic $p > 0$. Prove that M is triangularizable over $\Bbbk$ if and only if:

   (i) every regular $\mathscr{J}$-class is a subsemigroup;
   (ii) the maximal subgroups of $\langle E(M) \rangle$ are p-groups;
   (iii) each maximal subgroup G of M has normal p-Sylow subgroup N such that G/N is abelian and if n is the least common multiple of the orders of elements of G/N, then $x^n - 1$ splits over $\Bbbk$ into distinct linear factors.

11.14. Prove that a finite band is triangularizable over any field.

11.15. Let M be a finite monoid with trivial maximal subgroups, whose regular $\mathscr{J}$-classes are subsemigroups.

(a) Prove that $\Lambda(M) = \{MeM \mid e \in E(M)\}$.
(b) Prove that $\sigma \colon M \longrightarrow \Lambda(M)$ given by $\sigma(m) = Mm^\omega M$ is a surjective **LI**-morphism.

(c) Prove that if $\Bbbk$ is a field, then $\Bbbk M/\operatorname{rad}(\Bbbk M) \cong \Bbbk\Lambda(M)$.
(d) Prove that M is triangularizable over any field $\Bbbk$.

11.16. Let $\Bbbk$ be an algebraically closed field and A a finite dimensional $\Bbbk$-algebra. Let $\{P_1, \ldots, P_s\}$ be a complete set of representatives of the isomorphism classes of projective indecomposable A-modules and put $V = P_1 \oplus \cdots \oplus P_s$.

(a) Prove that $B = \operatorname{End}_A(V)^{op}$ is a basic finite dimensional $\Bbbk$-algebra.
(b) Prove that B is Morita equivalent to A.
(c) Prove that every basic algebra Morita equivalent to A is isomorphic to B.

11.17. Let $\Bbbk$ be an algebraically closed field of characteristic p and let C_p be a cyclic group of order p with generator a. Let $M = \{1\} \cup ([2] \times C_p \times [2])$ where 1 is the identity element and where

$$(i, g, j)(i', g', j') = \begin{cases} (i, gg', j'), & \text{if } (j, i') \neq (2, 2) \\ (i, gag', j'), & \text{if } (j, i') = (2, 2). \end{cases}$$

(a) Prove that M is not a rectangular monoid.
(b) Prove that $\Bbbk M$ is basic.

Part VI

Applications

12

Zeta Functions of Languages and Dynamical Systems

In this chapter, we apply the character theory of finite monoids to provide a proof of a theorem of Berstel and Reutenauer on the rationality of zeta functions of cyclic regular languages [BR90]. This generalizes the rationality of zeta functions of sofic shifts [Man71,LM95], an important result in symbolic dynamics. Background on free monoids, formal languages, and automata can be found in [Eil74, Eil76, Lot97, Lot02, BPR10, BR11].

12.1 Zeta functions

If A is a finite set, then A^* will denote the *free monoid* on the set A. Elements of A^* are (possibly empty) words in the alphabet A and the product is concatenation. We write $|w|$ for the length of a word w and write 1 for the empty word. A subset $L \subseteq A^*$ is called a *(formal) language* (since it consists of a collection of words).

The *zeta function* ζ_L of a language $L \subseteq A^*$ is the power series defined by

$$\zeta_L(t) = \exp\left(\sum_{n=1}^{\infty} a_n \frac{t^n}{n}\right)$$

where a_n is the number of words of length n in L. Notice that ζ_L does not keep track of whether $1 \in L$, that is, $\zeta_L = \zeta_{L \cup \{1\}}$. Our goal will be to show that ζ_L is rational if L is a cyclic regular language.

The set of *regular languages* over an alphabet A is the smallest collection of subsets of A^* that contains the finite subsets and is closed under union, product, and generation of submonoids. Regular languages play a fundamental role in theoretical computer science and have also entered into group theory, cf. [ECH⁺92]. A classical theorem of Kleene states that L is regular if and

© Springer International Publishing Switzerland 2016   177
B. Steinberg, *Representation Theory of Finite Monoids*,
Universitext, DOI 10.1007/978-3-319-43932-7_12

only if it is accepted by a finite state automaton, if and only if there is a homomorphism $\eta\colon A^* \longrightarrow M$ with M a finite monoid such that $\eta^{-1}(\eta(L)) = L$. Replacing M by $\eta(M)$, we may assume without loss of generality that η is surjective. There is, in fact, a unique (up to isomorphism) minimal cardinality choice for M, called the *syntactic monoid* of M. The syntactic monoid can be effectively computed from any of the standard ways of presenting a regular language (via an automaton or a regular expression). We shall use in this book only the formulation of regularity of a language in terms of finite monoids. The algebraic theory of regular languages uses finite monoids to classify regular languages [Eil74, Eil76, Pin86, Str94].

A language $L \subseteq A^*$ is called *cyclic* if it satisfies the following two conditions for all $u, v \in A^*$:

(a) $uv \in L \iff vu \in L$;
(b) for all $n \geq 1$, $u \in L \iff u^n \in L$.

One example of a cyclic regular language is the following. Let M be a finite monoid with a zero element z and let A be a generating set for M. Let $\varphi\colon A^* \longrightarrow M$ be the canonical surjection and let $L = \{u \in A^* \mid \varphi(u)^\omega \neq z\}$. Then L is a cyclic regular language.

Another example comes from symbolic dynamics. The set $A^{\mathbb{Z}}$ is a topological space, homeomorphic to the Cantor set, if we endow it with the topology of a product of discrete spaces. We view elements of $A^{\mathbb{Z}}$ as bi-infinite words $\cdots a_{-2}a_{-1}.a_0a_1 \cdots$ where the decimal point is placed immediately to the left of the image of the origin under the corresponding map $f\colon \mathbb{Z} \longrightarrow A$ given by $f(n) = a_n$. The *shift map* $T\colon A^{\mathbb{Z}} \longrightarrow A^{\mathbb{Z}}$ is defined by

$$T(\cdots a_{-2}a_{-1}.a_0a_1 \cdots) = \cdots a_{-2}a_{-1}a_0.a_1 \cdots .$$

A *symbolic dynamical system*, or *shift*, is a nonempty, closed, shift-invariant subspace $\mathscr{X} \subseteq A^{\mathbb{Z}}$ for some finite set A. The *zeta function* of $\mathscr{X}$ is

$$\zeta_{\mathscr{X}}(t) = \exp\left(\sum_{n=1}^{\infty} a_n \frac{t^n}{n}\right)$$

where a_n is the number of points $x \in \mathscr{X}$ with $T^n(x) = x$, i.e., the number of fixed points of T^n. Such a fixed point is of the form $\cdots ww.www \cdots$ where $|w| = n$, and hence a_n is finite. In fact, if

$$P(\mathscr{X}) = \{w \in A^* \mid \cdots ww.www \cdots \in \mathscr{X}\} \cup \{1\},$$

then $P(\mathscr{X})$ is a cyclic language and $\zeta_{P(\mathscr{X})} = \zeta_{\mathscr{X}}$. If $\mathscr{X}$ is a sofic shift, then $P(\mathscr{X})$ will be regular. We shall use here the original definition of Weiss [Wei73] of a sofic shift.

A shift $\mathscr{X} \subseteq A^{\mathbb{Z}}$ is a *sofic shift* if there is a finite monoid M with a zero element z and a homomorphism $\varphi\colon A^* \longrightarrow M$ such that $x \in \mathscr{X}$ if and only if $\varphi(w) \neq z$ for every finite factor (i.e., subword) $w \in A^*$ of $x \in A^{\mathbb{Z}}$. It is then straightforward to see that $P(\mathscr{X}) = \varphi^{-1}(T)$ where

$$T = \{m \in M \mid m^n \neq z, \, \forall n \geq 0\} = \{m \in M \mid m^\omega \neq z\}.$$

Therefore, $P(\mathscr{X})$ is a cyclic regular language. The reader is referred to [BR90, LM95] for more details.

Example 12.1. As an example, let $\Gamma = (V, E)$ be a simple undirected finite graph with vertex set V and edge set E, which is not a tree. We choose an orientation for E thereby allowing us to identify paths in Γ with certain words over the alphabet $E \cup E^{-1}$. Let $\mathscr{X}_\Gamma \subseteq (E \cup E^{-1})^{\mathbb{Z}}$ be the subspace of bi-infinite reduced paths in the graph Γ (that is, paths with no subpaths of the form ee^{-1} or $e^{-1}e$ with $e \in E$). Then one can verify that $\mathscr{X}_\Gamma$ is a sofic shift (in fact, it is what is called a shift of finite type).

Indeed, let $M_\Gamma = \{1, z\} \cup (E \cup E^{-1})^2$ where 1 is the identity, z is a zero element and

$$(x, y)(u, v) = \begin{cases} (x, v), & \text{if } yu \text{ is a reduced path} \\ z, & \text{else} \end{cases}$$

for $x, y, u, v \in E \cup E^{-1}$. Define $\eta \colon (E \cup E^{-1})^* \longrightarrow M_\Gamma$ by $\eta(x) = (x, x)$ for $x \in E \cup E^{-1}$. Then it is straightforward to verify that M_Γ is a monoid and $x \in (E \cup E^{-1})^{\mathbb{Z}}$ is reduced if and only if $\eta(w) \neq z$ for each finite factor w of x.

The zeta function $\zeta_{\mathscr{X}_\Gamma}$ is known as the *Ihara zeta function* of Γ [Ter99]. So the rationality of zeta functions of cyclic regular languages recovers the well-known fact that the Ihara zeta function is rational.

The zeta function of a cyclic language admits an Euler product expansion, due to Berstel and Reutenauer [BR90]. Two words in $u, v \in A^*$ are said to be *conjugate* if there exist $r, s \in A^*$ such that $u = rs$ and $v = sr$. This is an equivalence relation on A^*. A word $w \in A^*$ is *primitive* if it is not a proper power. Each conjugate of a primitive word is primitive. Indeed, if $rs = z^n$ with $n \geq 2$, then we can factor $z = uv$ with $r = z^k u$ and $s = vz^m$ where $k, m \geq 0$ and $n = k + m + 1$. Then $sr = vz^m z^k u = v(uv)^{k+m} u = (vu)^{k+m} vu = (vu)^n$ and so every conjugate of a proper power is a proper power. A primitive word w of length n has exactly n distinct conjugates; in fact, a word w of length n has exactly n conjugates if and only if it is primitive. This statement relies on the well-known fact that two words commute if and only if they are powers of a common element, cf. [Lot97, Proposition 1.3.2]. Every word is a power of a unique primitive word, called its *primitive root*. By definition, a cyclic language is closed under taking conjugates, primitive roots, and positive powers.

Theorem 12.2. *Let $L \subseteq A^*$ be a cyclic language. Then we have the Euler product expansion*

$$\zeta_L(t) = \exp\left(\sum_{n=1}^{\infty} a_n \frac{t^n}{n}\right) = \prod_{k=1}^{\infty} \frac{1}{(1-t^k)^{b_k}} \qquad (12.1)$$

where a_n is the number of words of length n in L and b_k is the number of conjugacy classes of primitive words of length k intersecting L. Consequently, $\zeta_L(t)$ has integer coefficients when expanded as a power series.

Proof. First we note that if we count words in L by their primitive roots, then

$$a_n = \sum_{k|n} b_k k. \qquad (12.2)$$

We now compute logarithmic derivatives of both sides of the second equality of (12.1) and show that they are equal. Since both sides evaluate to 1 at $t = 0$, it will then follow that (12.1) is valid. Let $f(t) = \log \zeta_L(t)$. Then we have, using (12.2), that

$$tf'(t) = \sum_{n=1}^{\infty} a_n t^n = \sum_{k=1}^{\infty} \sum_{d=1}^{\infty} b_k k t^{kd} = \sum_{k=1}^{\infty} b_k \frac{k t^k}{1-t^k}$$

$$= t\frac{d}{dt}\left[\sum_{k=1}^{\infty} -\log(1-t^k)^{b_k}\right] = t\frac{d}{dt}\log\left(\prod_{k=1}^{\infty} \frac{1}{(1-t^k)^{b_k}}\right).$$

This completes the proof.   □

In the case of a shift, conjugacy classes of primitive words of length k correspond to periodic orbits of size k and the Euler product expansion of Theorem 12.2 is well known [LM95]. The corresponding Euler product expansion for Ihara zeta functions is also well known. Here conjugacy classes of primitive words correspond to primitive cyclically reduced closed paths in the graph, taken up to cyclic conjugacy.

12.2 Rationality of the zeta function of a cyclic regular language

Let us slightly generalize the definition of a zeta function of a language. Let $f\colon A^* \longrightarrow \mathbb{C}$ be a mapping. Define the *zeta function* of f by

$$\zeta_f(t) = \exp\left(\sum_{n=1}^{\infty} \sum_{|w|=n} f(w)\frac{t^n}{n}\right).$$

The zeta function of a language $L \subseteq A^*$ is then the zeta function of the indicator function δ_L of L. As usual, a virtual character of A^* is a difference $\chi_\rho - \chi_\psi$ of characters of finite dimensional representations ρ and ψ of A^*. Note that one has that $\zeta_{f-g} = \zeta_f/\zeta_g$ for mappings $f, g\colon A^* \longrightarrow \mathbb{C}$.

Lemma 12.3. Let $f: A^* \longrightarrow \mathbb{C}$ be a virtual character. Assume that $f = \chi_\rho - \chi_\psi$ where $\rho: A^* \longrightarrow M_r(\mathbb{C})$ and $\psi: A^* \longrightarrow M_s(\mathbb{C})$ are representations. Let $S = \sum_{a \in A} \rho(a)$ and $T = \sum_{a \in A} \psi(a)$. Then the equality

$$\zeta_f(t) = \frac{\det(I - tT)}{\det(I - tS)}$$

holds and hence ζ_f is a rational function.

Proof. First observe that if B is a $k \times k$ matrix over $\mathbb{C}$, then

$$\exp\left(\sum_{n=1}^{\infty} \operatorname{Tr}(B^n) \frac{t^n}{n}\right) = \frac{1}{\det(I - tB)}. \tag{12.3}$$

Indeed, if $\lambda_1, \ldots, \lambda_k$ are the eigenvalues of B with multiplicities, then

$$\begin{aligned}
\exp\left(\sum_{n=1}^{\infty} \operatorname{Tr}(B^n) \frac{t^n}{n}\right) &= \exp\left(\sum_{n=1}^{\infty} \frac{(\lambda_1^n + \cdots + \lambda_k^n)t^n}{n}\right) \\
&= \prod_{i=1}^{k} \exp\left(\sum_{n=1}^{\infty} \frac{(\lambda_i t)^n}{n}\right) \\
&= \prod_{i=1}^{k} \exp\left(-\log(1 - \lambda_i t)\right) \\
&= \prod_{i=1}^{k} \frac{1}{1 - \lambda_i t} \\
&= \frac{1}{\det(I - tB)}
\end{aligned}$$

as required.

From the equality $\zeta_f = \zeta_{\chi_\rho}/\zeta_{\chi_\psi}$, it suffices to prove $\zeta_{\chi_\rho}(t) = 1/\det(I - tS)$ and $\zeta_{\chi_\psi}(t) = 1/\det(I - tT)$. Since the proofs are identical, we handle only the first case. Observe that

$$\sum_{|w|=n} \rho(w) = \left[\sum_{a \in A} \rho(a)\right]^n = S^n$$

and, therefore,

$$\sum_{|w|=n} \chi_\rho(w) = \sum_{|w|=n} \operatorname{Tr}(\rho(w)) = \operatorname{Tr}\left(\sum_{|w|=n} \rho(w)\right) = \operatorname{Tr}(S^n).$$

The result now follows by applying (12.3) with $B = S$. $\qquad\square$

To establish that the zeta function of a cyclic regular language $L \subseteq A^*$ is rational, we shall prove that the indicator function δ_L is a virtual character and apply Lemma 12.3.

Proposition 12.4. *Let $L \subseteq A^*$ be a cyclic regular language and $\eta \colon A^* \longrightarrow M$ a surjective homomorphism with M finite and $L = \eta^{-1}(\eta(L))$. Set $X = \eta(L)$.*

(i) $\delta_L = \delta_X \circ \eta$.
(ii) X enjoys the following two properties for all $m, m' \in M$.
  (a) $mm' \in X \iff m'm \in X$.
  (b) for all $n \geq 1$, $m \in X \iff m^n \in X$.
(iii) If $e \in E(M)$, then $X \cap G_e = \emptyset$ or $G_e \subseteq X$.
(iv) The indicator function $\delta_X \colon M \longrightarrow \mathbb{C}$ is a virtual character.

Proof. The first item is clear because $w \in L$ if and only if $\eta(w) \in X$. To prove (ii), let $m = \eta(w)$ and $m' = \eta(w')$. Then we have that

$$mm' \in X \iff \eta(ww') \in X \iff ww' \in L \iff w'w \in L$$
$$\iff \eta(w'w) \in X \iff m'm \in X,$$

establishing (a). Similarly, we have that, for all $n \geq 1$,

$$m \in X \iff \eta(w) \in X \iff w \in L \iff w^n \in L$$
$$\iff \eta(w^n) \in X \iff m^n \in X,$$

yielding (b).

We deduce (iii) from (ii) because if $g \in X \cap G_e$ and $n = |G_e|$, then we have $e = g^n \in X$. But if $h \in G_e$, we also have $h^n = e \in X$ and so $h \in X$. Therefore, $G_e \subseteq X$.

To prove (iv), we first note that δ_X is a class function because $\delta_X(mm') = \delta_X(m'm)$ by (ii)(a) and $\delta_X(m) = \delta_X(m^{\omega+1})$ by (ii)(b). Moreover, if G_e is a maximal subgroup, then by (iii), $(\delta_X)|_{G_e}$ is either identically 0 or 1. In the former case, $(\delta_X)|_{G_e}$ is the character of the zero $\mathbb{C}G_e$-module and in the latter it is the character of the trivial $\mathbb{C}G_e$-module. We conclude that δ_X is a virtual character by Corollary 7.16. □

If $\eta \colon A^* \longrightarrow M$ is a homomorphism and $\rho \colon M \longrightarrow M_r(\mathbb{C})$ is a representation, then $\rho \circ \eta$ is a representation of A^* and $\chi_{\rho \circ \eta} = \chi_\rho \circ \eta$. Consequently, if f is a virtual character of M, then $f \circ \eta$ is a virtual character of A^*. Proposition 12.4(i) and (iv), in conjunction with Lemma 12.3, then yield the following theorem of Berstel and Reutenauer.

Theorem 12.5. *Let $L \subseteq A^*$ be a cyclic regular language. Then the indicator function $\delta_L \colon A^* \longrightarrow \mathbb{C}$ is a virtual character and hence ζ_L is a rational function.*

We remark that the proof of Theorem 12.5 is constructive, but not very explicit, because the proof of Corollary 7.16 involves inverting the decomposition matrix (cf. Theorem 6.5).

As a corollary of Theorem 12.5, we obtain the rationality of zeta functions of sofic shifts [Man71, LM95].

Corollary 12.6 (Manning). *The zeta function of a sofic shift is rational.*

12.3 Computing the zeta function

In this section, we compute some zeta functions of sofic shifts to show that the methods are applicable.

12.3.1 Edge shifts

Let Q be a (finite) quiver (i.e., directed graph) with vertex set Q_0 and edge set Q_1. The *edge shift* $\mathscr{X}_Q \subseteq Q_1^{\mathbb{Z}}$ associated with Q is the shift consisting of all bi-infinite paths in Q. As in Chapter 8, we shall concatenate the edges in a path from right to left. We assume that Q is not acyclic so that $\mathscr{X}_Q$ is nonempty. Let A be the *adjacency matrix* of Q: so the rows and columns of A are indexed by Q_0 and A_{rs} is the number of edges from s to r.

Let us consider the inverse monoid

$$B_Q = \{1, z\} \cup (Q_0 \times Q_0)$$

where 1 is an identity, z is a zero element, and the remaining products are defined by

$$(u, v)(r, s) = \begin{cases} (u, s), & \text{if } v = r \\ z, & \text{else.} \end{cases} \tag{12.4}$$

Note that $E(B_Q) = \{1, z\} \cup \{(v, v) \mid v \in Q_0\}$. Also note that the inverse monoid involution is given by $1^* = 1$, $z^* = z$ and $(u, v)^* = (v, u)$.

Define $\varphi: Q_1^* \longrightarrow B_Q$ by $\varphi(a) = (r, s)$ whenever $a: s \longrightarrow r$ is an edge. Then it is straightforward to verify that a bi-infinite word x over Q_1 is a path if and only if $\varphi(w) \neq z$ for all finite subwords w of x. Therefore, $\mathscr{X}_Q$ is a sofic shift. Also we have that $P(\mathscr{X}_Q) = \varphi^{-1}(F)$ where $F = E(B_Q) \setminus \{z\}$. Note that φ is onto if and only if Q is strongly connected. Nonetheless, we will be able to express δ_F as a virtual character $\chi_V - \chi_W$ of B_Q, which is enough to compute the zeta function of the edge shift. Although it is easy to find this representation in an *ad hoc* manner, we follow the systematic method. We fix a vertex q_0 and put $e = (q_0, q_0)$. It is straightforward to check that $L_e = \{(r, q_0) \mid r \in Q_0\}$ and $J_e = Q_0 \times Q_0$.

We take as our $\mathscr{J}$-classes representatives $\{z, e, 1\}$. The corresponding maximal subgroups, G_z, G_e, and G_1, are trivial. Let S_z, S_e, and S_1 be the trivial modules for these three maximal subgroups over $\mathbb{C}$. Then, retaining the notation of Theorem 5.5, $S_z^\sharp$ is the trivial $\mathbb{C}B_Q$-module, $S_e^\sharp = \mathbb{C}L_e$ and $S_1^\sharp$ affords the representation sending 1 to 1 and $B_Q \setminus \{1\}$ to 0. We compute the decomposition matrix L, i.e., the matrix of

$$\text{Res: } G_0(\mathbb{C}B_Q) \longrightarrow G_0(\mathbb{C}G_z) \times G_0(\mathbb{C}G_e) \times G_0(\mathbb{C}G_1)$$

with respect to the ordered bases $\{S_z^\sharp, S_e^\sharp, S_1^\sharp\}$ and $\{S_z, S_e, S_1\}$. As L is lower triangular and unipotent by Theorem 6.5 the only computations to make are that $[fS_z^\sharp] = [S_f]$ for $f = \{e, 1\}$ and $[1S_e^\sharp] = |Q_0| \cdot [S_1]$ as $\dim S_e^\sharp = |L_e| = |Q_0|$. Therefore, we have

$$L = \begin{bmatrix} 1 & 0 & 0 \\ 1 & 1 & 0 \\ 1 & |Q_0| & 1 \end{bmatrix}.$$

Now δ_F is a class function and restricts to the zero mapping on G_z and to the trivial character on G_e and G_1. Hence it is a virtual character and, using the notation of Theorem 7.15, we have $\delta_F = \Delta_{B_Q}(\text{Res}^{-1}(0, [S_e], [S_1]))$. That is, we must solve $Lx = (0, 1, 1)^T$. The solution is $(0, 1, 1 - |Q_0|)^T$ and so

$$\delta_F = \chi_{S_e^\sharp} - (|Q_0| - 1) \cdot \chi_{S_1^\sharp}.$$

Let $\rho: B_Q \longrightarrow M_{|Q_0|}(\mathbb{C})$ be the representation afforded by $S_e^\sharp$ and let $\psi: B_q \longrightarrow \mathbb{C}$ be the representation afforded by $(|Q_0| - 1) \cdot S_1^\sharp$. Let $S = \sum_{a \in Q_1} \rho(\varphi(a))$ and $T = \sum_{a \in Q_1} \psi(\varphi(a))$. Then by Lemma 12.3, we have that

$$\zeta_{\mathscr{X}_Q} = \frac{\det(I - tT)}{\det(I - tS)}.$$

Since ψ vanishes on $B_Q \setminus \{1\}$, it follows that $T = 0$. On the other hand, $S_e^\sharp$ has basis (s, q_0) with $s \in Q_0$. Using (12.4) we compute that

$$\sum_{a \in Q_1} \varphi(a)(s, q_0) = \sum_{r \in Q_0} A_{rs}(r, q_0)$$

where A is the adjacency matrix of Q. Therefore, $S = A$ and we have arrived at the following classical result of Bowen and Lanford [LM95].

Theorem 12.7. *Let Q be a finite quiver that is not acyclic and let $\mathscr{X}_Q$ be the corresponding edge shift. Then the zeta function of $\mathscr{X}_Q$ is given by*

$$\zeta_{\mathscr{X}_Q}(t) = \frac{1}{\det(I - tA)}$$

where A is the adjacency matrix of Q.

Up to conjugacy (i.e., isomorphism of dynamical systems), edge shifts associated with quivers are precisely the shifts of finite type [LM95]. We remark that the zeta function is a conjugacy invariant [LM95].

12.3.2 The even shift

Next we consider a sofic shift which is not of finite type. Let $A = \{a, b\}$ and let $\mathscr{X}$ consist of all bi-infinite words $x \in A^{\mathbb{Z}}$ such that between any two consecutive occurrences of b, there are an even number of occurrences of a. In other words, $\mathscr{X}$ consists of all labels of bi-infinite paths in the automaton in Figure 12.1. One easily checks that $\mathscr{X}$ is a shift, called the *even shift* (cf. [LM95]).

Fig. 12.1. The even shift

Let I_2 be the symmetric inverse monoid on $\{1, 2\}$ and put $z = 1_\emptyset$, the zero element of I_2. Define $\varphi \colon A^* \longrightarrow I_2$ by $\varphi(a) = (1\ 2)$ and $\varphi(b) = 1_{\{2\}}$. Note that φ is surjective. A straightforward computation shows that $\varphi(w) = z$ if and only if w contains an odd number of occurrences of a between some consecutive pair of occurrences of b and hence $\mathscr{X}$ is a sofic shift. Also $P(\mathscr{X}) = \varphi^{-1}(F)$ where $F = S_2 \cup E(I_2) \setminus \{1_\emptyset\}$. Again, we must express δ_F as a virtual character.

Let us choose $1, e = 1_{\{2\}}, z$ as our set of $\mathscr{J}$-class representatives. Note that G_z and G_e are trivial, whereas $G_1 = S_2$ is the symmetric group on two symbols. Denote by S_z, S_e, and S_1 the trivial modules for these three maximal subgroups over $\mathbb{C}$ and let V be the sign representation of $G_1 = S_2$. Then $S_z^\sharp$ is the trivial $\mathbb{C}I_2$-module, $S_e^\sharp = \mathbb{C}L_e$, $S_1^\sharp$ affords the representation sending S_2 to 1 and $I_2 \setminus S_2$ to 0 and $V^\sharp$ affords the representation which restricts to S_2 as the sign representation and annihilates $I_2 \setminus S_2$.

We compute the matrix L of

$$\mathrm{Res}\colon G_0(\mathbb{C}I_2) \longrightarrow G_0(\mathbb{C}G_z) \times G_0(\mathbb{C}G_e) \times G_0(\mathbb{C}G_1)$$

with respect to the ordered bases $\{S_z^\sharp, S_e^\sharp, S_1^\sharp, V^\sharp\}$ and $\{S_z, S_e, S_1, V\}$. We shall compute it by hand, although one could also apply Theorem 9.16 (or, rather, its proof). Given that L is lower triangular and unipotent by Theorem 6.5, it remains to observe that $[fS_z^\sharp] = [S_f]$ for $f = \{e, 1\}$ and that the restriction of $S_e^\sharp$ to G_1 is the regular representation. Indeed, L_e consists of the two rank one partial injective mappings of $\{1, 2\}$ with domain $\{2\}$ and the left action by the transposition $(1\ 2)$ interchanges these two maps. Therefore, $[1S_e^\sharp] = [S_1] + [V]$ and hence

$$L = \begin{bmatrix} 1 & 0 & 0 & 0 \\ 1 & 1 & 0 & 0 \\ 1 & 1 & 1 & 0 \\ 0 & 1 & 0 & 1 \end{bmatrix}.$$

The mapping δ_F is a class function and restricts to the zero mapping on G_z and to the trivial character on both G_1 and G_e. Therefore, retaining the notation of Theorem 7.15, we have $\delta_F = \Delta_{I_2}(\mathrm{Res}^{-1}(0, [S_e], [S_1], 0))$. That is, we must solve $Lx = (0, 1, 1, 0)^T$. The solution is $(0, 1, 0 - 1)^T$ and so

$$\delta_F = \chi_{S_e^\sharp} - \chi_{V^\sharp}.$$

Let $\rho: I_2 \longrightarrow M_2(\mathbb{C})$ be the representation afforded by $S_e^\sharp$ and let $\psi: I_2 \longrightarrow \mathbb{C}$ be the representation afforded by $V^\sharp$. Put $S = \rho(\varphi(a)) + \rho(\varphi(b))$ and $T = \psi(\varphi(a)) + \psi(\varphi(b))$. Lemma 12.3 yields that

$$\zeta_x = \frac{\det(I - tT)}{\det(I - tS)}.$$

Since $\psi(\varphi(a)) = -1$ and $\psi(\varphi(b)) = 0$, we have $\det(I - tT) = 1 + t$. On the other hand, $S_e^\sharp$ has basis the two rank one maps with domain $\{2\}$. The transposition $\rho(a)$ swaps them and $\rho(b)$ fixes one and annihilates the other. Therefore, we have in an appropriate basis that

$$S = \begin{bmatrix} 0 & 1 \\ 1 & 0 \end{bmatrix} + \begin{bmatrix} 1 & 0 \\ 0 & 0 \end{bmatrix} = \begin{bmatrix} 1 & 1 \\ 1 & 0 \end{bmatrix}$$

and so $\det(I - tS) = 1 - t - t^2$. We have thus proved the following result.

Theorem 12.8. *The zeta function of the even shift is* $\dfrac{1+t}{1-t-t^2}$.

12.3.3 The Ihara zeta function

Let $\Gamma = (V, E)$ be a simple undirected graph which is not a tree and fix an orientation on Γ. Put $E^\pm = E \cup E^{-1}$. Recall that $\mathscr{X}_\Gamma \subseteq (E^\pm)^{\mathbb{Z}}$ is the space of bi-infinite reduced paths in Γ and that $\zeta_{\mathscr{X}_\Gamma}$ is known as the *Ihara zeta function* of Γ.

Define a matrix $T: E^\pm \times E^\pm \longrightarrow \mathbb{C}$, called *Hashimoto's edge adjacency operator*, by

$$T(x, y) = \begin{cases} 1, & \text{if } xy \text{ is a reduced path in } \Gamma \\ 0, & \text{else.} \end{cases}$$

Theorem 12.9. *The Ihara zeta function of* $\Gamma = (V, E)$ *is given by*

$$\zeta_{\mathscr{X}_\Gamma}(t) = \frac{1}{\det(I - tT)}$$

where T is Hashimoto's edge adjacency operator.

Proof. Let $\eta\colon (E^{\pm})^* \longrightarrow M_\Gamma$ be as in Example 12.1. Note that

$$E(M_\Gamma) = \{1, z\} \cup \{(x, y) \in (E^{\pm})^2 \mid yx \text{ is a reduced path}\}$$

and that $P(\mathscr{X}_\Gamma) = \eta^{-1}(X)$ where $X = E(M_\Gamma) \setminus \{z\}$. We express δ_X as a virtual character of M_Γ. Make $\mathbb{C}E^{\pm}$ a $\mathbb{C}M_\Gamma$-module by putting $1u = u$, $zu = 0$, and

$$(x, y)u = \begin{cases} x, & \text{if } yu \text{ is reduced} \\ 0, & \text{else} \end{cases}$$

for $x, y, u \in E^{\pm}$. In particular, $(x, y)u = u$ if and only if $u = x$ and yx is reduced, that is, if and only if $x = u$ and $(x, y) \in E(M_\Gamma)$. Thus the character χ of $\mathbb{C}E^{\pm}$ is given by $\chi(1) = |E^{\pm}|$, $\chi(z) = 0$ and

$$\chi(x, y) = \begin{cases} 1, & \text{if } (x, y) \in E(M_\Gamma) \\ 0, & \text{else.} \end{cases}$$

We conclude that $\delta_X = \chi - (|E^{\pm}| - 1)\theta$ where θ is the character of the degree one representation sending 1 to 1 and $M_\Gamma \setminus \{1\}$ to 0.

Now $\sum_{x \in E^{\pm}} \theta(\eta(x)) = 0$, whereas

$$\sum_{x \in E^{\pm}} \eta(x)y = \sum_{x \in E^{\pm}} T(x, y)x$$

and so if ρ is the representation afforded by $\mathbb{C}E^{\pm}$ with respect to the basis $E^{\pm}$, then $\sum_{x \in E^{\pm}} \rho(\eta(x)) = T$. We conclude from Lemma 12.3 that

$$\zeta_{\mathscr{X}_\Gamma}(t) = \frac{1}{\det(I - tT)},$$

as was required.   □

12.4 Exercises

12.1. Prove that A^* is indeed a free monoid on A. That is, given a monoid M and a mapping $\varphi\colon A \longrightarrow M$, prove that there is a unique homomorphism $\Phi\colon A^* \longrightarrow M$ such that $\Phi|_A = \varphi$.

12.2. Verify the assertions in Example 12.1.

12.3. Let A be a finite set and $F \subseteq A^*$. Let $\mathscr{X}_F$ be the subspace of $A^{\mathbb{Z}}$ consisting of all bi-infinite words containing no finite factor $w \in F$.

  (i) Prove that if $\mathscr{X}_F$ is nonempty, then $\mathscr{X}_F$ is a shift.
  (ii) Prove that every shift $\mathscr{X} \subseteq A^{\mathbb{Z}}$ is of the form $\mathscr{X}_F$.
  (iii) Prove that if F is finite and $\mathscr{X}_F$ is nonempty, then $\mathscr{X}_F$ is a sofic shift. A shift of this sort is called a *shift of finite type*.

12.4. Let $n \geq 2$. Compute the zeta function of the shift $\mathcal{X}_n$ where $\mathcal{X}_n$ consists of all $x \in \{a, b\}^{\mathbb{Z}}$ such that the number of occurrences of a between two consecutive occurrences of b is divisible by n. (Hint: consider the inverse submonoid of I_n generated by $1_{\{n\}}$ and the cyclic permutation $(1\ 2 \cdots n)$.)

12.5. Compute the zeta function of the shift $\mathcal{X}$ consisting of all $x \in \{a, b\}^{\mathbb{Z}}$ containing no two consecutive occurrences of b. This is usually called the *Fibonacci shift* since the number of words of length n containing no two consecutive occurrences of b is a Fibonacci number.

12.6. Let $L \subseteq A^*$ be a language. Define an equivalence relation $\equiv_L$ on A^*, called the *syntactic congruence* of L, by $x \equiv_L y$ if and only if one has

$$uxv \in L \iff uyv \in L$$

for all $u, v \in A^*$.

(a) Prove that $\equiv_L$ is a congruence on A^*. The quotient $M_L = A^*/\equiv_L$ is called the *syntactic monoid* of L.
(b) Prove that if $\varphi \colon A^* \longrightarrow M_L$ is the canonical projection, then $L = \varphi^{-1}(\varphi(L))$.
(c) Prove that if $\eta \colon A^* \longrightarrow M$ is a homomorphism with $\eta^{-1}(\eta(L)) = L$, then $\ker \eta \subseteq \equiv_L$.
(d) Deduce that L is regular if and only if M_L is finite.
(e) Let G be a group and let $\eta \colon A^* \longrightarrow G$ be a surjective homomorphism. Prove that if $L = \eta^{-1}(1)$, then $G \cong M_L$. Deduce that L is regular if and only if G is finite.

12.7. Compute the Ihara zeta function of the graph below.

12.8. A finite *automaton* is a 5-tuple $\mathscr{A} = (Q, A, \delta, i, T)$ where Q is a finite set, A is a finite alphabet, $\delta \colon A \times Q \longrightarrow Q$, $i \in Q$, and $T \subseteq Q$. Define $\delta^* \colon A^* \times Q \longrightarrow Q$ recursively by

$$\delta^*(1, q) = q$$
$$\delta^*(wa, q) = \delta^*(w, \delta(a, q))$$

for $q \in Q$, $a \in A$ and $w \in A^*$. The language accepted by $\mathscr{A}$ is

$$L(\mathscr{A}) = \{w \in A^* \mid \delta^*(w, i) \in T\}.$$

Prove that $L \subseteq A^*$ is regular if and only if $L = L(\mathscr{A})$ for some finite automaton $\mathscr{A}$.

12.9. Prove that a language $L \subseteq A^*$ is regular if and only if the indicator function δ_L generates a finite dimensional submodule of $\mathbb{F}_2^{A^*}$ where $\mathbb{F}_2^{A^*}$ is given a right module structure via $(fv)(w) = f(vw)$ for $f \colon A^* \longrightarrow \mathbb{F}_2$ and $v, w \in A^*$.

12.10 (M.-P. Schützenberger). Let $L \subseteq A^*$ be a regular language and

$$f_L(t) = \sum_{n=0}^{\infty} a_n t^n$$

where a_n is the number of words of length n belonging to L. Prove that $f_L(t)$ is rational. (Hint: let $\eta \colon A^* \longrightarrow M$ be a surjective homomorphism with M a finite monoid and $\eta^{-1}(\eta(L)) = L$ and let $\rho \colon M \longrightarrow M_n(\mathbb{C})$ be the regular representation of M; put $B = \sum_{a \in A} \rho(\eta(a))$ and show that

$$f_L(t) = \sum_{m \in \eta(L)} \sum_{n=0}^{\infty} (Bt)^n_{m,1} = \sum_{m \in \eta(L)} (I - tB)^{-1}_{m,1};$$

then use the classical adjoint formula for the inverse of a matrix over a commutative ring.)

13

Transformation Monoids

In this chapter we shall use the representation theory of finite monoids to study finite monoids acting on finite sets. Such actions play an important role in automata theory and we provide here some applications in this direction. In particular, we study connections with the popular Černý conjecture [Č64]; see [Vol08] for a survey. This chapter is primarily based upon the paper [Ste10b].

13.1 Transformation monoids

In this section, we discuss transformation monoids. Much of this is folklore. Fix a finite monoid M for the section. The minimal ideal of M shall be denoted by $I(M)$. Let Ω be a finite set equipped with an action of M on the left. We say that M is *transitive* on Ω if $M\omega = \Omega$ for all $\omega \in \Omega$, that is, Ω has no proper, nonempty M-invariant subset. The *rank* of $m \in M$ is defined by

$$\mathrm{rk}(m) = |m\Omega|.$$

The rank is constant on $\mathscr{J}$-classes. More precisely, we have the following.

Proposition 13.1. *If $u, v, m \in M$, then $\mathrm{rk}(umv) \leq \mathrm{rk}(m)$.*

Proof. Trivially $|mv\Omega| \leq |m\Omega| = \mathrm{rk}(m)$. But the map $mv\Omega \longrightarrow umv\Omega$ given by $x \mapsto ux$ is surjective. Thus $|umv\Omega| \leq |mv\Omega| \leq \mathrm{rk}(m)$ and hence $\mathrm{rk}(umv) \leq \mathrm{rk}(m)$. $\qquad\square$

We now consider elements of minimum rank. Put

$$\mathrm{rk}(\Omega) = \min\{\mathrm{rk}(m) \mid m \in M\}$$

for a finite M-set Ω. We show that the minimum rank is achieved on the minimal ideal.

© Springer International Publishing Switzerland 2016
B. Steinberg, *Representation Theory of Finite Monoids*,
Universitext, DOI 10.1007/978-3-319-43932-7_13

Lemma 13.2. *Let M act on a finite set Ω. Let $r = \mathrm{rk}(\Omega)$. Then $\mathrm{rk}(m) = r$ for all $m \in I(M)$. If M acts faithfully on Ω, then conversely $\mathrm{rk}(m) = r$ implies that $m \in I(M)$.*

Proof. Let $\mathrm{rk}(u) = r$ and let $m \in I(M)$. Trivially, $\mathrm{rk}(m) \geq r$. But $m \in MuM$ and so $\mathrm{rk}(m) \leq r$ by Proposition 13.1. The first statement follows.

Suppose now that the action is faithful and let $\mathrm{rk}(m) = r$. We prove that $m \in I(M)$. First note that $r \leq |m^2\Omega| \leq |m\Omega| = r$ and so $m^2\Omega = m\Omega$. Thus m permutes $m\Omega$ and hence m^ω fixes $m\Omega$. Therefore, $m^\omega m = m$ by faithfulness of the action, whence $MmM = Mm^\omega M$. So without loss of generality, we may assume that m is an idempotent. Let $z \in I(M)$ and consider $e = (mzm)^\omega$, an element of $E(I(M))$. Note that $em = e$ by construction because $m \in E(M)$. Also, we have $e\Omega \subseteq m\Omega$ and both sets have the same size r, whence $e\Omega = m\Omega$. Since e, m are idempotent, they both fix their image set. Thus $m\alpha = e(m\alpha) = e\alpha$ for all $\alpha \in \Omega$. Therefore, $m = e \in I(M)$ by faithfulness. $\qquad\square$

The pair (M, Ω) is called a *transformation monoid* if M acts faithfully on the left of Ω. We write T_Ω for the *full transformation monoid* on Ω, that is, the monoid of all self-maps of Ω. Transformation monoids on Ω amount to submonoids of T_Ω. The previous lemma shows that the minimal ideal of a transformation monoid consists of the elements of minimal rank.

A useful fact about transitive actions is the following observation.

Proposition 13.3. *Let M act transitively on Ω. Then $I(M)\omega = \Omega$ for all $\omega \in \Omega$.*

Proof. Trivially, $I(M)\omega$ is a nonempty M-invariant subset. $\qquad\square$

A *congruence* $\equiv$ on an M-set Ω is an equivalence relation on Ω such that $\alpha \equiv \beta$ implies $m\alpha \equiv m\beta$ for all $m \in M$. In this case, $\Omega/{\equiv}$ is an M-set with action $m[\alpha]_\equiv = [m\alpha]_\equiv$. A transformation monoid (M, Ω) is said to be *primitive* if M is transitive on Ω and the only congruences on Ω are the universal equivalence relation and the equality relation. Note that in [Ste10b], transitivity is not assumed in the definition of primitivity.

13.2 Transformation modules

Fix a field $\Bbbk$ for this section and a finite monoid M acting on the left of a finite set Ω. Then $\Bbbk\Omega$ is a left $\Bbbk M$-module by extending the action of M on the basis Ω linearly. We call it the *transformation module* associated with the action. Also, we have that $\Bbbk^\Omega = \{f\colon \Omega \longrightarrow \Bbbk\}$ is a right $\Bbbk M$-module by putting $(fm)(\omega) = f(m\omega)$. We have a dual pairing

$$\langle \cdot, \cdot \rangle \colon \Bbbk^\Omega \times \Bbbk\Omega \longrightarrow \Bbbk$$

defined by $\langle f, \omega \rangle = f(\omega)$ for $\omega \in \Omega$. Note that $\langle fm, \omega \rangle = \langle f, m\omega \rangle$ for all $m \in M$.

If $X \subseteq \Omega$, put $\lfloor X \rfloor = \sum_{\alpha \in X} \alpha \in \Bbbk\Omega$.

Theorem 13.4. *Let (M, Ω) be a transitive transformation monoid. Let*

$$V = \Bbbk\{\lfloor m\Omega \rfloor - \lfloor n\Omega \rfloor \mid m, n \in I(M)\}.$$

Then V is a $\Bbbk M$-submodule. Moreover, we have the following.

(i) $\dim V \leq |\Omega| - \mathrm{rk}(\Omega)$.
(ii) $mV = 0$ *if and only if* $m \in I(M)$.

Proof. Let $k = |\Omega|$ and $r = \mathrm{rk}(\Omega)$. If $m \in M$ and $x \in I(M)$, then $|mx\Omega| = r = |x\Omega|$ and so $y \mapsto my$ is a bijection $x\Omega \longrightarrow mx\Omega$. Thus $m\lfloor x\Omega \rfloor = \lfloor mx\Omega \rfloor$. It now follows, since $I(M)$ is an ideal, that V is a $\Bbbk M$-submodule.

We turn to the proof of (i). Fix $e \in E(I(M))$ and, for $\alpha \in e\Omega$, put $P_\alpha = \{\beta \in \Omega \mid e\beta = \alpha\}$. Then $\{P_\alpha \mid \alpha \in e\Omega\}$ is a partition of Ω into r parts. We claim that $|m\Omega \cap P_\alpha| = 1$ for all $m \in I(M)$ and $\alpha \in e\Omega$. Indeed, if $|m\Omega \cap P_\alpha| > 1$ for some $\alpha \in e\Omega$, then $r = |em\Omega| < |m\Omega| = r$, a contradiction. Thus $|m\Omega \cap P_\alpha| \leq 1$ for each $\alpha \in e\Omega$. But then we have

$$r = |m\Omega| = \left| \bigcup_{\alpha \in e\Omega} (m\Omega \cap P_\alpha) \right| = \sum_{\alpha \in e\Omega} |m\Omega \cap P_\alpha| \leq |e\Omega| = r.$$

We conclude that $|m\Omega \cap P_\alpha| = 1$ for all $\alpha \in e\Omega$.

Let $\delta_{P_\alpha} \colon \Omega \longrightarrow \Bbbk$ be the indicator function of P_α for $\alpha \in e\Omega$. Then we compute, for $m, n \in I(M)$,

$$\begin{aligned}
\langle \delta_{P_\alpha}, \lfloor m\Omega \rfloor - \lfloor n\Omega \rfloor \rangle &= \langle \delta_{P_\alpha}, \lfloor m\Omega \rfloor \rangle - \langle \delta_{P_\alpha}, \lfloor n\Omega \rfloor \rangle \\
&= \sum_{\beta \in m\Omega} \delta_{P_\alpha}(\beta) - \sum_{\gamma \in n\Omega} \delta_{P_\alpha}(\gamma) \\
&= |m\Omega \cap P_\alpha| - |n\Omega \cap P_\alpha| \\
&= 0
\end{aligned}$$

by the claim. Thus $V \subseteq W^\perp$ where W is the span of the mappings δ_{P_α} with $\alpha \in e\Omega$. Since the mappings δ_{P_α} with $\alpha \in e\Omega$ have disjoint supports, they form a linearly independent set. It follows that $\dim W = r$ and hence $\dim V \leq \dim W^\perp = k - r$. This proves the first item.

Suppose now that $m \in I(M)$. Then, for all $x \in I(M)$, we have $mx\Omega \subseteq m\Omega$ and both sets have size r. Thus $mx\Omega = m\Omega$ and so $m\lfloor x\Omega \rfloor = \lfloor mx\Omega \rfloor = \lfloor m\Omega \rfloor$. It follows that $mV = 0$. Suppose that $m \notin I(M)$. Then $\mathrm{rk}(m) > r$ by Lemma 13.2. Let $e \in E(I(M))$. Then $me\Omega \subsetneq m\Omega$ because $|me\Omega| = r < |m\Omega|$. Let $\alpha \in m\Omega \setminus me\Omega$ and choose $\beta \in \Omega$ with $m\beta = \alpha$. Since $I(M)\beta = \Omega$ by Proposition 13.3, there exists $n \in I(M)$ such that $n\beta = \beta$. Then $m(\lfloor n\Omega \rfloor - \lfloor e\Omega \rfloor) = \lfloor mn\Omega \rfloor - \lfloor me\Omega \rfloor \neq 0$ because $\alpha \in mn\Omega \setminus me\Omega$. This completes the proof. $\qquad\square$

Next we consider the relationship between congruences and transformation modules.

Proposition 13.5. *Let* $\equiv$ *be a congruence on* Ω. *Then there is a surjective* $\Bbbk M$-*module homomorphism* $\eta\colon \Bbbk\Omega \longrightarrow \Bbbk[\Omega/\equiv]$ *given by* $\omega \mapsto [\omega]_\equiv$ *for* $\omega \in \Omega$. *Moreover,* $\ker\eta$ *is spanned by the differences* $\alpha - \beta$ *with* $\alpha \equiv \beta$.

Proof. It is clear that η is a surjective $\Bbbk M$-module homomorphism. Obviously each difference $\alpha - \beta$ with $\alpha \equiv \beta$ is in $\ker\eta$. Fix a set T of representatives for each $\equiv$-class and write $\overline{\alpha}$ for the representative in T of the class of α. We claim that the nonzero elements of the form $\alpha - \overline{\alpha}$ form a basis for $\ker\eta$. The number of such elements is $|\Omega| - |\Omega/\equiv| = \dim\ker\eta$ so it is enough to show that they span $\ker\eta$. Indeed, if $v = \sum_{\alpha \in \Omega} c_\alpha \cdot \alpha$ is in $\ker\eta$, then

$$0 = \eta(v) = \sum_{\beta \in T}\sum_{\overline{\alpha}=\beta} c_\alpha [\beta]_\equiv.$$

Thus for each $\beta \in T$, we have that $\sum_{\overline{\alpha}=\beta} c_\alpha = 0$ and so

$$v = \sum_{\beta \in T}\sum_{\overline{\alpha}=\beta} c_\alpha(\alpha - \beta).$$

This completes the proof. ☐

As a special case, we may consider the universal equivalence relation $\equiv$ on Ω. The corresponding map is $\eta\colon \Bbbk\Omega \longrightarrow \Bbbk$ sending each element of Ω to 1. This is usually called the *augmentation map* and hence we write $\ker\eta = \mathrm{Aug}(\Bbbk\Omega)$. It is spanned by the differences $\alpha - \beta$ with $\alpha,\beta \in \Omega$. Of course, $\dim\mathrm{Aug}(\Bbbk\Omega) = |\Omega| - 1$.

Theorem 13.6. *Let* (M, Ω) *be a transitive transformation monoid. Suppose that* $\mathrm{Aug}(\Bbbk\Omega)$ *is a simple* $\Bbbk M$-*module.*

(i) (M, Ω) *is primitive.*

(ii) *Either* (M, Ω) *is a permutation group or* $I(M)$ *is the set of all constant mappings on* Ω.

Proof. To prove (i), observe that if $\equiv$ is a congruence, then the span of the $\alpha - \beta$ with $\alpha \equiv \beta$ is a $\Bbbk M$-submodule of $\mathrm{Aug}(\Bbbk\Omega)$ of codimension $|\Omega/\equiv| - 1$ by Proposition 13.5. Hence $\equiv$ is either universal or the equality relation by simplicity of $\mathrm{Aug}(\Bbbk\Omega)$. We conclude that (M, Ω) is primitive.

Let $r = \mathrm{rk}(\Omega)$. Let $V = \Bbbk\{\lfloor m\Omega \rfloor - \lfloor n\Omega \rfloor \mid m,n \in I(M)\}$. Note that $V \subseteq \mathrm{Aug}(\Bbbk\Omega)$ since the value of the augmentation mapping on $\lfloor m\Omega \rfloor$ with $m \in I(M)$ is r. By Theorem 13.4, we have that V is a submodule of $\mathrm{Aug}(\Bbbk\Omega)$ of dimension at most $|\Omega| - r$. By simplicity, we conclude $V = \mathrm{Aug}(\Bbbk\Omega)$ or $V = 0$. In the first case, we must have $r = 1$ and so $I(M)$ consists of constant maps. As $I(M)\beta = \Omega$ for all $\beta \in \Omega$ by Proposition 13.3, in fact, $I(M)$ contains all constant maps. If $V = 0$, then we have $M = I(M)$ by Theorem 13.4 as M annihilates V. Thus $M = J_1 = G_1$ and so M is a group. This completes the proof. ☐

A permutation group $G \leq S_\Omega$ is called a *synchronizing group* if, for each mapping $f \in T_\Omega \setminus S_\Omega$, the monoid $\langle G, f \rangle$ contains a constant map. This notion first appeared in the paper [AS06] of Arnold and the author and in an unpublished note of J. Araújo from October 2006. Recent work includes [ABC13, AC14, ACS15].

Corollary 13.7. *If $G \leq S_\Omega$ is 2-transitive, then G is synchronizing.*

Proof. Let $f \in T_\Omega \setminus S_\Omega$ and $M = \langle G, f \rangle$. Proposition B.12 says that if G is 2-transitive, then $\mathrm{Aug}(\mathbb{C}\Omega)$ is a simple $\mathbb{C}G$-module and hence a simple $\mathbb{C}M$-module. Since M is not a permutation group, it follows that it contains a constant map by Theorem 13.6.   □

Remark 13.8. If $G \leq S_\Omega$ is 2-homogeneous, that is, acts transitively on unordered pairs of elements of Ω, then it is known that $\mathrm{Aug}(\mathbb{R}\Omega)$ is a simple $\mathbb{R}G$-module (cf. [ACS15]). Thus one can deduce as above that G is synchronizing.

The following theorem was first proved by Pin [Pin78]. The proof below is due to the author and Arnold [AS06].

Theorem 13.9. *Let $G \leq S_\Omega$ be transitive with $|\Omega| = p$ prime. Then G is synchronizing.*

Proof. First note that G contains a p-cycle g where $p = |G|$. Indeed, since G is transitive one has that $p \mid |G|$. But then G has an element g of order p, which is necessarily a p-cycle. Let $C = \langle g \rangle$. It suffices to prove that $\mathrm{Aug}(\mathbb{Q}\Omega)$ is a simple $\mathbb{Q}C$-module. The result then follows by the argument in the proof of Corollary 13.7.

Note that $\mathbb{Q}C \cong \mathbb{Q}[x]/(x^p - 1) \cong \mathbb{Q}(\zeta_p) \times \mathbb{Q}$ where ζ_p is a primitive p^{th}-root of unity. Hence $\mathbb{Q}C$ has two simple modules: the trivial module and one of dimension $p-1$. Since C is transitive, $\mathbb{Q}\Omega$ contains only one copy of the trivial module, spanned by $\lfloor \Omega \rfloor$ (cf. Proposition B.10). As $\mathrm{Aug}(\mathbb{Q}\Omega)$ has dimension $p-1$, we deduce that $\mathrm{Aug}(\mathbb{Q}\Omega)$ is the $(p-1)$-dimensional simple $\mathbb{Q}C$-module. This completes the proof.   □

The above result can also be deduced from the following lemma of P. Neumann [Neu09].

Lemma 13.10. *Let (M, Ω) be a transformation monoid with group of units G. Assume that G acts transitively on Ω. Let $m \in I(M)$. Then $|m^{-1}(m\alpha)| = |\Omega|/\mathrm{rk}(\Omega)$ for all $\alpha \in \Omega$, that is, the partition associated with m is uniform.*

Proof. Let $n = |\Omega|$ and $r = \mathrm{rk}(\Omega)$. First note that $MmM = Mm^\omega M$ implies that $m \in G_{m^\omega}$. Indeed, $mM = m^\omega M$ and $Mm = Mm^\omega$ by stability (Theorem 1.13). Thus $m \in m^\omega Mm^\omega \cap J_{m^\omega} = G_{m^\omega}$ (using Corollary 1.16). It follows that m and m^ω induce the same partition of Ω and so we may assume without loss of generality that $m \in E(M)$. Put $P_\alpha = m^{-1}(m\alpha)$. The proof of Theorem 13.4 shows that $|P_\alpha \cap x\Omega| = 1$ for all $x \in I(M)$ and hence $|P_\alpha \cap gm\Omega| = 1$ for all $g \in G$.

Let $\eta \colon \mathbb{C}\Omega \longrightarrow \mathbb{C}$ be the augmentation mapping and put

$$U = \frac{1}{|G|} \sum_{g \in G} g.$$

Then $U\beta = \frac{1}{n}\lfloor \Omega \rfloor$ for $\beta \in \Omega$ by Proposition B.10 and so $U\lfloor m\Omega \rfloor = \frac{r}{n}\lfloor \Omega \rfloor$. Therefore, we have

$$r\frac{|P_\alpha|}{n} = \langle \delta_{P_\alpha}, \frac{r}{n}\lfloor \Omega \rfloor \rangle = \langle \delta_{P_\alpha}, U\lfloor m\Omega \rfloor \rangle = \frac{1}{|G|}\sum_{g \in G}\langle \delta_{P_\alpha}, \lfloor gm\Omega \rfloor \rangle = 1$$

because $|P_\alpha \cap gm\Omega| = 1$ for all $g \in G$. This completes the proof. $\square$

It follows in the context of Lemma 13.10 that if $|\Omega|$ is prime and G is transitive, then either $I(M) = G$ or $I(M)$ consists of constant maps.

We end this section with a discussion of partial transformation modules. An *action* of a monoid M on a set Ω by partial transformations is a partial mapping $A \colon M \times \Omega \longrightarrow M$, written $A(m, \omega) = m\omega$, satisfying $1\omega = \omega$ for all $\omega \in \Omega$ and $m(n\omega) = (mn)\omega$ for all $m, n \in M$ and $\omega \in \Omega$, where equality means either both sides are defined and equal or both sides are undefined.

A typical example is that if Λ is an M-set and Δ is an M-invariant subset, then M acts on $\Lambda \setminus \Delta$ by partial transformations via restriction of the action. Conversely, if M acts on Ω by partial transformations and $\theta \notin \Omega$, then $\Omega \cup \{\theta\}$ is a left M-set via $m\theta = \theta$, for all $m \in M$, and by putting $m\alpha = \theta$ whenever $m\alpha$ is undefined. Then $\{\theta\}$ is an M-invariant subset and restricting the action to Ω recovers the original action by partial transformations.

For example, if $e \in E(M)$, then $Me \setminus L_e$ is an M-invariant subset of Me and so M acts on L_e by partial transformations via restriction of the action.

If M is a finite monoid acting on a finite set Ω by partial transformations and $\Bbbk$ is a field, then $\Bbbk\Omega$ becomes a $\Bbbk M$-module by putting

$$m \odot \alpha = \begin{cases} m\alpha, & \text{if } m\alpha \text{ is defined} \\ 0, & \text{else} \end{cases}$$

for $m \in M$ and $\alpha \in \Omega$. From now on we drop the notation "$\odot$." We call $\Bbbk\Omega$ a *partial transformation module*. For example, $\Bbbk L_e$ is a partial transformation module for $e \in E(M)$. We shall see later that if M is $\mathscr{R}$-trivial, then each projective indecomposable $\Bbbk M$-module is a partial transformation module. If $\Omega = \Lambda \setminus \Delta$ where Λ is an M-set and Δ is an M-invariant subset, then $\Bbbk\Omega \cong \Bbbk\Lambda/\Bbbk\Delta$.

13.3 The Černý conjecture

Let us say that $A \subseteq T_\Omega$ is *synchronizing* if $M_A = \langle A \rangle$ contains a constant mapping. Synchronizing subsets of T_Ω correspond to synchronizing automata with state set Ω in Computer Science. One can think of Ω as the set of states

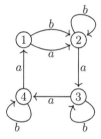

Fig. 13.1. The Černý example for $n = 4$

of the automaton and of A as the input alphabet. See Exercise 12.8 for the formal definition of an automaton. An automaton is synchronizing if some input sequence "resets" it to a fixed state, regardless of the original state of the automaton. A nice survey on synchronizing automata is [Vol08].

Let A^* denote the free monoid on A. A word $w \in A^*$ is called a *reset word* for $A \subseteq T_\Omega$ if its image under the canonical surjection $A^* \longrightarrow M_A$ is a constant map. The following conjecture of Černý [Č64] has been a popular open problem in automata theory for a half-century.

Conjecture 13.11 (Černý). Let $A \subseteq T_\Omega$ be synchronizing and let $n = |\Omega|$. Then there is a reset word for A of length at most $(n-1)^2$, that is, $A^{(n-1)^2}$ contains a constant mapping.

The current best upper bound for the length of a reset word is $(n^3 - n)/6$, due to Pin [Pin81]. The exercises sketch a simpler cubic upper bound. Černý himself provided the lower bound of $(n-1)^2$ with the following example.

Example 13.12. Let $a = (1\ 2\ \cdots\ n)$ and let b be the idempotent sending 1 to 2 and fixing all other elements. Then Černý [Č64] proved that $b(a^{n-1}b)^{n-2}$, which has length $(n-1)^2$, is the unique minimum length reset word for $A = \{a, b\}$. See Figure 13.1 for the case $n = 4$.

In this section we give several examples of how representation theory has been used to successfully attack special cases of the Černý conjecture.

Lemma 13.13. *Let (M, Ω) be a transformation monoid and $\Bbbk$ a field. Let $\varepsilon \colon \Bbbk^\Omega \longrightarrow \Bbbk$ be given by*

$$\varepsilon(f) = \langle f, \lfloor \Omega \rfloor \rangle = \sum_{\omega \in \Omega} f(\omega).$$

Then $\ker \varepsilon$ is a $\Bbbk M$-submodule of $\Bbbk^\Omega$ if and only if M is a group.

Proof. If $m \in M$ is a permutation, then $\varepsilon(fm) = \langle fm, \lfloor \Omega \rfloor \rangle = \langle f, m \lfloor \Omega \rfloor \rangle = \langle f, \lfloor \Omega \rfloor \rangle = \varepsilon(f)$. It follows that if M is a group, then ε is a homomorphism of right $\Bbbk M$-modules, where $\Bbbk$ is viewed as the trivial right $\Bbbk M$-module, and hence $\ker \varepsilon$ is a submodule. If $m \in M$ is not a permutation, choose $\alpha \in \Omega$ with $|m^{-1}\alpha| \geq 2$. Then

$$\varepsilon\left(\delta_\alpha - \frac{1}{|\Omega|}\delta_\Omega\right) = 0$$

but

$$\varepsilon\left(\delta_\alpha m - \frac{1}{|\Omega|}\delta_\Omega m\right) = \varepsilon\left(\delta_{m^{-1}\alpha} - \frac{1}{|\Omega|}\delta_\Omega\right) = |m^{-1}\alpha| - 1 > 0.$$

This completes the proof.   □

The following two lemmas are used frequently in conjunction to obtain bounds on the length of a reset word.

Lemma 13.14. *Let $A \subseteq T_\Omega$ be synchronizing. Suppose that there exists $k > 0$ such that, for each subset $S \subsetneq \Omega$ with $|S| \geq 2$, there exists $w \in A^*$ of length at most k such that $|w^{-1}S| > |S|$. Then there is a reset word for A of length at most $1 + k(n-2)$.*

Proof. Since A is synchronizing, there exists $a \in A$ which is not a permutation. Then there exists $\alpha \in \Omega$ with $|a^{-1}\alpha| \geq 2$. Repeated application of the hypothesis results in a sequence $w_1, \ldots, w_s$, with $0 \leq s \leq n-2$, of words of length at most k such that $|w_s^{-1}w_{s-1}^{-1} \cdots w_1^{-1}a^{-1}\alpha| = |\Omega|$. Then $aw_1 \cdots w_s$ is a reset word (with image α) of length at most $1 + k(n-2)$.   □

The next lemma is obvious, but important. It is the observation that a sum of non-positive numbers cannot be 0 unless all the summands are 0.

Lemma 13.15. *Let $A \subseteq T_\Omega$ and $S \subsetneq \Omega$. Let $X \subseteq A^*$ be finite. If*

$$\sum_{w \in X}(|w^{-1}S| - |S|) = 0$$

and $|u^{-1}S| - |S| \neq 0$ for some $u \in X$, then $|v^{-1}S| > |S|$ for some $v \in X$.

Motivated by the Černý examples (Example 13.12), it is natural to consider the conjecture in the special case that one of the transformations is a cyclic permutation. The following theorem is due to Pin [Pin78], which was the first result along this line. The proof here follows [AS06].

Theorem 13.16. *Let $|\Omega|$ be a prime p. Suppose that $a \in S_\Omega$ is a p-cycle and $B \subseteq T_\Omega$ with $B \nsubseteq S_\Omega$. Then $A = \{a\} \cup B$ is synchronizing and there exists a reset word of length at most $(p-1)^2$.*

Proof. Let $G = \langle a \rangle$ and $M = \langle A \rangle$. We know that A is synchronizing by Theorem 13.9. As $(p-1)^2 = 1 + p(p-2)$, it suffices to show by Lemma 13.14 that if $S \subsetneq \Omega$ with $|S| \geq 2$, there exists $w \in A^*$ of length at most p such that $|w^{-1}S| > |S|$.

Consider the right $\mathbb{Q}M$-module $\mathbb{Q}^\Omega$ and the mapping $\varepsilon \colon \mathbb{Q}^\Omega \longrightarrow \mathbb{Q}$ given by $\varepsilon(f) = \sum_{\alpha \in \Omega} f(\alpha)$. Then $V = \ker \varepsilon$ is a $\mathbb{Q}G$-submodule, by Lemma 13.13, of dimension $p-1$. Recall from the proof of Theorem 13.9 that $\mathbb{Q}G$ has two simple modules: the trivial module and one of dimension $p-1$. Clearly $(\delta_\alpha - \frac{1}{p}\delta_\Omega)a = \delta_{a^{-1}\alpha} - \frac{1}{p}\delta_\Omega$ and hence $\ker \varepsilon$ is not a direct sum of copies of the trivial module. Thus V is isomorphic to the nontrivial simple $\mathbb{Q}G$-module. Note that identifying $\mathbb{Q}G$ with

$$\mathbb{Q}[x]/(x^p - 1) \cong \mathbb{Q}[x]/(x-1) \times \mathbb{Q}[x]/(1 + x + \cdots + x^{p-1})$$

we see that the nontrivial simple $\mathbb{Q}G$-module is $\mathbb{Q}[x]/(1 + x + \cdots + x^{p-1})$. It follows that $1 + a + \cdots + a^{p-1}$ annihilates V. Since V is not a $\mathbb{Q}M$-submodule by Lemma 13.13, there exists $b \in B$ with $Vb \not\subseteq V$.

We claim that if $S \subsetneq \Omega$ with $|S| \geq 2$, then $|b^{-1}a^{-k}S| > |S|$ for some $0 \leq k \leq p-1$. Indeed, put

$$\gamma_S = \delta_S - \frac{|S|}{|\Omega|}\delta_\Omega \in V.$$

As V is a simple $\mathbb{Q}G$-module, we must have $V = \mathbb{Q}\{\gamma_S a^i \mid 0 \leq i \leq p-1\}$. Thus there exists $0 \leq i \leq p-1$ with $\gamma_S a^i b \notin V$. Now, observe that

$$\gamma_S a^i b = \delta_{b^{-1}a^{-i}S} - \frac{|S|}{|\Omega|}\delta_\Omega \notin V$$

and hence $0 \neq \varepsilon(\gamma_S a^i b) = |b^{-1}a^{-i}S| - |S|$.

Using that $1 + a + \cdots + a^{p-1}$ annihilates V we have that

$$0 = \varepsilon\left(\gamma_S \sum_{j=0}^{p-1} a^j b\right) = \sum_{j=0}^{p-1}\left(|b^{-1}a^{-j}S| - |S|\right).$$

Thus there exists $0 \leq k \leq p-1$ with $|b^{-1}a^{-k}S| > |S|$ by Lemma 13.15. This completes the proof in light of Lemma 13.14. $\qquad\square$

Černý's conjecture was proved in general for the case where some element of A is a cyclic permutation by Dubuc [Dub98].

The next result we present is due to Kari [Kar03]. We shall require a well-known combinatorial lemma about representations of free monoids.

Lemma 13.17. *Let $\Bbbk$ be a field and A a set. Suppose that V is a finite dimensional right $\Bbbk A^*$-module and that $W \subseteq U$ are subspaces with $WA^* \not\subseteq U$. Let $\{w_1, \ldots, w_k\}$ be a spanning set for W. Then there exist $x \in A^*$ of length at most $\dim U - \dim W + 1$ and $i \in \{1, \ldots, k\}$ such that $w_i x \notin U$.*

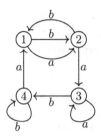

Fig. 13.2. An Eulerian automaton

Proof. Let $A^{\leq i} = \{x \in A^* \mid |x| \leq i|\}$ for $i \geq 0$ be the set of words in A^* of length at most i. Put $W_i = WA^{\leq i}$. Then we have a chain of subspaces

$$W = W_0 \subseteq W_1 \subseteq \cdots$$

whose union is $WA^* \nsubseteq U$. Moreover, if $W_i = W_{i+1}$, then W_i is invariant under A, and hence A^*, and so $W_i = WA^* \nsubseteq U$. Thus if j is maximal with $W_j \subseteq U$, then

$$W = W_0 \subsetneq W_1 \subsetneq \cdots \subsetneq W_j \subseteq U$$

and so $j \leq \dim U - \dim W$. But then there exists $x \in A^{\leq j+1}$ and w_i such that $w_i x \notin U$ because $W_{j+1} \nsubseteq U$. This completes the proof. $\square$

A subset $A \subseteq T_\Omega$ is called *Eulerian* if, for each $\alpha \in \Omega$, one has that

$$\sum_{a \in A} |a^{-1}\alpha| = |A| \tag{13.1}$$

and $\langle A \rangle$ is transitive on Ω. The reason for the terminology is that if we form a digraph by connecting α to $a\alpha$ for each $a \in A$, then condition (13.1) says that we obtain an Eulerian digraph (i.e., a digraph with a directed Euler circuit). See Figure 13.2.

Theorem 13.18. *Let $A \subseteq T_\Omega$ be Eulerian and synchronizing with $|\Omega| = n$. Then there is a reset word of length at most $n^2 - 3n + 3 \leq (n-1)^2$.*

Proof. Set $M = \langle A \rangle$, let $P = \sum_{a \in A} a \in \mathbb{Q}M$ and put $k = |A|$. Recall that $\varepsilon \colon \mathbb{Q}^\Omega \longrightarrow \mathbb{Q}$ is given by $\varepsilon(f) = \sum_{\omega \in \Omega} f(\omega) = \langle f, \lfloor \Omega \rfloor \rangle$. We compute that

$$P\lfloor \Omega \rfloor = \sum_{a \in A} \sum_{\omega \in \Omega} a\omega = \sum_{\alpha \in \Omega} \sum_{a \in A} |a^{-1}\alpha|\alpha = k\lfloor \Omega \rfloor$$

by (13.1). Thus, for any mapping $f \in \mathbb{Q}^\Omega$ and $m \geq 0$, we have

$$\varepsilon(fP^m) = \langle fP^m, \lfloor \Omega \rfloor \rangle = \langle f, P^m \lfloor \Omega \rfloor \rangle = k^m \langle f, \lfloor \Omega \rfloor \rangle = k^m \varepsilon(f). \tag{13.2}$$

We now claim that if $S \subsetneq \Omega$ with $|S| \geq 2$, then there exists a word v of length at most $n-1$ such that $|v^{-1}S| > |S|$. It will then follow by Lemma 13.14 that there is a reset word of length $1 + (n-1)(n-2) = n^2 - 3n + 3$, as required.

Let $U = \ker \varepsilon$ and put $\gamma_S = \delta_S - \frac{|S|}{n}\delta_\Omega \in U$. Let $W = \mathbb{Q}\gamma_S$. If $w \in A^*$ is a reset word with image contained in S (such exists by transitivity of M), then $w^{-1}S = \Omega$ and so

$$\gamma_S w = \delta_{w^{-1}S} - \frac{|S|}{n}\delta_\Omega = \left(1 - \frac{|S|}{n}\right)\delta_\Omega \notin \ker \varepsilon = U.$$

Applying Lemma 13.17, there is $u \in A^*$ with $|u| \leq \dim U - \dim W + 1 = n-1$ such that $\gamma_S u \notin U$. But then $0 \neq \varepsilon(\gamma_S u) = |u^{-1}S| - |S|$.

Put $j = |u|$. Then we have, by (13.2), that

$$0 = k^j \varepsilon(\gamma_S) = \varepsilon(\gamma_S P^j) = \sum_{|w|=j} \varepsilon(\gamma_S w) = \sum_{|w|=j} (|w^{-1}S| - |S|).$$

As $|u| = j$ and $0 \neq |u^{-1}S| - |S|$, we conclude that there exists v of length $j \leq n-1$ with $|v^{-1}S| > |S|$ by Lemma 13.15. This completes the proof. $\quad\square$

Our final application of representation theory to the Černý problem is a result from [AS09], generalizing an earlier result from [AMSV09].

Lemma 13.19. *Let M be an A-generated finite monoid all of whose regular $\mathscr{J}$-classes are subsemigroups and $\Bbbk$ a field. Let V be a simple $\Bbbk M$-module and suppose that $mV = 0$ for some $m \in M$. Then $aV = 0$ for some $a \in A$.*

Proof. Let $e \in E(M)$ be an apex for V. Then $I(e)$ is a prime ideal by Proposition 2.3 and $m \in I(e)$. If $m = a_1 \cdots a_r$ with $a_1, \ldots, a_r \in A$, then $a_i \in I(e)$ for some i because $I(e)$ is prime. But then $a_i V = 0$. $\quad\square$

As a consequence, one obtains the following generalization to arbitrary finite dimensional modules.

Theorem 13.20. *Let M be a finite A-generated monoid in which each regular $\mathscr{J}$-class is a subsemigroup and let $\Bbbk$ be a field. Let V be a $\Bbbk M$-module of length r and suppose $mV = 0$ for some $m \in M$. Then there is a word w over A of length at most r such that $wV = 0$.*

Proof. We induct on the length of V, the case $r = 1$ being handled by Lemma 13.19. Suppose that V has length $r > 1$. Then there is a maximal submodule W of V such that W has length $r - 1$ and V/W is simple. Since $m(V/W) = 0$, there exists $a \in A$ such that $a(V/W) = 0$, i.e., $aV \subseteq W$ by Lemma 13.19. Since $mW = 0$, there exists by induction a word u of length at most $r - 1$ such that $uW = 0$. Then $uaV \subseteq uW = 0$. As $|ua| \leq r$, this completes the proof. $\quad\square$

As a corollary, we obtain the following Černý-type result.

Corollary 13.21. *Let $A \subseteq T_\Omega$ be such that each regular $\mathcal{J}$-class of $M = \langle A \rangle$ is a subsemigroup. Let $r = \mathrm{rk}(\Omega)$ and $n = |\Omega|$. Then there is a word $w \in A^*$ of length at most $n - r$ such that w represents an element of M of rank r. In particular, if A is synchronizing, then there is a reset word of length at most $n - 1$.*

Proof. Assume first that M is transitive on Ω. Consider the $\mathbb{C}M$-module $\mathbb{C}\Omega$. Let $V = \mathbb{C}\{\lfloor m\Omega \rfloor - \lfloor n\Omega \rfloor \mid m, n \in I(M)\}$. Then V is a submodule of dimension at most $n - r$ and $mV = 0$ if and only if $\mathrm{rk}(m) = r$ by Theorem 13.4 and Lemma 13.2. As the length of a module is bounded by its dimension, Theorem 13.20 applied to V provides the required conclusion.

Next suppose that M is not transitive, but that Ω decomposes as a disjoint union $\Omega = \Omega_1 \cup \cdots \cup \Omega_s$ of transitive M-sets Ω_i for $i = 1, \ldots, s$. Put $r_i = r(\Omega_i)$ and $n_i = |\Omega_i|$. Then one easily checks that

$$n = n_1 + \cdots + n_s \quad \text{and} \quad r = r_1 + \cdots + r_s$$

because elements of $I(M)$ have minimal rank on each Ω_i by Lemma 13.2. Let M_i be the quotient of M that acts faithfully on Ω_i. By Exercise 1.26, each regular $\mathcal{J}$-class of M_i is a subsemigroup. By the previous case, we can find $w_i \in A^*$ of length at most $n_i - r_i$ with $|w_i \Omega_i| = r_i$ for $i = 1, \ldots, s$. Then putting $w = w_1 \cdots w_s$, we have that

$$|w| \leq (n_1 + \cdots + n_s) - (r_1 + \cdots + r_s) = n - r$$

and

$$|w\Omega| = \sum_{i=1}^{s} |w\Omega_i| = \sum_{i=1}^{s} r_i = r$$

where the penultimate equality uses Lemma 13.2.

Finally, we handle the general case. Let $\Omega' = I(M)\Omega$; note that Ω' is M-invariant. Observe that $r(\Omega) = r(\Omega')$ as a consequence of Lemma 13.2 because if $m \in I(M)$, then since $I(M)$ is a regular $\mathcal{J}$-class, there exists $a \in M$ with $mam = m$ and so $m\Omega = mam\Omega \subseteq m\Omega' \subseteq m\Omega$. Let $n' = |\Omega'|$. We claim that Ω' decomposes as a disjoint union of transitive actions of M. Indeed, if $\alpha \in \Omega'$ and $\beta = m\alpha$ with $m \in M$, then we can find $\gamma \in \Omega$ and $x \in I(M)$ with $x\gamma = \alpha$. Then $mx \in I(M)$ and so $mx \in L_x$ by stability (Theorem 1.13) because $I(M)$ is a $\mathcal{J}$-class. Thus there exists $m' \in M$ with $m'mx = x$ and so $m'\beta = m'm\alpha = m'mx\gamma = x\gamma = \alpha$. Therefore, $M\alpha$ is a transitive M-set for $\alpha \in \Omega'$. It follows that the distinct sets of the form $M\alpha$ with $\alpha \in \Omega'$ partition Ω' into transitive M-invariant subsets. Let M' be the quotient of M acting faithfully on Ω'. Then each regular $\mathcal{J}$-class of M' is a subsemigroup by Exercise 1.26. Thus, by the previous case, we can find $w \in A^*$ with $|w| \leq n' - r$ and $|w\Omega'| = r$.

Consider now the module $V = \mathbb{C}\Omega/\mathbb{C}\Omega'$. Then $xV = 0$ for $x \in I(M)$ by construction. Thus, by Theorem 13.20, there exists $u \in A^*$ with

$|u| \leq \dim V = n - n'$ such that $uV = 0$, that is, such that $u\Omega \subseteq \Omega'$. Therefore, we have that $r \leq |wu\Omega| \leq |w\Omega'| = r$ and $|wu| \leq n' - r + n - n' = n - r$. This completes the proof.   □

Let $1 \leq r \leq n$ and define a mapping $a \colon [n] \longrightarrow [n]$ by

$$a(x) = \begin{cases} x + 1, & \text{if } 1 \leq x < n \\ n - r + 1, & \text{if } x = n. \end{cases}$$

Consideration of $A = \{a\} \subseteq T_n$ shows that Corollary 13.21 is sharp.

Examples of monoids in which each regular $\mathscr{J}$-class is a subsemigroup include commutative monoids, $\mathscr{R}$-trivial monoids, bands, and, more generally, rectangular monoids.

13.4 Exercises

13.1. Suppose that M acts transitively on a set Ω and $e \in E(M)$. Prove that eMe acts transitively on $e\Omega$. Deduce that if $e \in E(I(M))$, then G_e acts transitively on $e\Omega$.

13.2. Suppose that (M, Ω) is a primitive transformation monoid and let $e \in E(I(M))$. Prove that $(G_e, e\Omega)$ is a primitive permutation group.

13.3. Let (M, Ω) be a finite transformation monoid. Prove that the following are equivalent.

(i) Ω is a union of transitive M-invariant subsets.
(ii) Ω is a disjoint union of transitive M-invariant subsets.
(iii) $\Omega = I(M)\Omega$.

13.4. Let M be a finite monoid, Ω a finite left M-set, and $\Bbbk$ a field. Prove that $\Bbbk^\Omega \cong \mathrm{Hom}_\Bbbk(\Bbbk\Omega, \Bbbk) = D(\Bbbk\Omega)$.

13.5. Let M be a finite monoid, Ω a finite left M-set, and $\Bbbk$ a field. Let $\sim$ be the least equivalence relation $\sim$ on Ω such that $\omega \sim m\omega$ for all $m \in M$ and $\omega \in \Omega$. Let r be the number of $\sim$-classes. Prove that $[\mathrm{soc}(\Bbbk^\Omega) : \Bbbk] = r$ where, as usual, $\Bbbk$ denotes the trivial $\Bbbk M$-module.

13.6. Let M be a finite monoid and $\Bbbk$ an algebraically closed field. Then $\Bbbk^M$ has the structure of a right $\Bbbk M$-module by viewing M as a left M-set. Suppose that $S_1, \ldots, S_r$ form a complete set of representatives of the isomorphism classes of simple $\Bbbk M$-modules and that S_k affords the representation $\varphi^{(k)} \colon M \longrightarrow M_{n_k}(\Bbbk)$ with $\varphi^{(k)}(m) = (\varphi_{ij}^{(k)}(m))$. Prove that the mappings

$$\varphi_{ij}^{(k)} \colon M \longrightarrow \Bbbk$$

with $1 \leq k \leq r$ and $1 \leq i, j \leq n_k$ form a basis for $\mathrm{soc}(\Bbbk^M)$.

13.7. Let (M, Ω) be a transitive finite transformation monoid containing a constant mapping and $\Bbbk$ a field.

(a) Prove that $\Bbbk\Omega$ is a projective indecomposable module with $\mathrm{rad}(\Bbbk\Omega) = \mathrm{Aug}(\Bbbk\Omega)$ and $\Bbbk\Omega/\mathrm{rad}(\Bbbk\Omega)$ isomorphic to the trivial module.
(b) Prove that $\Bbbk^{\Omega}$ is an injective indecomposable right module with simple socle $\Bbbk\delta_\Omega$, which is isomorphic to the trivial module.

13.8. Give a direct proof, avoiding representation theory, that a 2-homogeneous group is synchronizing.

13.9. Prove that a synchronizing group is primitive.

13.10. Let $A \subseteq T_\Omega$ with $|\Omega| = n$.

(a) Show that A is synchronizing if and only if, for any pair of distinct elements $\alpha, \beta \in \Omega$, there exists $w \in A^*$ with $w\alpha = w\beta$.
(b) Prove that if A is synchronizing, then one can always choose w in (a) so that $|w| \leq \binom{n}{2}$. (Hint: consider the action of A^* by partial transformations on the set of pairs of elements of Ω and use the pigeonhole principle.)
(c) Deduce that there is a reset word of length at most $1 + (n-2)\binom{n}{2}$ for A.

13.11. Prove that $b(a^{n-1}b)^{n-2}$ is the unique minimal length reset word for Černý's example (Example 13.12).

13.12. Prove that the Černý conjecture is true if and only if it is true for all synchronizing $A \subseteq T_\Omega$ with $M_A = \langle A \rangle$ transitive on Ω.

13.13. Let us say that $A \subseteq T_\Omega$ is pseudo-Eulerian if $\langle A \rangle$ is transitive and there is a strictly positive mapping $f: A \longrightarrow \mathbb{R}$ such that

$$\sum_{a \in A} f(a)\lfloor \Omega \rfloor = \lambda\lfloor \Omega \rfloor$$

for some $\lambda \in \mathbb{R}$.

(a) Prove that if A is Eulerian, then A is pseudo-Eulerian.
(b) Prove that if $A \subseteq T_\Omega$ is pseudo-Eulerian and synchronizing with $|\Omega| = n$, then there is a reset word of length at most $n^2 - 3n + 3$.
(c) Give an example of a synchronizing subset $A \subseteq T_\Omega$ that is pseudo-Eulerian but not Eulerian for some set Ω.

13.14. Let G be a finite group of order n and let $G = \langle B \rangle$. Let $A \subseteq T_G$ be such that $B \subseteq A$ and A is synchronizing, where we view $G \subseteq S_G$ via the regular representation. Prove that there is a reset word of length at most $2n^2 - 6n + 5$. (Hint: use that $\sum_{g \in G} g$ annihilates $\ker \varepsilon$ (defined as per Lemma 13.13) and that each element of G can be represented by a word in B of length at most $n - 1$.)

13.15 (Rystsov). Let G be a subgroup of S_n and let $e \in E(T_n)$ have rank $n - 1$ where $e(i) = e(j)$ with $i < j$. Prove that $G \cup \{e\}$ is synchronizing if and only if the graph Γ with vertex set $[n]$ and edge set consisting of all pairs $\{g(i), g(j)\}$ with $g \in G$ is connected.

14

Markov Chains

A Markov chain is a stochastic process on a finite state space such that the system evolves from one state to another according to a prescribed probabilistic law. For example, card shuffling can be modeled via a Markov chain. The state space is all 52! orderings of a deck of cards. Each step of the Markov chain corresponds to performing a riffle shuffle to the deck.

The analysis of Markov chains using group representation theory was pioneered to a large extent by Diaconis, and the reader is encouraged to look at his beautiful lecture notes [Dia88]. The reader may also consult [CSST08] or [Ste12a, Chapter 11]. The usage of techniques from monoid theory to study Markov chains is a relatively new subject. Work in this area began in a seminal paper of Bidigare, Hanlon, and Rockmore [BHR99] on hyperplane walks, and was continued by Brown, Diaconis, Björner, Athanasiadis, Chung, and Graham, amongst others [BD98, BBD99, Bro00, Bro04, Bjö08, Bjö09, AD10, CG12, Sal12]. Most of these works concern left regular bands, that is, von Neumann regular $\mathscr{R}$-trivial monoids. In his 1998 ICM lecture [Dia98], Diaconis highlighted these developments and in Section 4.1, entitled *What is the ultimate generalization?*, essentially asked how far can the monoid techniques be taken.

Recently, the author [Ste06, Ste08], Ayyer, Klee, and Schilling [AKS14a, AKS14b], and the author with Ayyer, Schilling, and Thiéry in [ASST15a, ASST15b] have extended the techniques beyond left regular bands.

In this chapter we only touch on the basics of this topic. In particular, we focus on the computation of the eigenvalues for the transition matrix of the Markov chain and diagonalizability since it is in this aspect that representation theory most directly intervenes. A good reference on finite state Markov chain theory is [LPW09].

© Springer International Publishing Switzerland 2016
B. Steinberg, *Representation Theory of Finite Monoids*,
Universitext, DOI 10.1007/978-3-319-43932-7_14

14.1 Markov Chains

A *probability* on a finite set X is a mapping $P\colon X \longrightarrow \mathbb{R}$ such that:

(i) $P(x) \geq 0$ for all $x \in X$;

(ii) $\displaystyle\sum_{x \in X} P(x) = 1$.

The *support* of P is $\operatorname{supp}(P) = \{x \in X \mid P(x) \neq 0\}$. We shall systematically identify P with the element $\sum_{x \in X} P(x)x \in \mathbb{C}X$.

A *Markov chain* $\mathscr{M} = (T, \Omega)$ consists of a finite set Ω, called the *state space*, and a *transition matrix* $T\colon \Omega \times \Omega \longrightarrow \mathbb{R}$ satisfying the following properties:

(i) $T(\alpha, \beta) \geq 0$ for all $\alpha, \beta \in \Omega$;

(ii) $\displaystyle\sum_{\alpha \in \Omega} T(\alpha, \beta) = 1$ for all $\beta \in \Omega$.

One interprets $T(\alpha, \beta)$ to be the probability of going from state β to state α in one step of the Markov chain[1]. We view T as an operator on $\mathbb{C}\Omega$ via the usual rule

$$T\beta = \sum_{\alpha \in \Omega} T(\alpha, \beta)\alpha$$

for $\beta \in \Omega$. Then one can interpret $T^n(\alpha, \beta)$ as the probability of going from state β to state α in exactly n steps of the chain. Since we are interested in the dynamics of the operator T, it is natural to try and compute the eigenvalues of T. We shall use representation theory as a tool.

If $\mathscr{M} = (T, \Omega)$ is a Markov chain and ν is a probability on Ω, which we think of as an initial distribution, then $T\nu$ is another probability on Ω, as the reader easily checks. One says that ν is a *stationary distribution* if $T\nu = \nu$. Stationary distributions always exist and under mild hypotheses there is a unique stationary distribution π and $T^n\mu$ converges to π for any initial distribution μ. One is typically interested in the rate of convergence. For example, the Markov chain might be a model of shuffling a deck of cards and the rate of convergence is how quickly the deck mixes. We shall not, however, focus on this topic here. There are some monoid theoretic techniques for bounding mixing times for monoid random walks in [ASST15b, Section 3], but they do not rely so much on representation theory. In the case of groups, there are Fourier analytic techniques for bounding mixing times, cf. [Dia88].

---

[1] In probability theory, it is traditional to use the transpose of what we are calling the transition matrix. However, because we are using left actions and left modules, it is more natural for us to use this formulation.

14.2 Random walks

Let M be a finite monoid, P a probability on M, and Ω a finite left M-set. The *random walk* of M on Ω *driven* by P is the Markov chain $\mathcal{M} = (T, \Omega)$ with transition matrix $T\colon \Omega \times \Omega \longrightarrow \mathbb{R}$ given by

$$T(\alpha, \beta) = \sum_{\{m \in M \mid m\beta = \alpha\}} P(m).$$

In other words, $T(\alpha, \beta)$ is the probability that acting upon β with a randomly chosen element of M, distributed according to P, results in α.

Proposition 14.1. *The pair (T, Ω) constructed above is a Markov chain.*

Proof. Trivially, $T(\alpha, \beta) \geq 0$ for all $\alpha, \beta \in \Omega$. Fix $\beta \in \Omega$. Then we compute

$$\sum_{\alpha \in \Omega} T(\alpha, \beta) = \sum_{\alpha \in \Omega} \sum_{\{m \in M \mid m\beta = \alpha\}} P(m) = \sum_{m \in M} P(m) = 1$$

because, for each $m \in M$, there is exactly one $\alpha \in \Omega$ with $\alpha = m\beta$. We conclude that (T, Ω) is a Markov chain. $\qquad\square$

It is worth noting that every Markov chain (T, Ω) can be obtained as a random walk of the full transformation monoid T_Ω on Ω with respect to an appropriate probability on T_Ω. This is a reformulation of the fact that the extreme points of the polytope of stochastic matrices are the column monomial matrices. We do not give a proof here, but refer the reader instead to [ASST15b, Theorem 2.3]. It is important to note that a Markov chain can be realized as a group random walk if and only if it has a uniform stationary distribution by the Birkhoff-von Neumann theorem [Zie95], which states that the permutation matrices are the extreme points of the polytope of doubly stochastic matrices. Thus monoid theory has a wider potential for applications.

If Ω is a left M-set, then $\mathbb{C}\Omega$ is a $\mathbb{C}M$-module as per Chapter 13. If we identify a probability P on M with the element

$$\sum_{m \in M} P(m)m$$

of $\mathbb{C}M$, then P is an operator on $\mathbb{C}\Omega$. It turns out that the transition matrix T of the random walk of M on Ω driven by P is the matrix of the operator P with respect to the basis Ω. This is what underlies the applications of representation theory to Markov chains: $\mathbb{C}M$-submodules of $\mathbb{C}\Omega$ are T-invariant subspaces.

Proposition 14.2. *Let P be a probability on M and Ω a left M-set. Then the transition matrix T of the random walk of M on Ω driven by P is the matrix of $P = \sum_{m \in M} P(m)m$ acting on $\mathbb{C}\Omega$ with respect to the basis Ω.*

Proof. If $\beta \in \Omega$, then we compute

$$P\beta = \sum_{m \in M} P(m)m\beta = \sum_{\alpha \in \Omega} \sum_{\{m \in M | m\beta = \alpha\}} P(m)\alpha = \sum_{\alpha \in \Omega} T(\alpha, \beta)\alpha$$

as required.   □

14.3 Examples

In this section we describe several famous Markov chains and discuss how to model them as monoid random walks. In Section 14.6, we shall reprise these examples and perform a detailed analysis using the theory that we shall develop. The examples in this section will all be left regular band walks. The theory of left regular band walks was developed by Brown [Bro00, Bro04], following the groundbreaking work of Bidigare, Hanlon, and Rockmore [BHR99].

We recall that a left regular band is an $\mathscr{R}$-trivial monoid M in which each element is regular. In this case each element of M is idempotent and

$$\Lambda(M) = \{MmM \mid m \in M\} = \{Mm \mid m \in M\}$$

is a lattice with intersection as the meet. The mapping $\sigma \colon M \longrightarrow \Lambda(M)$ given by $\sigma(m) = MmM$ is a monoid homomorphism. See Proposition 2.9 for details.

14.3.1 The Tsetlin library

The *Tsetlin library* is a well-studied Markov chain [Hen72, Pha91, DF95, FH96, Fil96, BHR99]. This was the first Markov chain analyzed via left regular band techniques. Imagine that you have a shelf of books that you wish to keep organized, but no time to do it. A simple self-organizing system is to always replace a book at the front of the shelf after using it. In this way, over time, your favorite books will be at the front of the shelf and your least favorite books at the back. Therefore, if you know the relative frequency with which you use a book, then you will know roughly where to find it.

Formally speaking, one has a set B (of books) and a probability P on B. The states of the Markov chain are the $|B|!$ linear orderings of the set of books B, i.e., the set of possible orderings of the books on the shelf. At each step of the chain, a book b is chosen with probability $P(b)$ and placed at the front of the shelf.

Let us remark that this Markov chain has a card shuffling interpretation in the special case that P is the uniform distribution, that is, $P(b) = 1/|B|$ for all $b \in B$. One can then think of B as the set of cards in a deck and the state space as the set of all possible orderings of the cards. One shuffles the deck by randomly choosing a card and placing it at the top. This is sometimes called the *random-to-top shuffle*. It is the time reversal of the *top-to-random shuffle* where you shuffle a deck by inserting the top card into a random position. The transition matrix of the top-to-random shuffle is the transpose of that of the random-to-top shuffle and these chains have the same eigenvalues and mixing time. See [LPW09, Section 4.6] for details.

Let us now model the Tsetlin library as a left regular band random walk. To do this we shall introduce the *free left regular band* $F(B)$ on the set B. Let us begin by motivating the construction. The left regular band identity $xyx = xy$ says that one can always remove repetitions from any product. Since the free monoid B^* consists of all words over the alphabet B, including the empty one, it follows that the free left regular band should consist of all words $w \in B^*$ with no repeated letters. So let $F(B)$ be the set of repetition-free words over the alphabet B. A product on $F(B)$ is given by $u \odot v = \overline{uv}$ where if $w \in B^*$, then $\overline{w}$ is the result of removing all repetitions of letters as you scan w from left to right. For example, if $B = \{1, 2, 3, 4\}$, then $3 \odot 1342 = 3142$. Notice that left multiplying a word containing all the letters by a generator moves that generator to the front. We shall prove that $F(B)$ is a left regular band; its freeness will be left to the exercises. Associativity of this multiplication is a straightforward exercise.

Let $\mathscr{P}(B)$ be the power set of B, viewed as a monoid via the operation of union. In other words, we view $\mathscr{P}(B)$ as a lattice with respect to the ordering $\supseteq$. Then there is a monoid homomorphism $c \colon F(B) \longrightarrow \mathscr{P}(B)$ sending a word w to its set $c(w)$ of letters, called the *content* of w. For example, $c(312) = \{1, 2, 3\}$. Let $\Omega_B = \{w \in F(B) \mid c(w) = B\}$ be the set of words with full content.

Proposition 14.3. *Let B be a finite set.*

(i) $uF(B) \subseteq vF(B)$ if and only if v is a prefix of u.

(ii) $F(B)u \subseteq F(B)v$ if and only if $c(u) \supseteq c(v)$.

Proof. If v is a prefix of u, then $u = vx = \overline{vx}$ in B^* and hence $u = v \odot x$. Conversely, if $v \odot x = u$, then since v has no repeated letters, $u = \overline{vx} = vx'$ where x' is obtained from x by deleting some letters. This establishes (i).

To prove (ii), if $u = x \odot v$, then $c(u) = c(x) \cup c(v)$ and hence $c(u) \supseteq c(v)$. Conversely, if $c(u) \supseteq c(v)$, then $u \odot v = \overline{uv} = u$ because each letter of v is a repetition of a letter occurring in u. Thus $F(B)u \subseteq F(B)v$, as required. □

As a consequence we deduce that $F(B)$ is indeed a left regular band, and more.

Corollary 14.4. *Let B be a finite set. Then $F(B)$ is a left regular band and $(\Lambda(F(B)), \subseteq) \cong (\mathscr{P}(B), \supseteq)$ via $F(B)uF(B) \mapsto c(u)$. Consequently, Ω_B is the minimal ideal of $F(B)$.*

Proof. The first item of Proposition 14.3 implies that $F(B)$ is $\mathscr{R}$-trivial. Trivially, $u \odot u = \overline{uu} = u$ and so $F(B)$ is a left regular band. The identification of $\Lambda(F(B))$ with $\mathscr{P}(B)$ is immediate from the second item of Proposition 14.3 because $MmM = Mm$ in a left regular band M. The final statement follows from this identification.   □

We now verify that the Tsetlin library is the random walk of $F(B)$ on Ω_B driven by P.

Proposition 14.5. *The Tsetlin library Markov chain with set of books B and probability P on B is the random walk of $F(B)$ on Ω_B driven by P.*

Proof. Clearly, we can identify Ω_B with all linear orderings of B. The action of a generator $b \in B$ on a word $w \in \Omega_B$ is given by $b \odot w = \overline{bw}$. But $\overline{bw}$ is obtained from w by placing b at the front and removing the occurrence of b in w. The proposition then follows.   □

14.3.2 The inverse riffle shuffle

The standard way of shuffling a deck of cards, by dividing it roughly in half and then interleaving the two halves, is called *riffle shuffling*. The Gilbert-Shannon-Reeds model is a classical Markov chain model of this process (cf. [LPW09, Section 8.3]). In this model, the number of cards in the top half of the deck (after dividing in two) is binomially distributed and then all interleavings are considered equally likely. The time reversal of the riffle shuffle is the *inverse riffle shuffle*. Its transition matrix is the transpose of the transition matrix for the riffle shuffle and they have the same eigenvalues and mixing time. In fact, Bayer and Diaconis used the inverse riffle shuffle to analyze the mixing time of the riffle shuffle in their famous paper [BD92], which established the "7-shuffle" rule of thumb. Let us describe the inverse riffle shuffle in detail since it is a left regular band random walk [BHR99, Bro04].

One has a deck of n cards. The state space is the set of all $n!$ possible orderings of the cards. One shuffles by randomly choosing a subset A of cards (with all subsets equally likely) and moving the cards from A to the front while maintaining the relative ordering of the cards in A (and its complement). Note that if either $A = [n]$ or $A = \emptyset$, the result is to stay at the current state. You should think of A as the set of cards which formed the top half of the deck before interleaving to see this process as performing the riffle shuffle backwards in time. For example, if there are 4 cards in the order 1432 and we choose the subset $A = \{2, 4\}$, then the new ordering of the deck is 4213.

To analyze this Markov chain, we use a combinatorial model of the hyperplane face monoid used in Bidigare, Hanlon, and Rockmore [BHR99]. Let Σ_n be the set of all ordered set partitions of $[n] = \{1, \ldots, n\}$. In other words, an element of Σ_n is an r-tuple $(P_1, \ldots, P_r)$ of subsets of $[n]$ such that $\{P_1, \ldots, P_r\}$ is a set partition. We define a product by putting

$$(P_1,\ldots,P_r)(Q_1,\ldots,Q_s) = (P_1\cap Q_1,\ldots,P_1\cap Q_s,\cdots,P_r\cap Q_1,\ldots,P_r\cap Q_s)^\wedge$$

where if $R_1,\ldots,R_t$ are subsets of $[n]$, then $(R_1,\ldots,R_t)^\wedge$ is the result of removing all empty R_i from the list. For example, if $n=4$, then

$$(\{2,4\},\{1,3\})(\{1\},\{4\},\{3\},\{2\}) = (\{4\},\{2\},\{1\},\{3\}). \tag{14.1}$$

The reader should check that Σ_n is a monoid with identity $([n])$. Let Ω_n be the set of all ordered set partitions of $[n]$ into singletons. Notice that Ω_n can be identified with the set of linear orderings of $[n]$. The computation (14.1) indicates how the inverse riffle shuffle can be modeled as a random walk of Σ_n on Ω_n.

Let Π_n denote the lattice of set partitions of $[n]$. As usual, a partition $\{P_1,\ldots,P_r\}$ is smaller than a partition $\{Q_1,\ldots,Q_s\}$ if, for each i, there exists j (necessarily unique) with $P_i\subseteq Q_j$. The meet $\{P_1,\ldots,P_r\}\wedge\{Q_1,\ldots,Q_s\}$ is the partition whose blocks consist of all nonempty intersections of the form $P_i\cap Q_j$ with $1\le i\le r$, $1\le j\le s$.

We can define $c\colon \Sigma_n\longrightarrow\Pi_n$ by $c((P_1,\ldots,P_r))=\{P_1,\ldots,P_r\}$. From the above description of the meet in Π_n, it is immediate that c is a homomorphism.

Lemma 14.6. *Let $(P_1,\ldots,P_r)$ and $(Q_1,\ldots,Q_s)$ be elements of Σ_n. Then*

$$(P_1,\ldots,P_r)(Q_1,\ldots,Q_s) = (P_1,\ldots,P_r) \tag{14.2}$$

if and only if $c((P_1,\ldots,P_r))\le c((Q_1,\ldots,Q_s))$.

Proof. If (14.2) holds, then $c((P_1,\ldots,P_r))\wedge c((Q_1,\ldots,Q_s))=c((P_1,\ldots,P_r))$, giving the desired inequality. Conversely, if $c((P_1,\ldots,P_r))\le c((Q_1,\ldots,Q_s))$, then, for each i, either $P_i\cap Q_j=\emptyset$ or $P_i\cap Q_j=P_i$ and the second equality holds for exactly one index j. With this in mind, we easily deduce (14.2) from the definition of the product in Σ_n. □

We obtain as a consequence that Σ_n is a left regular band.

Corollary 14.7. *Let $n\ge 1$.*

(i) Σ_n is a left regular band.
(ii) One has

$$\Sigma_n(P_1,\ldots,P_r)\subseteq\Sigma_n(Q_1,\ldots,Q_s)$$

if and only if $c((P_1,\ldots,P_r))\le c((Q_1,\ldots,Q_s))$.
(iii) The map c induces an isomorphism $MmM\mapsto c(m)$ of $\Lambda(\Sigma_n)$ and Π_n.
(iv) Ω_n is the minimal ideal of Σ_n.

Proof. For (i) we check the identity $xyx=xy$ from Lemma 2.8. Since c is a homomorphism, it is obvious that if $x,y\in\Sigma_n$, then $c(xy)\le c(x)$. Therefore, $xyx=xy$ by Lemma 14.6. This proves (i). Lemma 14.6 is equivalent to (ii) because Σ_n is a left regular band. Since $MmM=Mm$ in a left regular band, (iii) follows from (ii). Finally, (iv) may be deduced from (iii) because the partition into singletons is the smallest element of Π_n. □

We are now prepared to prove that the inverse riffle shuffle is a random walk of Σ_n on Ω_n.

Proposition 14.8. *The inverse riffle shuffle Markov chain for a deck of n cards is the random walk of Σ_n on Ω_n driven by the probability P given by*

$$P((Q_1, \ldots, Q_s)) = \begin{cases} \frac{1}{2^{n-1}}, & \text{if } s = 1, \\ \frac{1}{2^n}, & \text{if } s = 2 \\ 0, & \text{else} \end{cases}$$

for $(Q_1, \ldots, Q_s) \in \Sigma_n$.

Proof. As was mentioned earlier, we can identify linear orderings of $[n]$ with Ω_n via

$$i_1 \cdots i_n \longmapsto (\{i_1\}, \ldots, \{i_n\}).$$

The random walk of Σ_n on Ω_n driven by P can be described as follows. If we are in state $(\{i_1\}, \ldots, \{i_n\})$, we choose $A \subseteq [n]$ with probability $1/2^n$ and we move to state

$$(A, B)^{\wedge}(\{i_1\}, \ldots, \{i_n\}) = (A \cap \{i_1\}, \ldots, A \cap \{i_n\}, B \cap \{i_1\}, \ldots, B \cap \{i_n\})^{\wedge}$$

where $B = [n] \setminus A$ (here we have used $(\emptyset, [n])^{\wedge} = ([n]) = ([n], \emptyset)^{\wedge}$). But the right-hand side of the above equation is the ordered partition into singletons in which the elements of A appear before the elements of B and within A and B are ordered as per $(\{i_1\}, \ldots, \{i_n\})$. This is exactly the result of moving the cards from A to the front and maintaining the relative ordering in A and B. The proposition follows. $\qquad \square$

14.3.3 Ehrenfest urn model

We consider here a variant of the *Ehrenfest urn model*. In this model there are n balls and two urns: A and B. In one of the standard versions of the model you randomly pick a ball and an urn and move that ball to the chosen urn. The state space is then $\{A, B\}^n$ where an X in position i means ball i is in urn X.

Alternatively, one can view this as the lazy hypercube random walk. Namely, the hypercube graph is the graph with vertex set $\{A, B\}^n$ in which two vertices are connected if they differ in a single entry. In the lazy random walk the state space consists of the vertices. With probability $1/2$ the state does not change; otherwise, the state changes to a random neighbor.

In fact, it is not any more difficult to work with the following more general model. We have a probability Q on $\{A, B\} \times [n]$. The state space is $\{A, B\}^n$. With probability $Q(X, i)$ we place ball i in urn X. The original Ehrenfest urn model is the case where $Q(X, i) = 1/2n$ for all $(X, i) \in \{A, B\} \times [n]$.

Let $L = \{1, A, B\}$ with the product defined so that 1 is the identity and $XY = X$ for $X, Y \in \{A, B\}$. Trivially, L is a left regular band and hence so is the direct product L^n. In fact, $\{A, B\}^n$ is the minimal ideal of L^n. To see this, define $c\colon L^n \longrightarrow \mathscr{P}([n])$ by

$$c(x_1, \ldots, x_n) = \{i \mid x_i \neq 1\}.$$

One easily checks that c is a homomorphism, where again $\mathscr{P}([n])$ is made a monoid via union, and hence ordered by reverse inclusion.

Proposition 14.9. *Let $n \geq 1$.*

(i) One has

$$L^n(x_1, \ldots, x_n) \subseteq L^n(y_1, \ldots, y_n)$$

  if and only if $c(x_1, \ldots, x_n) \supseteq c(y_1, \ldots, y_n)$.
(ii) $\Lambda(L^n) \cong \mathscr{P}([n])$ via $L^n m L^n \mapsto c(m)$.
(iii) $\{A, B\}^n$ is the minimal ideal of L^n.

Proof. Recall that for a left regular band M, one has that $Mm \subseteq Mn$ if and only if $mn = m$. We may then deduce (i) as follows. If $x_i = 1$, then $x_i y_i = x_i$ if and only if $y_i = 1$. If $x_i \neq 1$, then $x_i y_i = x_i$ holds independently of the value of y_i. Thus $xy = x$ if and only if $c(x) \supseteq c(y)$. The remaining items are immediate from the first. □

We are now prepared to realize the Ehrenfest urn model as a random walk of L^n on $\{A, B\}^n$.

Proposition 14.10. *The Ehrenfest urn Markov chain with n balls and probability Q on $\{A, B\} \times [n]$ is the random walk of L^n on $\{A, B\}^n$ driven by the probability P given by*

$$P(x_1, \ldots, x_n) = \begin{cases} Q(A, i), & \text{if } c(x_1, \ldots, x_n) = \{i\}, \ x_i = A \\ Q(B, i), & \text{if } c(x_1, \ldots, x_n) = \{i\}, \ x_i = B \\ 0, & \text{else} \end{cases}$$

for $(x_1, \ldots, x_n) \in L^n$.

Proof. For $(X, i) \in \{A, B\} \times [n]$, let $e_{X,i}$ be the element of L^n with X in position i and 1 in all other positions. If we are in state $(y_1, \ldots, y_n)$, then with probability $Q(X, i)$ we move to state $e_{X,i} \cdot (y_1, \ldots, y_n)$. But $e_{X,i} \cdot (y_1, \ldots, y_n)$ is obtained from $(y_1, \ldots, y_n)$ by replacing y_i by X. Thus we have recovered the Ehrenfest urn Markov chain. □

14.4 Eigenvalues

We return now to the development of the general theory. In this section, we compute the eigenvalues of the transition matrix of a random walk of a triangularizable monoid on a set. Recall that M is triangularizable over $\mathbb{C}$ if each simple $\mathbb{C}M$-module is one-dimensional. The following result was first stated in [Ste08, Section 8.3].

Theorem 14.11. *Let M be a triangularizable monoid over $\mathbb{C}$, Ω a left M-set and P a probability on M. Let $\chi_1, \ldots, \chi_s$ be the irreducible characters of M and let m_i be the multiplicity of the simple module affording χ_i as a composition factor of $\mathbb{C}\Omega$. Then the transition matrix T for the random walk of M on Ω driven by P has eigenvalues $\lambda_1, \ldots, \lambda_s$ where*

$$\lambda_i = \sum_{m \in M} P(m)\chi_i(m)$$

and λ_i has multiplicity m_i.

Proof. Since M is triangularizable each representation of M is equivalent to one by upper triangular matrices, as observed after Proposition 10.8. That is, there is a basis $\{v_1, \ldots, v_n\}$ for $\mathbb{C}\Omega$ such that $V_i = \mathbb{C}\{v_1, \ldots, v_i\}$ is a $\mathbb{C}M$-submodule, for $1 \leq i \leq n$, and

$$\mathbb{C}\Omega = V_n \supseteq V_{n-1} \supseteq \cdots \supseteq V_0 = 0$$

is a composition series.

Let θ_i be the character afforded by the one-dimensional simple module V_i/V_{i-1} for $i = 1, \ldots, n$. Then with respect to the basis $\{v_1, \ldots, v_n\}$, we have that $\mathbb{C}\Omega$ affords a representation of the form $\rho \colon M \longrightarrow M_n(\mathbb{C})$ with

$$\rho(m) = \begin{bmatrix} \theta_1(m) & * & \cdots & * \\ 0 & \theta_2(m) & \ddots & \vdots \\ \vdots & & \ddots & \ddots & * \\ 0 & \cdots & 0 & \theta_n(m) \end{bmatrix}.$$

Since T is the matrix of the operator P on $\mathbb{C}\Omega$ with respect to the basis Ω by Proposition 14.2, it follows that T is similar to the matrix

$$T' = \sum_{m \in M} P(m)\rho(m)$$

$$= \begin{bmatrix} \sum_{m \in M} P(m)\theta_1(m) & * & \cdots & * \\ 0 & \sum_{m \in M} P(m)\theta_2(m) & \ddots & \vdots \\ \vdots & & \ddots & * \\ 0 & & \cdots & 0 \;\; \sum_{m \in M} P(m)\theta_n(m) \end{bmatrix}.$$

Since χ_i appears among the θ_j exactly m_i times, the result follows. $\square$

We note that $m_i = 0$ and $\lambda_i = \lambda_j$ for $i \neq j$ can occur. Since triangularizable monoids are rectangular with abelian maximal subgroups by Theorem 11.16, the characters of a triangularizable monoid M can be explicitly computed using Proposition 11.17 and the character theory of abelian groups. Moreover, the multiplicities can be obtained by inverting the character table.

Alternatively, one can use that the semisimple quotient of M is the algebra of a commutative inverse monoid and apply Theorem 9.9 to compute multiplicities. Details can be found in [Ste08, Section 8.3] or Exercise 14.13. Let us specialize to the case of an $\mathscr{R}$-trivial monoid, where we can be more explicit. This result is a special case of results from [Ste06]. The reader should consult Appendix C for the definition of the Möbius function of a poset.

Theorem 14.12. *Let M be an $\mathscr{R}$-trivial monoid, Ω a left M-set, and P a probability on M. Let $\Lambda(M) = \{MeM \mid e \in E(M)\}$. Then the transition matrix T for the random walk of M on Ω driven by P has an eigenvalue*

$$\lambda_{MeM} = \sum_{\{m \in M \mid MmM \supseteq MeM\}} P(m) \qquad (14.3)$$

for each $MeM \in \Lambda(M)$. The multiplicity of λ_{MeM} is given by

$$m_{MeM} = \sum_{\{MfM \in \Lambda(M) \mid MfM \subseteq MeM\}} |f\Omega| \cdot \mu(MfM, MeM) \qquad (14.4)$$

where μ is the Möbius function of the lattice $\Lambda(M)$. In particular, each eigenvalue of T is nonnegative.

Proof. First of all, $\mathscr{R}$-trivial monoids are triangularizable by Corollary 10.9. The characters of M are indexed by $\Lambda(M)$ and given by

$$\chi_{MeM}(m) = \begin{cases} 1, & \text{if } MmM \supseteq MeM \\ 0, & \text{else} \end{cases}$$

by Corollary 5.7. Thus the eigenvalue λ_{MeM} associated with χ_{MeM} according to Theorem 14.11 is as per (14.3).

The character θ afforded by $\mathbb{C}\Omega$ is given by

$$\theta(m) = |\{\alpha \in \Omega \mid m\alpha = \alpha\}|.$$

In particular, if $f \in E(M)$, then $\theta(f) = |f\Omega|$. We can then deduce (14.4) from Corollary 7.21. □

Theorem 14.12 has so far been the principal tool in computing eigenvalues of transition matrices of Markov chains using monoids that are not groups.

14.5 Diagonalizability

In this section, we give a proof of Brown's theorem on the diagonalizability of the transition matrix of a left regular band random walk [Bro00]. The treatment here is from the author's unpublished note [Ste10a]; a more sophisticated version of this argument, which works for some more general classes of $\mathscr{R}$-trivial monoids, appears in [ASST15b, Theorem 4.3].

Let M be a left regular band, that is, a monoid satisfying the identity $xyx = xy$. Then there is a homomorphism $\sigma\colon M \longrightarrow \Lambda(M)$ where $\Lambda(M)$ is a lattice such that $\sigma(m) \leq \sigma(n)$ if and only if $mn = m$. Concretely, one can take $\Lambda(M) = \{MmM \mid m \in M\}$ to be the lattice of principal ideals and $\sigma\colon M \longrightarrow \Lambda(M)$ to be given by $\sigma(m) = MmM$ where $\Lambda(M)$ is equipped with the meet operation (which turns out to be intersection for a left regular band, although we do not use this fact here). The property that $\sigma(m) \leq \sigma(n)$ if and only if $mn = m$ follows from Corollary 2.7 since each $m \in M$ is idempotent.

Brown's diagonalizability theorem is, in fact, about elements of left regular band algebras over any field. For the case of probability distributions it suffices to work over $\mathbb{R}$, but there is no added difficulty in tackling the general case. So let $\Bbbk$ be a field and let

$$w = \sum_{m \in M} w_m m \in \Bbbk M. \tag{14.5}$$

For $X \in \Lambda(M)$, define

$$\lambda_X = \sum_{\sigma(m) \geq X} w_m. \tag{14.6}$$

Brown [Bro00,Bro04] showed that the commutative algebra $\Bbbk[w]$ is *split semisimple* (i.e., the minimal polynomial of w splits over $\Bbbk$ into distinct linear factors) provided that $X > Y$ implies $\lambda_X \neq \lambda_Y$. We prove this by showing that if $\lambda_1, \ldots, \lambda_k$ are the distinct elements of $\{\lambda_X \mid X \in \Lambda(M)\}$, then

$$0 = \prod_{i=1}^{k}(w - \lambda_i). \tag{14.7}$$

This immediately implies that the minimal polynomial of w has distinct roots and hence $\Bbbk[w]$ is split semisimple. Everything is based on the following formula for mw. So let us assume from now on that $X > Y$ implies that $\lambda_X \neq \lambda_Y$.

Lemma 14.13. *Let $m \in M$. Then*

$$mw = \lambda_{\sigma(m)}m + \sum_{\sigma(n) \not\geq \sigma(m)} w_n mn$$

and moreover, $\sigma(m) > \sigma(mn)$ for all n with $\sigma(n) \not\geq \sigma(m)$.

Proof. Using that $\sigma(n) \geq \sigma(m)$ implies $mn = m$, we compute

$$mw = \sum_{\sigma(n) \geq \sigma(m)} w_n mn + \sum_{\sigma(n) \not\geq \sigma(m)} w_n mn$$

$$= \sum_{\sigma(n) \geq \sigma(m)} w_n m + \sum_{\sigma(n) \not\geq \sigma(m)} w_n mn$$

$$= \lambda_{\sigma(m)}m + \sum_{\sigma(n) \not\geq \sigma(m)} w_n mn.$$

It remains to observe that $\sigma(n) \not\geq \sigma(m)$ implies that $\sigma(mn) = \sigma(m)\sigma(n) < \sigma(m)$.   □

The proof of (14.7) proceeds via an induction along $\Lambda(M)$. Let us write $\widehat{1}$ for the maximum element of $\Lambda(M)$. If $X \in \Lambda(M)$, put

$$\Phi_X = \{\lambda_Y \mid Y \leq X\} \text{ and } \Phi'_X = \{\lambda_Y \mid Y < X\}.$$

Our hypothesis says exactly that $\Phi_X = \{\lambda_X\} \uplus \Phi'_X$ (disjoint union). Define polynomials $p_X(z)$ and $q_X(z)$, for $X \in \Lambda(M)$, by

$$p_X(z) = \prod_{\lambda_i \in \Phi_X} (z - \lambda_i)$$

$$q_X(z) = \prod_{\lambda_i \in \Phi'_X} (z - \lambda_i) = \frac{p_X(z)}{z - \lambda_X}.$$

Notice that, for $X > Y$, we have $\Phi_Y \subseteq \Phi'_X$, and hence $p_Y(z)$ divides $q_X(z)$, because $\lambda_X \notin \Phi_Y$ by assumption. Also observe that

$$p_{\widehat{1}}(z) = \prod_{i=1}^{k}(z - \lambda_i)$$

and hence establishing (14.7) is equivalent to proving $p_{\widehat{1}}(w) = 0$.

Lemma 14.14. *If $m \in M$, then $m \cdot p_{\sigma(m)}(w) = 0$.*

Proof. The proof is by induction on $\sigma(m)$ in the lattice $\Lambda(M)$. Assume that the lemma holds for all $m' \in M$ with $\sigma(m') < \sigma(m)$. Then by Lemma 14.13

$$m \cdot p_{\sigma(m)}(w) = m \cdot (w - \lambda_{\sigma(m)}) \cdot q_{\sigma(m)}(w) = \sum_{\sigma(n) \not\geq \sigma(m)} w_n mn \cdot q_{\sigma(m)}(w) = 0.$$

The last equality follows because $\sigma(n) \not\geq \sigma(m)$ implies $\sigma(m) > \sigma(mn)$ and so $p_{\sigma(mn)}(z)$ divides $q_{\sigma(m)}(z)$, whence induction yields $mn \cdot q_{\sigma(m)}(w) = 0$. $\square$

Applying the lemma with m the identity element of M yields $p_{\widehat{1}}(w) = 0$ and hence we have proved the following theorem.

Theorem 14.15. *Let M be a left regular band and $\Bbbk$ a field. Suppose that $w \in \Bbbk M$ is as in (14.5) and λ_X is as in (14.6) for $X \in \Lambda(M)$. If $X > Y$ implies $\lambda_X \neq \lambda_Y$, then $\Bbbk[w]$ is split semisimple, that is, the minimal polynomial of w splits into distinct linear factors over $\Bbbk$.*

As a corollary, we may deduce that transition matrices of left regular band random walks are diagonalizable.

Theorem 14.16. *Let M be a left regular band, Ω a left M-set, and P a probability on M. Then the transition matrix T of the random walk of M on Ω driven by P is diagonalizable over $\mathbb{R}$.*

Proof. First note that if we replace M by the submonoid generated by the support of P, we obtain the same Markov chain (T, Ω). Thus we may assume without loss of generality that the support of P generates M. The transition matrix T is the matrix of the operator $P = \sum_{m \in M} P(m)m$ on $\mathbb{R}\Omega$ with respect to the basis Ω by Proposition 14.2. Therefore, the algebra $\mathbb{R}[T]$ is a homomorphic image of the algebra $\mathbb{R}[P]$ via a map sending P to T and hence the minimal polynomial of T divides the minimal polynomial of P. It thus suffices to prove that $\mathbb{R}[P]$ is split semisimple. We shall apply the criterion of Theorem 14.15.

Suppose that $X, Y \in \Lambda(M)$ with $X > Y$. Then we have that

$$\lambda_Y = \sum_{\sigma(m) \geq Y} P(m) \geq \sum_{\sigma(m) \geq X} P(m) = \lambda_X.$$

Moreover, the inequality is strict because M is generated by the support of P, $\{m \in M \mid \sigma(m) \geq X\}$ is a submonoid disjoint from $\sigma^{-1}(Y)$ and elements of $\sigma^{-1}(Y)$ can only be expressed as products of elements $m \in \operatorname{supp}(P)$ with $\sigma(m) \geq Y$. Thus there must exist $m \in M$ with $P(m) > 0$, $\sigma(m) \geq Y$ and $\sigma(m) \not\geq X$. Therefore, $X > Y$ implies $\lambda_Y > \lambda_X$ and so $\mathbb{R}[P]$ is split semisimple by Theorem 14.15. □

14.6 Examples: revisited

It is now time to apply the results of the previous sections to the Markov chains from Section 14.3 in order to provide some idea of their scope. We shall retain throughout the notation from Section 14.3 and it is suggested that the reader review that section before reading this one.

14.6.1 The Tsetlin library

Let B be a finite set (of books). We identify $\Lambda(F(B))$ with $(\mathscr{P}(B), \supseteq)$ via the mapping c as per Section 14.3.1. One then has that $\sigma(w) \geq MmM$ if and only if $c(w) \subseteq c(m)$.

To describe the multiplicities of the eigenvalues of the transition matrix for the Tsetlin library Markov chain, we need to recall the notion of derangement numbers. A *derangement* is a fixed-point-free permutation. Let d_n be the number of derangements of an n-element set. Note that $d_0 = 1$ and $d_1 = 0$.

Theorem 14.17. *Let T be the transition matrix of the Tsetlin library Markov chain with set of books B and probability P on B. Then there is an eigenvalue*

$$\lambda_X = \sum_{b \in X} P(b)$$

of T for each $X \subseteq B$. The multiplicity of λ_X is given by the derangement number $d_{|B|-|X|}$. Moreover, T is diagonalizable.

Proof. We view (T, Ω) as the random walk of $F(B)$ on Ω_B driven by P. One has that $c(b) \subseteq X$ if and only if $b \in X$ for $X \subseteq B$. Thus the eigenvalue associated with the element of $\Lambda(F(B))$ corresponding to X is λ_X by Theorem 14.12. It remains to compute the multiplicity of λ_X.

Let $|B| = n$. Fix a word w_X with content X for each $X \subseteq B$. Note that $w_X \Omega_B$ consists of all words of content B with w_X as a prefix. Hence $|w_X \Omega_B| = (n - |X|)! = |S_{B \setminus X}|$. We can identify $S_{B \setminus X}$ with the subgroup G_X of S_n fixing X pointwise. If we count elements of G_X by their fixed-point sets, then we obtain

$$|w_X \Omega_B| = |G_X| = \sum_{Y \supseteq X} d_{n - |Y|}$$

for all X. Applying Möbius inversion (Theorem C.6), we conclude that

$$d_{n - |X|} = \sum_{Y \supseteq X} |w_Y \Omega_B| \cdot \mu(Y, X) \tag{14.8}$$

where μ is the Möbius function of $(\mathscr{P}(B), \supseteq)$. But the right-hand side of (14.8) is the multiplicity of λ_X by Theorem 14.12.

The diagonalizability of T follows from Theorem 14.16. This completes the proof.   □

Specializing to the case where P is the uniform distribution, and hence the Tsetlin library chain is the time reversal of the top-to-random shuffle, we obtain the following result.

Corollary 14.18. *Let T be the transition matrix of the top-to-random shuffle Markov chain for a deck of n cards. Then the eigenvalues for T are $\lambda_j = j/n$ with $j = 0, \ldots, n$. The multiplicity of λ_j is $\binom{n}{j} d_{n-j}$. Furthermore, T is diagonalizable.*

Proof. We retain the notation of Theorem 14.17 with $B = [n]$ and $P(i) = 1/n$ for all $i \in [n]$. Note that T is the transpose of the transition matrix of the Tsetlin library with these parameters. If $X \subseteq [n]$, then $\lambda_X = |X|/n = \lambda_{|X|}$. Since there are $\binom{n}{j}$ subsets of size j, the result follows.   □

We remark that since $d_1 = 0$, it follows from Corollary 14.18 that $(n-1)/n$ is not actually an eigenvalue of the top-to-random shuffle (i.e., it occurs with multiplicity zero).

Remark 14.19. One can model the top-to-random shuffle as a symmetric group random walk. From that point of view, it is not at all obvious, *a priori*, that the eigenvalues should be nonnegative, while this is immediate from the monoid viewpoint.

14.6.2 The inverse riffle shuffle

We identify $\Lambda(\Sigma_n)$ with Π_n as in Section 14.3.2. We shall need the following lemma. Let us fix some notation. If $\pi \in \Pi_n$, then let O_π be the set of all permutations in S_n with orbit partition π of $[n]$.

Lemma 14.20. *Let $\pi = \{P_1, \ldots, P_r\}$ be a set partition of $[n]$. Put $\pi! = |P_1|! \cdots |P_r|!$. Then the equality*

$$\pi! = \sum_{\pi' \leq \pi} |O_{\pi'}|$$

holds.

Proof. The left-hand side is the cardinality of the subgroup

$$S_\pi \cong S_{P_1} \times \cdots \times S_{P_r}$$

of S_n that leaves the blocks of π invariant. The orbit partition of any element of S_π must refine π and hence the desired equality follows by counting elements of S_π by their orbit partition. $\qquad\Box$

Now we can prove the main result. We remind the reader that the transition matrix of the riffle shuffle is the transpose of the transition matrix of the inverse riffle shuffle and so they have the same eigenvalues and diagonalizability properties (cf. [BD92, LPW09]).

Theorem 14.21. *Let T be the transition matrix of the inverse riffle shuffle Markov chain for a deck of n cards. Then the eigenvalues of T are*

$$\lambda_k = \frac{1}{2^{n-k}}$$

with $1 \leq k \leq n$. The multiplicity of λ_k is the number of permutations of $[n]$ with k orbits. Moreover, T is diagonalizable.

Proof. We use the probability P from Proposition 14.8. Let $\pi \in \Pi_n$ with $\pi = \{P_1, \ldots, P_k\}$ and $A \subseteq [n]$. Put $B = [n] \setminus A$. Then one has that $c((A, B)^\wedge) \geq \pi$ if and only if A is the union of some subset of $\{P_1, \ldots, P_k\}$. Thus there are 2^k choices for A. It follows from Theorem 14.12 that the eigenvalue λ_π associated with the element of $\Lambda(\Sigma_n)$ corresponding to π is $2^k/2^n = 1/2^{n-k} = \lambda_k$. Notice that all set partitions with k blocks give the eigenvalue λ_k.

We now compute the multiplicity of λ_π for $\pi \in \Pi_n$. Fix a preimage $e_\pi \in \Sigma_n$ of each element $\pi \in \Pi_n$. We claim $|e_\pi \Omega_n| = \pi!$. Indeed, if $e_\pi = (P_1, \ldots, P_r)$, then $e_\pi \Omega_n$ can be identified with those linear orderings on $[n]$ in which the elements of P_1 appear before those of P_2, and so forth. Thus there are $|P_1|! \cdots |P_r|! = \pi!$ such elements. By Lemma 14.20 and Möbius inversion (Theorem C.6) we have that

$$|O_\pi| = \sum_{\pi' \leq \pi} \pi'! \cdot \mu(\pi', \pi) = \sum_{\pi' \leq \pi} |e_{\pi'} \Omega_n| \cdot \mu(\pi', \pi)$$

where μ is the Möbius function of Π_n. Therefore, λ_π has multiplicity $|O_\pi|$ by Theorem 14.12. As all partitions π with k blocks provide the same eigenvalue λ_k, we deduce that the multiplicity of λ_k is the number of permutations of $[n]$ with k orbits.

The diagonalizability of T is immediate from Theorem 14.16. This establishes the theorem. □

We remark that permutations with k orbits are those permutations with k cycles in their cycle decomposition, including cycles of length 1.

14.6.3 The Ehrenfest urn model

We identify $\Lambda(L^n)$ with $(\mathscr{P}([n]), \supseteq)$ via c as per Section 14.3.3.

Theorem 14.22. *Let T be the transition matrix for the Ehrenfest urn model with probability Q on $\{A, B\} \times [n]$. Then T has an eigenvalue*

$$\lambda_S = \sum_{i \in S} (Q(A, i) + Q(B, i))$$

for each $S \subseteq [n]$ with multiplicity one. Furthermore, T is diagonalizable.

Proof. Let P be the probability from Proposition 14.10. Put $\Omega = \{A, B\}^n$. As before, let $e_{X,i}$ be the element of L^n with $X \in \{A, B\}$ in position $i \in [n]$ and 1 in all other positions. Then $c(e_{X,i}) \subseteq S$ if and only if $i \in S$. Therefore, λ_S is the eigenvalue associated by Theorem 14.12 to the element of $\Lambda(L^n)$ corresponding to S.

In order to compute the multiplicity of λ_S, let $e_S = (x_1, \ldots, x_n)$ with

$$x_i = \begin{cases} A, & \text{if } i \in S \\ 1, & \text{else.} \end{cases}$$

Then $c(e_S) = S$ and we have that $e_S \Omega$ consists of all elements of Ω with A in each position belonging to S. Thus

$$|e_S \Omega| = 2^{n-|S|} = \sum_{T \supseteq S} 1.$$

Möbius inversion (Theorem C.6) then yields

$$1 = \sum_{T \supseteq S} |e_T \Omega| \cdot \mu(T, S)$$

where μ is the Möbius function of $\mathscr{P}([n])$. It follows that λ_S has multiplicity one by Theorem 14.12.

The diagonalizability of T again is a consequence of Theorem 14.16. □

Specializing to the case where Q is the uniform distribution, we obtain the following result for the classical Ehrenfest urn model or, equivalently, for the lazy hypercube random walk.

Corollary 14.23. *Let T be the transition matrix of the classical Ehrenfest urn model with n balls or, equivalently, of the lazy n-hypercube random walk. Then T has an eigenvalue $\lambda_k = k/n$ with multiplicity $\binom{n}{k}$ for $0 \leq k \leq n$. Moreover, T is diagonalizable.*

Proof. This follows easily from Theorem 14.22. In this case,

$$\lambda_S = \sum_{i \in S} \left(\frac{1}{2n} + \frac{1}{2n} \right) = \frac{|S|}{n} = \lambda_{|S|}.$$

As λ_S has multiplicity one and there are $\binom{n}{k}$ subsets of $[n]$ of cardinality k, we deduce that λ_k has multiplicity $\binom{n}{k}$, as required.   □

We remark that Corollary 14.23 can also be proved by modeling the lazy n-hypercube random walk as a random walk of $(\mathbb{Z}/2\mathbb{Z})^n$ on itself and using the character theory of abelian groups [Dia88, CSST08].

14.7 Exercises

14.1. Prove that if $T: \Omega \times \Omega \longrightarrow \mathbb{R}$ is the transition matrix of a Markov chain, then $T\nu$ is a probability for any probability ν on Ω.

14.2. Let M be a finite monoid acting transitively on a set Ω. Let P be a probability on M whose support generates M. Prove that the stationary distribution for the random walk of M on Ω driven by P is unique. (Hint: prove that the left eigenspace of the transition matrix associated with the eigenvalue 1 is one-dimensional.)

14.3. Compute the transition matrix for the Tsetlin library with 3 books with respect to the uniform distribution on the set of books.

14.4. Prove that if A is a set, M is a left regular band, and $\varphi: A \longrightarrow M$ is a mapping, then there.is a unique homomorphism $\Phi: F(A) \longrightarrow M$ such that $\Phi|_A = \varphi$, that is, $F(A)$ is, indeed, a free left regular band.

14.5. Prove that the nth-derangement number d_n is given by

$$d_n = n! \sum_{k=0}^{n} \frac{(-1)^k}{k!} = \sum_{k=0}^{n} (-1)^k \binom{n}{k} (n-k)!$$

for all $n \geq 0$.

14.6. Verify that Σ_n is a monoid.

14.7. A coupon collector wishes to collect all the coupons from a set B of coupons. Let P be a probability on B. The probability that the next coupon that he draws is $b \in B$ is $P(b)$. So the state space consists of all possible sets of coupons. If we are in state $X \subseteq B$, then with probability $P(b)$ we transition to state $X \cup \{b\}$.

(a) Model the coupon collector Markov chain as a random walk on $(\mathscr{P}(B), \cup)$.
(b) Compute the eigenvalues of the transition matrix.
(c) Prove that the transition matrix is diagonalizable.

14.8. Let B be a finite set (of say books) with $n \geq 1$ elements and let $k \geq 1$. Let P be a probability on B. Imagine that you have k indistinguishable copies of each book. The states of the Markov chain are all possible orderings of the nk books on a shelf. With probability $P(b)$ the last copy of book b on the shelf is moved to the front of the shelf. Model this Markov chain as a random walk of an $\mathscr{R}$-trivial monoid and compute the eigenvalues of the transition matrix with multiplicities.

14.9. Let G be a finite abelian group and P a probability on G. Let T be the transition matrix of the random walk of G on itself driven by P. Prove that there is an eigenvalue λ_χ of T for each character $\chi \colon G \longrightarrow \mathbb{C}$ given by

$$\lambda_\chi = \sum_{g \in G} P(g)\chi(g)$$

with multiplicity one and that T is diagonalizable.

14.10. Use Exercise 14.9 to give another proof of Corollary 14.23.

14.11 (D. Grinberg). Let M be a finite left regular band and $\Bbbk$ a field. Let $w \in \Bbbk M$ be as in (14.5) and define λ_X as in (14.6) for $X \in \Lambda(M)$. Let $p(z)$ denote the polynomial

$$\mathrm{lcm} \left\{ \prod_{i=1}^{r} (z - \lambda_{X_i}) \;\middle|\; X_1 < X_2 < \cdots < X_r \text{ is an increasing chain in } \Lambda(M) \right\}.$$

Prove that $p(w) = 0$.

14.12. Let $\mathcal{H} = \{H_1, \ldots, H_n\}$ be a finite set of linear hyperplanes in $\mathbb{R}^d$ and $f_1, \ldots, f_n$ linear forms on $\mathbb{R}^d$ with $H_i = \{x \in \mathbb{R}^d \mid f_i(x) = 0\}$. If $t \in \mathbb{R}$, put

$$\mathrm{sgn}(t) = \begin{cases} +, & \text{if } t > 0 \\ -, & \text{if } t < 0 \\ 0, & \text{if } t = 0. \end{cases}$$

Let $L = \{0, +, -\}$ with the monoid structure given by making 0 the identity and putting $xy = x$ for $x, y \in \{+, -\}$; this monoid is isomorphic to the one we called L previously and is a left regular band. Define $\sigma \colon \mathbb{R}^d \longrightarrow L^n$ by

$$\sigma(x) = (\mathrm{sgn}(f_1(x)), \ldots, \mathrm{sgn}(f_n(x))).$$

(a) Show that the image of σ is a submonoid of L^n, called the *hyperplane face monoid* $\mathcal{F}(\mathcal{H})$ associated with $\mathcal{H}$. (Hint: show that $\sigma(x)\sigma(y) = \sigma(z)$ where z is an element of the line segment from x to y sufficiently close to x.)

(b) Show that the fibers of σ are convex subsets of $\mathbb{R}^d$ (called *faces*).

(c) Show that $F, G \in \mathcal{F}(\mathcal{H})$ generate the same left ideal if and only if $\sigma^{-1}(F)$ and $\sigma^{-1}(G)$ span the same subspace of $\mathbb{R}^d$.

(d) The *intersection lattice* $\mathcal{L}(\mathcal{H})$ is the set of subspaces of $\mathbb{R}^d$ that are intersections of a (possibly empty) subset of $\mathcal{H}$, ordered by reverse inclusion. Prove that $\mathcal{L}(\mathcal{H}) \cong \Lambda(\mathcal{F}(\mathcal{H}))$.

(e) Show that the minimal ideal of $\mathcal{F}(\mathcal{H})$ consists of those elements of the form $\sigma(x)$ with $x \in \mathbb{R}^d \setminus \bigcup_{i=1}^n H_i$ and that the fibers of σ over the minimal ideal of $\mathcal{F}(\mathcal{H})$ are precisely the connected components (called *chambers*) of $\mathbb{R}^d \setminus \bigcup_{i=1}^n H_i$.

(f) Show that if $\mathcal{H}$ consists of the coordinate hyperplanes $H_1, \ldots, H_n$ in $\mathbb{R}^n$, where H_i is defined by the equation $x_i = 0$, then $\mathcal{F}(\mathcal{H}) \cong L^n$. This arrangement is called the *boolean arrangement*.

(g) Show that if $\mathcal{H}$ is the hyperplane arrangement in $\mathbb{R}^n$ given by the hyperplanes $H_{ij} = \{x \in \mathbb{R}^n \mid x_i = x_j\}$ for $1 \le i < j \le n$, then $\mathcal{F}(\mathcal{H}) \cong \Sigma_n$. This arrangement is called the *braid arrangement*. (Hint: correspond to each ordered set partition $(P_1, \ldots, P_r)$ the image under σ of those $x \in \mathbb{R}^d$ with $x_i = x_j$ if i, j are in the same block and $x_i < x_j$ if the block containing i precedes the block containing j.)

14.13. Let M be a finite monoid triangularizable over $\mathbb{C}$ and $e_1, \ldots, e_s$ form a complete set of idempotent representatives of the regular $\mathscr{J}$-classes of M. Let $\Lambda(M) = \{Me_iM \mid i = 1, \ldots, s\}$; it is a lattice when ordered by inclusion by Proposition 2.3, Theorem 2.4, and Proposition 11.14.

(a) Show that $N = \bigcup_{i=1}^s G_{e_i}$, equipped with the product $\odot$ defined as follows, is a commutative inverse monoid. If $g \in G_{e_i}$ and $h \in G_{e_j}$, then

$$g \odot h = e_k g h e_k = e_k g e_k g e_k$$

where $Me_k M = Me_i M \wedge Me_j M$ in $\Lambda(M)$. (Hint: you may find Exercise 11.5 useful.)

(b) Show that $\theta \colon M \longrightarrow N$ given by $\theta(m) = e_i m e_i$ where $Me_i M = Mm^\omega M$ is a surjective homomorphism.

(c) Prove that $\ker \theta$ is the Rhodes radical congruence $\equiv_{\mathbb{C}}$.

(d) Let P be a probability on M and Ω a finite left M-set. Let T be the transition matrix of the random walk of M on Ω driven by P and χ an irreducible character of the abelian group G_{e_i}.

(i) Prove that the eigenvalue of T associated with χ is

$$\lambda_\chi = \sum_{MmM \supseteq Me_i M} P(m)\chi(e_i me_i).$$

(ii) Prove that the multiplicity of λ_{χ_i} is given by

$$m_\chi = \sum_{g \in G_{e_i}} \chi(g) \sum_{Me_j M \subseteq Me_i M} |\mathrm{Fix}(e_j g e_j)| \cdot \mu(Me_j M, Me_i M)$$

where $\mathrm{Fix}(m)$ is the set of fixed points of m on Ω and μ is the Möbius function of the lattice $\Lambda(M)$. (Hint: use that every character of a $\mathbb{C}M$-module is the character of a semisimple $\mathbb{C}M$-module and that all completely reducible representations of M factor through θ; then apply Theorem 9.9.)

Part VII

Advanced Topics

15

Self-injective, Frobenius, and Symmetric Algebras

In this chapter we characterize regular monoids with a self-injective algebra. It turns out that the algebra of such a monoid is a product of matrix algebras over group algebras and hence is a symmetric algebra. The results of this chapter are new to the best of our knowledge.

15.1 Background on self-injective algebras

The reader is referred to [Ben98, Section 1.6] for the material of this section. Fix a field $\Bbbk$ and a finite dimensional $\Bbbk$-algebra A for the section. The algebra A is said to be *self-injective* if the regular A-module A is an injective module. This is equivalent to each projective module being injective and to each projective indecomposable module being injective. A part of [Ben98, Proposition 1.6.2] is that injective and projective modules coincide for a self-injective finite dimensional algebra A.

Proposition 15.1. *Let A be a finite dimensional self-injective $\Bbbk$-algebra. Then a finite dimensional A-module V is projective if and only if it is injective.*

The algebra A is said to be a *Frobenius algebra* if there is a linear map $\lambda \colon A \longrightarrow \Bbbk$ such that $\ker \lambda$ contains no nonzero left or right ideal of A. If, in addition, $\lambda(ab) = \lambda(ba)$ for all $a, b \in A$, then A is called a *symmetric algebra*. The following proposition is Exercise 4 of [Ben98, Section 1.6].

Proposition 15.2. *Let A be a symmetric algebra. Then $M_n(A)$ is a symmetric algebra for all $n \geq 1$.*

The reader should also verify that a direct product of symmetric algebras is symmetric (cf. Exercise 15.2). We recall that if V is a right A-module, then $D(V) = \mathrm{Hom}_{\Bbbk}(V, \Bbbk)$ is a left A-module. See Section A.4 for details on the standard duality D.

© Springer International Publishing Switzerland 2016
B. Steinberg, *Representation Theory of Finite Monoids*,
Universitext, DOI 10.1007/978-3-319-43932-7_15

Frobenius algebras are self-injective. The following result combines [Ben98, Proposition 1.6.2] and [Ben98, Theorem 1.6.3].

Theorem 15.3. *Let A be a Frobenius algebra. Then $A \cong D(A)$ and hence A is self-injective. Moreover, if A is symmetric, then $P/\operatorname{rad}(P) \cong \operatorname{soc}(P)$ for all projective indecomposable A-modules P.*

The primary example of a symmetric algebra is a group algebra [Ben98, Proposition 3.1.2].

Theorem 15.4. *Let G be a finite group and $\Bbbk$ a field. Define $\lambda \colon \Bbbk G \longrightarrow \Bbbk$ by*

$$\lambda(g) = \begin{cases} 1, & \text{if } g = 1 \\ 0, & \text{else} \end{cases}$$

for $g \in G$. Then $\Bbbk G$ is a symmetric algebra with respect to the functional λ.

15.2 Regular monoids with self-injective algebras

Our goal is to generalize Theorem 5.19 to a result about self-injectivity. First we note that Corollary 9.4, Proposition 15.2, Exercise 15.2, and Theorem 15.4 easily imply the following result, which can be viewed as a warmup to our main result.

Corollary 15.5. *Let M be a finite inverse monoid and $\Bbbk$ a field. Then $\Bbbk M$ is a symmetric algebra.*

Now we state and prove the main theorem of this chapter.

Theorem 15.6. *Let M be a finite regular monoid and $\Bbbk$ a field. Let $e_1, \ldots, e_s$ form a complete set of idempotent representatives of the $\mathcal{J}$-classes of M and let n_i be the number of $\mathcal{L}$-classes in J_{e_i}. Then the following are equivalent.*

(i) $\Bbbk M$ is self-injective.
(ii) $\Bbbk M$ is a Frobenius algebra.
(iii) $\Bbbk M$ is a symmetric algebra.
(iv) $\Bbbk M \cong \prod_{i=1}^{s} M_{n_i}(\Bbbk G_{e_i})$.
(v) The sandwich matrix $P(e_i)$ is invertible over $\Bbbk G_{e_i}$ for $1 \leq i \leq s$.
(vi) The natural map $\varphi_{\Bbbk G_{e_i}} \colon \operatorname{Ind}_{G_{e_i}}(\Bbbk G_{e_i}) \longrightarrow \operatorname{Coind}_{G_{e_i}}(\Bbbk G_{e_i})$ is an isomorphism for $1 \leq i \leq s$.
(vii) $\operatorname{Ind}_{G_{e_i}}(\Bbbk G_{e_i}) \cong \operatorname{Coind}_{G_{e_i}}(\Bbbk G_{e_i})$ for $1 \leq i \leq s$.

Proof. The implication (iv) implies (iii) follows from Proposition 15.2, Exercise 15.2, and Theorem 15.4. The implications (iii) implies (ii) implies (i) hold for finite dimensional algebras in general. The equivalence of (v) and (vi) is a consequence of Lemma 5.20. The equivalence of (vi) and (vii) is the content of Exercise 4.3. It remains to show that (vii) implies (iv) and (i) implies (vii).

We proceed in a similar fashion to the proof of Theorem 5.19. Recall that, for $e \in E(M)$, we have

$$\mathrm{Ind}_{G_e}(\Bbbk G_e) = \Bbbk L_e \otimes_{\Bbbk G_e} \Bbbk G_e \cong \Bbbk L_e.$$

Consider a principal series

$$\emptyset = I_0 \subsetneq I_1 \subsetneq \cdots \subsetneq I_s = M$$

for M and without loss of generality assume that $e_k \in J_k = I_k \setminus I_{k-1}$, which is a regular $\mathscr{J}$-class (cf. Proposition 1.20), for $k = 1, \ldots, s$. Then we have

$$\Bbbk I_k / \Bbbk I_{k-1} \cong \Bbbk J_k \cong n_k \cdot \Bbbk L_{e_k} \tag{15.1}$$

by Lemma 5.18.

Suppose first that (vii) holds. We claim that

$$\Bbbk M / \Bbbk I_{k-1} \cong \Bbbk J_k \oplus \Bbbk M / \Bbbk I_k \tag{15.2}$$

as a $\Bbbk M$-module for $1 \leq k \leq s$. Indeed, let $A = \Bbbk M / \Bbbk I_{k-1}$. Then we have that $e_k A e_k \cong \Bbbk G_{e_k}$ (by Corollary 1.16), $\mathrm{Ind}_{e_k}(\Bbbk G_{e_k}) = \mathrm{Ind}_{G_{e_k}}(\Bbbk G_{e_k})$ and $\mathrm{Coind}_{e_k}(\Bbbk G_{e_k}) = \mathrm{Coind}_{G_{e_k}}(\Bbbk G_{e_k})$. As $\Bbbk G_{e_k}$ is self-injective, $\mathrm{Coind}_{G_{e_k}}(\Bbbk G_{e_k})$ is an injective A-module by Proposition 4.3. Since $\Bbbk L_e \cong \mathrm{Ind}_{G_{e_k}}(\Bbbk G_{e_k}) \cong \mathrm{Coind}_{G_{e_k}}(\Bbbk G_e)$, we conclude that $\Bbbk L_{e_k}$ is an injective A-module. From (15.1) we then conclude that $\Bbbk J_k$ is an injective A-module. Thus the exact sequence of A-modules

$$0 \longrightarrow \Bbbk J_k \longrightarrow \Bbbk A \longrightarrow \Bbbk M / \Bbbk I_k \longrightarrow 0$$

splits, establishing (15.2).

Applying (15.1) and (15.2) repeatedly yields

$$\Bbbk M \cong \bigoplus_{k=1}^{s} \Bbbk J_k \cong \bigoplus_{k=1}^{s} n_k \cdot \Bbbk L_{e_k}. \tag{15.3}$$

We claim that if $i \neq j$, then $\mathrm{Hom}_{\Bbbk M}(\Bbbk L_{e_i}, \Bbbk L_{e_j}) = 0$. Let us for the moment assume that this claim holds. Then from (15.3) we shall have that

$$\Bbbk M^{op} \cong \mathrm{End}_{\Bbbk M}(\Bbbk M) \cong \prod_{k=1}^{s} M_{n_k}(\mathrm{End}_{\Bbbk M}(\Bbbk L_{e_k})).$$

But by Proposition 4.6, we have that

$$\mathrm{End}_{\Bbbk M}(\Bbbk L_{e_k}) = \mathrm{End}_{A_{e_k}}(\mathrm{Ind}_{e_k}(\Bbbk G_{e_k})) \cong \mathrm{End}_{\Bbbk G_{e_k}}(\Bbbk G_{e_k}) \cong \Bbbk G_{e_k}^{op}$$

(where $A_{e_k} = \Bbbk M / \Bbbk I(e_k)$) and so (iv) will follow from the claim.

Assume first that $Me_jM \not\subseteq Me_iM$ and set $I = Me_iM \setminus J_i$. Putting $A = \Bbbk M/\Bbbk I$, we have that $\Bbbk L_{e_i}$ and $\Bbbk L_{e_j}$ are both A-modules and $Ae_i \cong \Bbbk L_{e_i}$ by Theorem 1.13. Therefore, we have

$$\mathrm{Hom}_{\Bbbk M}(\Bbbk L_{e_i}, \Bbbk L_{e_j}) = \mathrm{Hom}_A(Ae_i, \Bbbk L_{e_j}) \cong e_i \Bbbk L_{e_j} = 0$$

where the last equality follows because $e_i \in I(e_j)$.

Next assume that $Me_jM \subsetneq Me_iM$. Let $A = \Bbbk M/\Bbbk I(e_j)$, and so $e_j Ae_j \cong \Bbbk G_{e_j}$. Then $\Bbbk L_{e_i}$ and $\Bbbk L_{e_j}$ are both A-modules and $\Bbbk L_{e_j} \cong \mathrm{Ind}_{\Bbbk G_{e_j}}(\Bbbk G_{e_j}) \cong \mathrm{Coind}_{\Bbbk G_{e_j}}(\Bbbk G_{e_j}) = \mathrm{Coind}_{e_j}(\Bbbk G_{e_j})$. Proposition 4.1 then yields

$$\begin{aligned}
\mathrm{Hom}_{\Bbbk M}(\Bbbk L_{e_i}, \Bbbk L_{e_j}) &= \mathrm{Hom}_A(\Bbbk L_{e_i}, \mathrm{Coind}_{e_j}(\Bbbk G_{e_j})) \\
&\cong \mathrm{Hom}_{\Bbbk G_{e_j}}(e_j \Bbbk L_{e_i}, \Bbbk G_{e_j}) \\
&= 0
\end{aligned}$$

as $e_j \Bbbk L_{e_i} = 0$ if $Me_jM \subsetneq Me_iM$. This completes the proof of the claim. We conclude that (vii) implies (iv).

Next assume that (i) holds and we prove (vii). First we prove by induction on k that $\Bbbk M/\Bbbk I_k$ is a projective $\Bbbk M$-module. The base case that $k = 0$ is trivial because $\Bbbk M/\Bbbk I_0 \cong \Bbbk M$. Assume that $\Bbbk M/\Bbbk I_{k-1}$ is a projective $\Bbbk M$-module with $k > 0$ and let $A = \Bbbk M/\Bbbk I_{k-1}$. Then since A is projective as a $\Bbbk M$-module, it follows that every projective A-module is a projective $\Bbbk M$-module. But $\Bbbk L_{e_k} \cong Ae_k$ and hence is a projective A-module. We conclude that $\Bbbk J_k \cong n_k \cdot \Bbbk L_{e_k}$ (by (15.1)) is a projective A-module and hence a projective $\Bbbk M$-module. Therefore, $\Bbbk J_k \cong \Bbbk I_k/\Bbbk I_{k-1}$ is an injective $\Bbbk M$-module by self-injectivity of $\Bbbk M$ and so the exact sequence of $\Bbbk M$-modules

$$0 \longrightarrow \Bbbk I_k/I_{k-1} \longrightarrow \Bbbk M/\Bbbk I_{k-1} \longrightarrow \Bbbk M/\Bbbk I_k \longrightarrow 0$$

splits. We deduce that

$$\Bbbk M/\Bbbk I_{k-1} \cong \Bbbk I_k/I_{k-1} \oplus \Bbbk M/\Bbbk I_k \cong \Bbbk J_k \oplus \Bbbk M/\Bbbk I_k \qquad (15.4)$$

and hence $\Bbbk M/\Bbbk I_k$ is a projective $\Bbbk M$-module, being a direct summand in the projective $\Bbbk M$-module $\Bbbk M/\Bbbk I_{k-1}$. This completes the induction.

Repeated application of (15.4) shows that

$$\Bbbk M = \bigoplus_{k \geq 1}^{s} \Bbbk J_k \cong \bigoplus_{k=1}^{s} n_k \cdot \Bbbk L_{e_k} \qquad (15.5)$$

and hence each $\Bbbk L_{e_k}$ is a projective $\Bbbk M$-module for $1 \leq k \leq s$.

Let $\Bbbk G_{e_k} = \bigoplus_{r=1}^{m_k} P_{k,r}$ be the decomposition of $\Bbbk G_{e_k}$ into projective indecomposable modules. Then observe that

$$\Bbbk L_{e_k} \cong \mathrm{Ind}_{G_{e_k}}(\Bbbk G_{e_k}) \cong \bigoplus_{r=1}^{m_k} \mathrm{Ind}_{G_{e_k}}(P_{k,r})$$

is a direct sum decomposition into indecomposable modules, necessarily projective, by Corollary 4.10. Thus each projective indecomposable $\Bbbk M$-module is isomorphic to one of the form $\operatorname{Ind}_{G_{e_k}}(P)$ where P is a projective indecomposable $\Bbbk G_{e_k}$-module and $1 \leq k \leq s$ by (15.5).

Now let $A = \Bbbk M / \Bbbk I_{k-1}$ (and so $e_k A e_k \cong \Bbbk G_{e_k}$). Then A is projective, and hence injective, as a $\Bbbk M$-module and therefore is injective as an A-module. Thus A is self-injective. Let P be a projective indecomposable $\Bbbk G_{e_k}$-module. Then P is injective because $\Bbbk G_{e_k}$ is self-injective by Theorem 15.4. Thus $\operatorname{Coind}_{e_k}(P) = \operatorname{Coind}_{G_{e_k}}(P)$ is an injective indecomposable A-module by Proposition 4.3 and Corollary 4.10, and hence is a projective indecomposable A-module by Proposition 15.1. Therefore, since A is a projective $\Bbbk M$-module, we conclude that $\operatorname{Coind}_{G_{e_k}}(P)$ is a projective indecomposable $\Bbbk M$-module. We then deduce that $\operatorname{Coind}_{G_{e_k}}(P) \cong \operatorname{Ind}_{G_{e_i}}(Q)$ for some $1 \leq i \leq s$ and projective indecomposable $\Bbbk G_{e_i}$-module Q. But since $I(e_k)$ annihilates $\operatorname{Coind}_{G_{e_k}}(P)$ and $I(e_i)$ annihilates $\operatorname{Ind}_{G_{e_i}}(Q)$ we deduce that $i = k$ and $P \cong e_k \operatorname{Coind}_{G_{e_k}}(P) \cong e_k \operatorname{Ind}_{G_{e_k}}(Q) \cong Q$. Thus $\operatorname{Coind}_{G_{e_k}}(P) \cong \operatorname{Ind}_{G_{e_k}}(P)$ for each projective indecomposable $\Bbbk G_{e_k}$-module P. It follows that

$$\operatorname{Ind}_{G_{e_k}}(\Bbbk G_{e_k}) \cong \bigoplus_{r=1}^{m_k} \operatorname{Ind}_{G_{e_k}}(P_{k,r}) \cong \bigoplus_{r=1}^{m_k} \operatorname{Coind}_{G_{e_k}}(P_{k,r}) \cong \operatorname{Coind}_{G_{e_k}}(\Bbbk G_{e_k})$$

completing the proof that (i) implies (vii). This establishes the theorem. □

A corollary of Theorem 5.31 is the following.

Corollary 15.7. *Let $\mathbb{F}$ be a finite field and let $\Bbbk$ be a field of characteristic different than that of $\mathbb{F}$. Then $\Bbbk M_n(\mathbb{F})$ is a symmetric algebra (and hence self-injective).*

Unlike the case of semisimplicity, a monoid algebra $\Bbbk M$ can be a symmetric algebra without M being regular.

Example 15.8. Let $M = \{1, x, x^2\}$ where $x^2 = x^3$. Notice that M, and hence $\Bbbk M$ is commutative. Also M is not regular. We claim that $\Bbbk M$ is a symmetric algebra for any field $\Bbbk$. Indeed, define $\lambda \colon \Bbbk M \longrightarrow \Bbbk$ by $\lambda(1) = 2 = \lambda(x)$ and $\lambda(x^2) = 1$. Then $\ker \lambda$ has basis by $1 - x$ and $1 - 2x^2$. It is routine to verify that ker λ contains no nonzero ideal of $\Bbbk M$.

15.3 Exercises

15.1. Prove Proposition 15.2.

15.2. Prove that a direct product of symmetric algebras is symmetric.

15.3. Prove Theorem 15.4.

15.4. Verify that $\ker \lambda$ from Example 15.8 contains no nonzero ideals.

15.5. Let $\mathbb{F}$ be a finite field of characteristic p and let $\Bbbk$ be any field of characteristic p. Prove that $\Bbbk M_n(\mathbb{F})$ is not self-injective. (Hint: prove the sandwich matrix for the $\mathscr{J}$-class of rank 1 matrices is not invertible over the group algebra of the group of nonzero elements of $\mathbb{F}$.)

16

Global Dimension

An important homological invariant of a finite dimensional algebra A is its global dimension. It measures the maximum length of a minimal projective resolution of an A-module. Nico obtained a bound on the global dimension of $\Bbbk M$ for a regular monoid M in "good" characteristic [Nic71, Nic72]. Essentially, Nico discovered that $\Bbbk M$ is a quasi-hereditary algebra in the sense of Cline, Parshall, and Scott [CPS88], more than 15 years before the notion was defined, and proved the corresponding bound on global dimension. Putcha was the first to observe that $\Bbbk M$ is quasi-hereditary [Put98]. Here we provide a direct approach to Nico's theorem, without introducing the machinery of quasi-hereditary algebras, although it is hidden under the surface (compare with [DR89]). Our approach follows the ideas in [MS11] and uses the results of Auslander, Platzeck, and Todorov [APT92] on idempotent ideals in algebras. In this chapter, we do not hesitate to assume familiarity with notions from homological algebra, in particular, with properties of the Ext-functor. Standard references for homological algebra are [Wei94, HS97, CE99], although everything we shall need can be found in [Ben98] or the appendix of [ASS06].

16.1 Idempotent ideals and homological algebra

In this section $\Bbbk$ is a field and A is a finite dimensional $\Bbbk$-algebra. The *global dimension* of A is defined by

$$\text{gl. dim } A = \sup\{n \mid \text{Ext}_A^n(M, N) \neq 0, \text{for some } M, N \in A\text{-mod}\}. \quad (16.1)$$

Note that gl. dim $A = \infty$ is possible. For example, gl. dim $A = 0$ if and only if each finite dimensional A-module is injective, if and only if A is semisimple. Indeed, every A-module is injective if and only if every submodule has a complement. An algebra A has global dimension at most 1 if and only if each

© Springer International Publishing Switzerland 2016
B. Steinberg, *Representation Theory of Finite Monoids*,
Universitext, DOI 10.1007/978-3-319-43932-7_16

submodule of a finite dimensional projective module is projective. One can readily show that this is equivalent to each left ideal of A being a projective module. An algebra of global dimension at most 1 is called a *hereditary algebra*.

Lemma 16.1. *Let $M, N \in A$-mod. Then $\mathrm{Ext}^n_A(M, N) = 0$ if either of the following two conditions hold.*

(i) $\mathrm{Ext}^n_A(M, S) = 0$ for each composition factor S of N.
(ii) $\mathrm{Ext}^n_A(S', N) = 0$ for each composition factor S' of M.

Proof. We just handle the case of (i) as the other argument is dual. We proceed by induction on the number of composition factors (i.e., length) of N. The case that N has one composition factor is just $N = S$ and so there is nothing to prove. Assume that it is true for A-modules with k composition factors and suppose that N has $k+1$ composition factors. Let L be a maximal submodule of N and $S = N/L$. Notice that L has k composition factors, all of which are composition factors of N. Thus $\mathrm{Ext}^n_A(M, L) = 0$ by induction. By assumption, $\mathrm{Ext}^n_A(M, S) = 0$. From the exact sequence

$$0 \longrightarrow L \longrightarrow N \longrightarrow S \longrightarrow 0$$

and the long exact sequence for Ext, we obtain an exact sequence

$$0 = \mathrm{Ext}^n_A(M, L) \longrightarrow \mathrm{Ext}^n_A(M, N) \longrightarrow \mathrm{Ext}^n_A(M, S) = 0$$

and so $\mathrm{Ext}^n_A(M, N) = 0$, as required. □

As a consequence, it follows that the sup in (16.1) is achieved by a pair of simple modules.

Corollary 16.2. *Let A be a finite dimensional $\Bbbk$-algebra. Then*

$$\mathrm{gl.\,dim}\, A = \sup\{n \mid \mathrm{Ext}^n_A(S, S') \neq 0, \text{ for some } S, S' \text{ simple}\}$$

holds.

Proof. Suppose that $\mathrm{Ext}^n_A(S, S') = 0$ for all finite dimensional simple A-modules. Let M, N be finite dimensional A-modules. Then $\mathrm{Ext}^n_A(S', N) = 0$ for all simple A-modules S' by Lemma 16.1(i). Therefore, $\mathrm{Ext}^n_A(M, N) = 0$ by Lemma 16.1(ii). The corollary follows. □

The next result is a special case of a very general result of Adams and Rieffel [AR67]. The reader should refer to Chapter 4 for the definitions of Ind_e and Coind_e when e is an idempotent of an algebra A.

Theorem 16.3. *Let A be a finite dimensional $\Bbbk$-algebra and let $e \in A$ be an idempotent. Assume that Ae is a flat right eAe-module, respectively, eA is a projective left eAe-module. Then there are isomorphisms*

$$\mathrm{Ext}^n_A(\mathrm{Ind}_e(V), W) \cong \mathrm{Ext}^n_{eAe}(V, \mathrm{Res}_e(W)) \tag{16.2}$$

$$\mathrm{Ext}^n_A(W, \mathrm{Coind}_e(V)) \cong \mathrm{Ext}^n_{eAe}(\mathrm{Res}_e(W), V) \tag{16.3}$$

respectively, for all $n \geq 0$, $V \in eAe$-mod and $W \in A$-mod.

Proof. We just handle (16.2), as the other isomorphism is proved in a dual fashion. Let $P_\bullet \longrightarrow V$ be a projective resolution by finite dimensional eAe-modules. Then $\mathrm{Ind}_e(P_n)$ is a projective A-module, for all $n \geq 0$, by Proposition 4.3. Also the functor Ind_e is exact by Proposition 4.2. Therefore, $\mathrm{Ind}_e(P_\bullet) \longrightarrow \mathrm{Ind}_e(V)$ is a projective resolution. We conclude using the adjointness of Ind_e and Res_e (Proposition 4.1) that

$$\begin{aligned}
\mathrm{Ext}_A^n(\mathrm{Ind}_e(V), W) &\cong H^n(\mathrm{Hom}_A(\mathrm{Ind}_e(P_\bullet), W)) \\
&\cong H^n(\mathrm{Hom}_{eAe}(P_\bullet, \mathrm{Res}_e(W))) \\
&\cong \mathrm{Ext}_{eAe}^n(V, \mathrm{Res}_e(W))
\end{aligned}$$

as required.   □

An ideal I of A is said to be *idempotent* if $I^2 = I$. Idempotent ideals enjoy a number of nice homological properties.

Lemma 16.4. *Let I be an idempotent ideal of A. Then $\mathrm{Ext}_A^1(A/I, N) = 0$ for all A/I-modules N. If, in addition, I is a projective left A-module, then $\mathrm{Ext}_A^n(A/I, N) = 0$ for all $n > 0$.*

Proof. Since A is a projective module, the exact sequence

$$0 \longrightarrow I \longrightarrow A \longrightarrow A/I \longrightarrow 0 \tag{16.4}$$

gives rise to an exact sequence

$$\mathrm{Hom}_A(I, N) \longrightarrow \mathrm{Ext}_A^1(A/I, N) \longrightarrow \mathrm{Ext}_A^1(A, N) = 0.$$

But $IN = 0$ and so if $\varphi \in \mathrm{Hom}_A(I, N)$, then $\varphi(I) = \varphi(II) = I\varphi(I) = 0$. Therefore, $\mathrm{Hom}_A(I, N) = 0$ and so we conclude that $\mathrm{Ext}_A^1(A/I, N) = 0$.

Assume now that I is projective and $n > 1$. Then the exact sequence (16.4) gives rise to an exact sequence

$$0 = \mathrm{Ext}_A^{n-1}(I, N) \longrightarrow \mathrm{Ext}_A^n(A/I, N) \longrightarrow \mathrm{Ext}_A^n(A, N) = 0$$

and so $\mathrm{Ext}_A^n(A/I, N) = 0$, as required.   □

As a corollary, we show that under suitable hypotheses on the idempotent ideal I, we can compute the Ext-functor between A/I-modules over either A or A/I and obtain the same result.

Theorem 16.5. *Let A be a finite dimensional $\Bbbk$-algebra and let I be an idempotent ideal. Let M, N be finite dimensional A/I-modules. Then*

$$\mathrm{Ext}_A^1(M, N) \cong \mathrm{Ext}_{A/I}^1(M, N)$$

holds. If, in addition, I is a projective left A-module, then

$$\mathrm{Ext}_A^n(M, N) \cong \mathrm{Ext}_{A/I}^n(M, N) \tag{16.5}$$

for all $n \geq 0$.

Proof. Choose a presentation of M as an A/I-module of the form

$$0 \longrightarrow L \xrightarrow{\ f\ } r(A/I) \longrightarrow M \longrightarrow 0$$

with $r \geq 0$. Since $\text{Ext}^1_{A/I}(r(A/I), N) = 0$ by freeness of $r(A/I)$ and $\text{Ext}^1_A(r(A/I), N) = 0$ by Lemma 16.4 we have the following two exact sequences

$$\text{Hom}_{A/I}(r(A/I), N) \xrightarrow{\ f^*\ } \text{Hom}_{A/I}(L, N) \longrightarrow \text{Ext}^1_{A/I}(M, N) \longrightarrow 0$$

and

$$\text{Hom}_A(r(A/I), N) \xrightarrow{\ f^*\ } \text{Hom}_A(L, N) \longrightarrow \text{Ext}^1_A(M, N) \longrightarrow 0$$

where f^* is the map induced by f. But the functors $\text{Hom}_A(-, N)$ and $\text{Hom}_{A/I}(-, N)$ on A/I-mod are isomorphic. Thus both $\text{Ext}^1_{A/I}(M, N)$ and $\text{Ext}^1_A(M, N)$ can be identified with the cokernel of the morphism

$$\text{Hom}_{A/I}(r(A/I), N) \xrightarrow{\ f^*\ } \text{Hom}_{A/I}(L, N)$$

thereby establishing the first isomorphism.

The second isomorphism is proved simultaneously for all finite dimensional A/I-modules M by induction on n. The case $n = 0$ is clear and the case $n = 1$ has already been handled. So assume that (16.5) is true for all M and we prove the corresponding statement for $n+1$ with $n \geq 1$. We retain the above notation. Then, for all $k \geq 1$, we have $\text{Ext}^k_{A/I}(r(A/I), N) = 0$ by projectivity of $r(A/I)$ and $\text{Ext}^k_A(r(A/I), N) = 0$ by Lemma 16.4. Therefore, we have the following two exact sequences

$$0 \longrightarrow \text{Ext}^n_{A/I}(L, N) \longrightarrow \text{Ext}^{n+1}_{A/I}(M, N) \longrightarrow 0$$

and

$$0 \longrightarrow \text{Ext}^n_A(L, N) \longrightarrow \text{Ext}^{n+1}_A(M, N) \longrightarrow 0$$

By the inductive hypothesis, we have $\text{Ext}^n_{A/I}(L, N) \cong \text{Ext}^n_A(L, N)$ and so $\text{Ext}^{n+1}_{A/I}(M, N) \cong \text{Ext}^{n+1}_A(M, N)$. This completes the proof. $\qquad\square$

16.2 Global dimension and homological properties of regular monoids

In this section we prove Nico's result on the global dimension of the algebra of a regular monoid. The key ingredient is that we can always work modulo an ideal of the monoid. The following lemma is from [MS11].

Lemma 16.6. *Let M be a regular monoid and $\Bbbk$ a field. Let I be an ideal of M. Then*

$$\operatorname{Ext}^n_{\Bbbk M}(V, W) \cong \operatorname{Ext}^n_{\Bbbk M/\Bbbk I}(V, W)$$

for any $\Bbbk M/\Bbbk I$-modules V, W and all $n \geq 0$.

Proof. For the purposes of this proof, we allow $I = \emptyset$ and we proceed by induction on the number of $\mathscr{J}$-classes of M contained in I. If $I = \emptyset$, then there is nothing to prove. So assume that the lemma is true for ideals with $k - 1$ $\mathscr{J}$-classes and that I has $k \geq 1$ $\mathscr{J}$-classes. Let J be a maximal $\mathscr{J}$-class of I, that is, MJM is not contained in $MJ'M$ for any $\mathscr{J}$-class $J' \subseteq I$. Then $I' = I \setminus J$ is an ideal of M with $k - 1$ $\mathscr{J}$-classes. Clearly, V, W are $\Bbbk M/\Bbbk I'$-modules and so

$$\operatorname{Ext}^n_{\Bbbk M}(V, W) \cong \operatorname{Ext}^n_{\Bbbk M/\Bbbk I'}(V, W) \tag{16.6}$$

for all $n \geq 0$ by induction.

Note that $\Bbbk I/\Bbbk I'$ is an idempotent ideal of $\Bbbk M/\Bbbk I'$ (idempotent as a consequence of Proposition 1.23) and $\Bbbk M/\Bbbk I \cong (\Bbbk M/\Bbbk I')/(\Bbbk I/\Bbbk I')$. Let $e \in E(J)$. If r is the number of $\mathscr{L}$-classes contained in J, then we have

$$\Bbbk I/\Bbbk I' \cong \Bbbk J \cong r \cdot \Bbbk L_e \cong r \cdot [(\Bbbk M/\Bbbk I')e]$$

by Lemma 5.18 and Theorem 1.13. Thus $\Bbbk I/\Bbbk I'$ is projective as a $\Bbbk M/\Bbbk I'$-module. Applying Theorem 16.5 and (16.6) yields

$$\operatorname{Ext}^n_{\Bbbk M}(V, W) \cong \operatorname{Ext}^n_{\Bbbk M/\Bbbk I'}(V, W) \cong \operatorname{Ext}^n_{\Bbbk M/\Bbbk I}(V, W)$$

completing the induction and the proof of the lemma. $\qquad\square$

If M is a regular monoid and $e \in E(M)$, we define the *height* of e by

$$\operatorname{ht}(e) = \max\{k \mid MeM = Me_0M \subsetneq Me_1M \subsetneq \cdots \subsetneq Me_kM = M\}$$

with $e_0, \ldots, e_k \in E(M)$. Notice that $\operatorname{ht}(e) = 0$ if and only if $e = 1$. The maximum height is achieved on the minimal ideal. The key step to proving Nico's theorem is the following lemma bounding the vanishing degree of Ext between simples in terms of the heights of their apexes.

Lemma 16.7. *Let M be a regular monoid and $\Bbbk$ a field whose characteristic is 0 or does not divide the order of any maximal subgroup of M. Let S, S' be simple $\Bbbk M$-modules with respective apexes e, e'. Then $\operatorname{Ext}^r_{\Bbbk M}(S, S') = 0$ for all $r > \operatorname{ht}(e) + \operatorname{ht}(e')$.*

Proof. We induct on the quantity $\operatorname{ht}(e) + \operatorname{ht}(e')$. If $0 = \operatorname{ht}(e) + \operatorname{ht}(e')$, then $e = 1 = e'$ and so S, S' are modules over $\Bbbk M/\Bbbk I(1) \cong \Bbbk G_1$. Lemma 16.6 then yields

$$\operatorname{Ext}^r_{\Bbbk M}(S, S') \cong \operatorname{Ext}^r_{\Bbbk G_1}(S, S') = 0$$

for $r > 0$ because $\Bbbk G_1$ is a semisimple algebra. Assume now that the result holds for all smaller values of $\mathrm{ht}(e) + \mathrm{ht}(e')$ and that $\mathrm{ht}(e) + \mathrm{ht}(e') > 0$. Note that $r > 1$ from now on.

We handle two cases. Assume first that $Me'M \not\subseteq MeM$. Let $I = MeM \setminus J_e$; this is an ideal. Let $A = \Bbbk M / \Bbbk I$. Then $eAe \cong \Bbbk G_e$ by Corollary 1.16 and $Ae \cong \Bbbk L_e$ by Theorem 1.13. It follows that $\mathrm{Ind}_e(V) = \mathrm{Ind}_{G_e}(V)$ for any $\Bbbk G_e$-module V, where we view $\mathrm{Ind}_e \colon eAe\text{-mod} \longrightarrow A\text{-mod}$. Let us assume that $S = V^\sharp$ with V a simple $\Bbbk G_e$-module where we have retained the notation of Theorem 5.5. Note that S' is an A-module because $e \in I(e')$ and hence $I \subseteq I(e')$. Also, we have that $\mathrm{Ind}_e(V)$ is a projective A-module by Proposition 4.3 because V is a projective $\Bbbk G_e$-module by semisimplicity of $\Bbbk G_e$. Therefore, the exact sequence of A-modules

$$0 \longrightarrow \mathrm{rad}(\mathrm{Ind}_e(V)) \longrightarrow \mathrm{Ind}_e(V) \longrightarrow S \longrightarrow 0$$

yields an exact sequence

$$0 \longrightarrow \mathrm{Ext}_A^{r-1}(\mathrm{rad}(\mathrm{Ind}_e(V)), S') \longrightarrow \mathrm{Ext}_A^r(S, S') \longrightarrow 0.$$

Lemma 16.6 then implies that $\mathrm{Ext}_{\Bbbk M}^{r-1}(\mathrm{rad}(\mathrm{Ind}_e(V)), S') \cong \mathrm{Ext}_{\Bbbk M}^r(S, S')$. By Theorem 5.5 every composition factor L of $\mathrm{rad}(\mathrm{Ind}_e(V)) = \mathrm{rad}(\mathrm{Ind}_{G_e}(V))$ has apex f with $MeM \subsetneq MfM$. Therefore, by induction $\mathrm{Ext}_{\Bbbk M}^{r-1}(L, S') = 0$. We conclude that $\mathrm{Ext}_{\Bbbk M}^{r-1}(\mathrm{rad}(\mathrm{Ind}_e(V)), S') = 0$ by Lemma 16.1. Thus $\mathrm{Ext}_{\Bbbk M}^r(S, S') = 0$, as was required.

Next we assume that $Me'M \subseteq MeM$. Let $A = \Bbbk M / \Bbbk I(e')$. Then both S, S' are A-modules. Let $S' = W^\sharp$ with W a simple $\Bbbk G_{e'}$-module. Then W is an injective $\Bbbk G_{e'}$-module by semisimplicity of $\Bbbk G_{e'}$, whence $\mathrm{Coind}_{G_{e'}}(W) = \mathrm{Coind}_{e'}(W)$ is an injective A-module by Proposition 4.3, and $S' = \mathrm{soc}(\mathrm{Coind}_{G_{e'}}(W))$. Put $N = \mathrm{Coind}_{G_e}(W) / S'$. Then the exact sequence of A-modules

$$0 \longrightarrow S' \longrightarrow \mathrm{Coind}_{G_{e'}}(W) \longrightarrow N \longrightarrow 0$$

leads to an exact sequence

$$0 \longrightarrow \mathrm{Ext}_A^{r-1}(S, N) \longrightarrow \mathrm{Ext}_A^r(S, S') \longrightarrow 0.$$

Lemma 16.6 then yields that $\mathrm{Ext}_{\Bbbk M}^{r-1}(S, N) \cong \mathrm{Ext}_{\Bbbk M}^r(S, S')$. But each composition factor L of N has apex f with $Me'M \subsetneq MfM$ by Theorem 5.5. Induction then implies that $\mathrm{Ext}_{\Bbbk M}^{r-1}(S, L) = 0$. It follows that $\mathrm{Ext}_{\Bbbk M}^{r-1}(S, N) = 0$ by Lemma 16.1 and hence $\mathrm{Ext}_{\Bbbk M}^r(S, S') = 0$, as desired. This completes the proof of the lemma. □

We now prove Nico's theorem [Nic71, Nic72].

Theorem 16.8. *Let M be a finite regular monoid and $\Bbbk$ a field. Then* gl. dim $\Bbbk M$ *is finite if and only if the characteristic of $\Bbbk$ is 0 or does not* *divide the order of any maximal subgroup of M. If these conditions hold, then* gl. dim $\Bbbk M \leq 2k$ *where k is the length of the longest chain*

$$Mm_0M \subsetneqq Mm_1M \subsetneqq \cdots \subsetneqq Mm_kM = M$$

of principal ideals in M.

Proof. Suppose first that the characteristic of $\Bbbk$ divides the order of the maximal subgroup G_e. Then it is well known that gl. dim $\Bbbk G_e = \infty$ (cf. Exercise 16.6). Therefore, we can find, for any $n \geq 0$, a pair V, W of finite dimensional $\Bbbk G_e$-modules with $\operatorname{Ext}^n_{\Bbbk G_e}(V, W) \neq 0$. Let $A = \Bbbk M / \Bbbk I(e)$. Note that $Ae \cong \Bbbk L_e$ is a free right module over $eAe \cong \Bbbk G_e$ by Proposition 1.10. Therefore, Lemma 16.6 and Theorem 16.3 yield

$$\operatorname{Ext}^n_{\Bbbk M}(\operatorname{Ind}_{G_e}(V), \operatorname{Coind}_{G_e}(W)) \cong \operatorname{Ext}^n_A(\operatorname{Ind}_{G_e}(V), \operatorname{Coind}_{G_e}(W))$$
$$\cong \operatorname{Ext}^n_{eAe}(V, \operatorname{Res}_e(\operatorname{Coind}_e(W)))$$
$$\cong \operatorname{Ext}^n_{\Bbbk G_e}(V, W) \neq 0$$

where we have used that $\operatorname{Res}_e \circ \operatorname{Coind}_e$ is isomorphic to the identity functor by Proposition 4.5. We conclude that gl. dim $\Bbbk M = \infty$.

If $\Bbbk$ has characteristic 0 or its characteristic divides the order of no maximal subgroup of M, then gl. dim $\Bbbk M \leq 2k$ by Lemma 16.7 as each idempotent has height at most k. □

Computing Ext^1 between simple modules is important for computing the quiver of an algebra. The following proposition is then helpful.

Proposition 16.9. *Let M be a regular monoid and $\Bbbk$ a field such that the characteristic of $\Bbbk$ is 0 or does not divide the order of any maximal subgroup of M. Let S, S' be simple modules with apexes e, e', respectively. Then one has*

$$\operatorname{Ext}^1_{\Bbbk M}(S, S') = 0$$

if $MeM = Me'M$ or if MeM and $Me'M$ are incomparable.

Proof. Suppose that $MeM = Me'M$ or that MeM and $Me'M$ are incomparable. Let $I = MeM \setminus J_e$ and $A = \Bbbk M / \Bbbk I$. Then S, S' are A-modules and $eAe \cong \Bbbk G_e$ by Corollary 1.16. Suppose that $S = V^\sharp$ with V a simple $\Bbbk G_e$-module. Then V is projective by semisimplicity of $\Bbbk G_e$. Therefore, $\operatorname{Ind}_{G_e}(V) = \operatorname{Ind}_e(V)$ is a projective A-module by Proposition 4.3. The exact sequence of A-modules

$$0 \longrightarrow \operatorname{rad}(\operatorname{Ind}_e(V)) \longrightarrow \operatorname{Ind}_e(V) \longrightarrow S \longrightarrow 0$$

then yields an exact sequence

$$\operatorname{Hom}_A(\operatorname{rad}(\operatorname{Ind}_e(V)), S') \longrightarrow \operatorname{Ext}^1_A(S, S') \longrightarrow \operatorname{Ext}^1_A(\operatorname{Ind}_e(V), S') = 0.$$

But every composition factor of $\mathrm{rad}(\mathrm{Ind}_e(V)) = \mathrm{rad}(\mathrm{Ind}_{G_e}(V))$ has apex f with $MfM \supsetneq MeM$ by Theorem 5.5. By assumption on e', we conclude that $\mathrm{Hom}_A(\mathrm{rad}(\mathrm{Ind}_e(V)), S') = 0$ and hence $\mathrm{Ext}_A^1(S, S') = 0$. But $\mathrm{Ext}_{\Bbbk M}^1(S, S') = \mathrm{Ext}_A^1(S, S') = 0$ by Lemma 16.6. This completes the proof. $\square$

We now provide a homological proof of the sufficiency of the conditions for semisimplicity in Theorem 5.19. This proof is novel to the text.

Corollary 16.10. *Let M be a finite regular monoid and $\Bbbk$ a field. Suppose that the characteristic of $\Bbbk$ does not divide the order of G_e for any $e \in E(M)$ and that the natural homomorphism*

$$\varphi_{\Bbbk G_e} \colon \mathrm{Ind}_{G_e}(\Bbbk G_e) \longrightarrow \mathrm{Coind}_{G_e}(\Bbbk G_e)$$

is an isomorphism for all $e \in E(M)$. Then $\Bbbk M$ is semisimple.

Proof. The algebra $\Bbbk M$ is semisimple if and only if gl. $\dim \Bbbk M = 0$. We just are required to show, by Corollary 16.2 and Exercise 16.3, that $\mathrm{Ext}_{\Bbbk M}^1(S, S') = 0$ for all simple $\Bbbk M$-modules S, S'. Let S have apex e and S' have apex f. By Proposition 16.9 it suffices to consider the cases that $MeM \subsetneq MfM$ and $MfM \subsetneq MeM$. The algebras $\Bbbk G_e$ and $\Bbbk G_f$ are semisimple under the hypotheses on $\Bbbk$ by Maschke's theorem.

Assume first that $MeM \subsetneq MfM$. Let $A = \Bbbk M/\Bbbk I(e)$ and note that S, S' are A-modules, $eAe \cong \Bbbk G_e$, which is semisimple, and $\mathrm{Ind}_{G_e}(V) = \mathrm{Ind}_e(V)$ for any $\Bbbk G_e$-module V. By Corollary 4.22, we have that $\mathrm{Ind}_e(\Bbbk G_e)$ is a semisimple A-module. On the other hand, it is a projective A-module by Proposition 4.3. Suppose that $S = V^\sharp$ with $V \in \mathrm{Irr}_\Bbbk(G_e)$. Then V is a direct summand in $\Bbbk G_e$ (as $\Bbbk G_e$ is a semisimple algebra) and hence $\mathrm{Ind}_e(V)$ is a direct summand in $\mathrm{Ind}_e(\Bbbk G_e)$. It follows that $\mathrm{Ind}_e(V)$ is both semisimple and projective as an A-module. But then $\mathrm{Ind}_e(V)$ is, in fact, simple by Theorem 4.23 and $S = \mathrm{Ind}_e(V)$. Therefore, $\mathrm{Ext}_{\Bbbk M}^1(S, S') \cong \mathrm{Ext}_A^1(S, S') = 0$, where the first isomorphism follows from Lemma 16.6 and the second because S is a projective A-module.

Next assume that $MfM \subsetneq MeM$ and let $A = \Bbbk M/\Bbbk I(f)$. Then S, S' are A-modules, $fAf \cong \Bbbk G_f$, which is semisimple, and $\mathrm{Coind}_{G_f}(W) = \mathrm{Coind}_f(W)$ for any $\Bbbk G_f$-module W. Corollary 4.22 implies that $\mathrm{Coind}_f(\Bbbk G_f)$ is a semisimple A-module. But $\mathrm{Coind}_f(\Bbbk G_f)$ is an injective A-module by Proposition 4.3 (because $\Bbbk G_f$ is an injective $\Bbbk G_f$-module as the algebra $\Bbbk G_f$ is semisimple). Suppose that $S' = W^\sharp$ with $W \in \mathrm{Irr}_\Bbbk(G_f)$. Then W is a direct summand in $\Bbbk G_f$ (because $\Bbbk G_f$ is a semisimple algebra) and, therefore, $\mathrm{Coind}_f(W)$ is a direct summand in $\mathrm{Coind}_f(\Bbbk G_f)$. We conclude that $\mathrm{Coind}_f(W)$ is both a semisimple and injective A-module. However, it then follows that $\mathrm{Coind}_f(W)$ is simple by Theorem 4.23 and so $S' = \mathrm{Coind}_f(W)$. Therefore, $\mathrm{Ext}_{\Bbbk M}^1(S, S') \cong \mathrm{Ext}_A^1(S, S') = 0$, where the first isomorphism follows from Lemma 16.6 and the second because S' is an injective A-module. This completes the proof. $\square$

16.3 Exercises

16.1. Prove that the following are equivalent for a finite dimensional algebra A.

(i) gl. dim $A \leq 1$.
(ii) Each submodule of a finite dimensional projective A-module is projective.
(iii) Each left ideal of A is projective.
(iv) rad(A) is projective.
(v) rad(P) is projective for each projective indecomposable A-module P.

16.2. Let A be a finite dimensional $\Bbbk$-algebra and V, W be finite dimensional A-modules. Prove that $\operatorname{Ext}_A^n(V, W) \cong \operatorname{Ext}_{A^{op}}^n(D(W), D(V))$. Deduce that gl. dim $A =$ gl. dim A^{op}.

16.3. Assume that $\operatorname{Ext}_A^{n+1}(V, W) = 0$ for all finite dimensional A-modules V, W for a finite dimensional algebra A. Prove that gl. dim $A \leq n$.

16.4. Suppose that A is a finite dimensional $\Bbbk$-algebra and that

$$\cdots \longrightarrow P_n \xrightarrow{d_n} P_{n-1} \xrightarrow{d_{n-1}} \cdots \xrightarrow{d_1} P_0 \xrightarrow{d_0} V \longrightarrow 0$$

is a projective resolution of a finite dimensional A-module V. Prove that $\operatorname{Ext}_A^{n+k}(V, W) \cong \operatorname{Ext}_A^k(d_n(P_n), W)$ for all $n \geq 0$, $k \geq 1$ and finite dimensional A-modules W.

16.5. Let V be a finite dimensional A-module for a finite dimensional algebra A. Prove that $\operatorname{Ext}_A^{n+1}(V, W) = 0$ for all finite dimensional A-modules W if and only if V admits a projective resolution

$$0 \longrightarrow P_n \xrightarrow{d_n} P_{n-1} \xrightarrow{d_{n-1}} \cdots \xrightarrow{d_1} P_0 \xrightarrow{d_0} V \longrightarrow 0.$$

(Hint: use Exercise 16.4 to cut down an arbitrary projective resolution to one of this length.)

16.6. Let A be a finite dimensional self-injective $\Bbbk$-algebra and let V be a finite dimensional A-module. Prove that if V is not projective, then there is no finite projective resolution of V. Deduce that A is either semisimple or gl. dim $A = \infty$. Deduce, in particular, that if G is a finite group and the characteristic of $\Bbbk$ divides $|G|$, then gl. dim $\Bbbk G = \infty$. (Hint: if V is not projective and has a finite projective resolution, then consider a minimal length one and derive a contradiction using that projectives are injectives.)

16.7. Let M be a finite regular monoid and $\Bbbk$ a field whose characteristic divides the order of no maximal subgroup of M. Prove that if $e \in E(M)$, V is a $\Bbbk G_e$-module, and S is a simple $\Bbbk M$-module with apex f, then

$\operatorname{Ext}_{\Bbbk M}^n(\operatorname{Ind}_{G_e}(V), S) = 0$ unless $MfM \subseteq MeM$ and $n \leq \operatorname{ht}(e) - \operatorname{ht}(f)$. Deduce that if $\operatorname{Ind}_{G_e}(V)$ is simple for each $e \in E(M)$ and simple $\Bbbk G_e$-module V, then gl. dim $\Bbbk M$ is bounded by the length of the longest chain

$$Mm_0M \subsetneq Mm_1M \subsetneq \cdots \subsetneq Mm_kM = M$$

of principal ideals in M. Formulate and prove a corresponding result for $\operatorname{Coind}_{G_e}(V)$.

16.8. Let M be a finite regular monoid and $\Bbbk$ a field whose characteristic divides the order of no maximal subgroup of M. Prove that the following are equivalent.

(i) $\operatorname{Coind}_{G_e}(V)$ is simple for each simple $\Bbbk G_e$-module and $e \in E(M)$.
(ii) $\operatorname{Ind}_{G_e}(V)$ is a projective indecomposable module for each simple $\Bbbk G_e$-module and $e \in E(M)$.
(iii) The natural homomorphism $\varphi_{\Bbbk G_e}: \operatorname{Ind}_{G_e}(\Bbbk G_e) \longrightarrow \operatorname{Coind}_e(\Bbbk G_e)$ is surjective for each $e \in E(M)$.

(Hint: use Exercise 16.7.)

16.9. Let M be a finite regular monoid and $\Bbbk$ a field whose characteristic divides the order of no maximal subgroup of M. Prove that the following are equivalent.

(i) $\operatorname{Ind}_{G_e}(V)$ is simple for each simple $\Bbbk G_e$-module and $e \in E(M)$.
(ii) $\operatorname{Coind}_{G_e}(V)$ is an injective indecomposable module for each simple $\Bbbk G_e$-module and $e \in E(M)$.
(iii) The natural homomorphism $\varphi_{\Bbbk G_e}: \operatorname{Ind}_{G_e}(\Bbbk G_e) \longrightarrow \operatorname{Coind}_e(\Bbbk G_e)$ is injective for each $e \in E(M)$.

(Hint: use Exercise 16.7.)

16.10. Let $G \leq S_n$ and let $M = G \cup C$ where C is the set of constant mappings on $[n]$. Let $\Bbbk$ be a field of characteristic 0. Prove that $\Bbbk M$ is hereditary.

17

Quivers of Monoid Algebras

In this chapter we provide a computation of the quiver of a left regular band algebra, a result of Saliola [Sal07], and of a $\mathscr{J}$-trivial monoid algebra, a result of Denton, Hivert, Schilling, and Thiéry [DHST11]. These are special cases of the results of Margolis and the author [MS12a], computing the quiver of an arbitrary rectangular monoid algebra. However, the latter result is much more technical and beyond the scope of this text. We also describe the projective indecomposable modules for $\mathscr{R}$-trivial monoid algebras as partial transformation modules. This result is from the paper of Margolis and the author [MS12a] and again generalizes earlier results of Saliola [Sal07] for left regular bands and of Denton, Hivert, Schilling, and Thiéry [DHST11] for $\mathscr{J}$-trivial monoids.

17.1 Quivers of algebras

We define in this section the Ext-quiver (or Gabriel quiver) of a finite dimensional algebra over an algebraically closed field. This notion was introduced by Gabriel [Gab72]. Good references for this material are [DK94, ARS97, Ben98, ASS06].

If A is a finite dimensional algebra over an algebraically closed field $\Bbbk$, then the *Ext-quiver* or, more simply, *quiver* of A is the directed graph $Q(A)$ with vertex set $Q(A)_0$ the set of isomorphism classes of simple A-modules and with edge set $Q(A)_1$ described as follows. The number of arrows from $[S]$ to $[S']$ is $\dim \operatorname{Ext}_A^1(S, S')$. There are a number of alternative descriptions of $Q(A)$ that can be found in the literature [DK94, ARS97, Ben98, ASS06], but we shall not use them here. It is well known that $\operatorname{Ext}_A^1(S, S')$ classifies short exact sequences (or *extensions*, whence the name) of the form

$$0 \longrightarrow S' \longrightarrow E \longrightarrow S \longrightarrow 0 \tag{17.1}$$

© Springer International Publishing Switzerland 2016
B. Steinberg, *Representation Theory of Finite Monoids*,
Universitext, DOI 10.1007/978-3-319-43932-7_17

where (17.1) is *equivalent* to an extension

$$0 \longrightarrow S' \longrightarrow E' \longrightarrow S \longrightarrow 0$$

if there is a commutative diagram

$$
\begin{array}{ccccccccc}
0 & \longrightarrow & S' & \longrightarrow & E & \longrightarrow & S & \longrightarrow & 0 \\
  &   & \| &   & \downarrow &   & \| &   &   \\
0 & \longrightarrow & S' & \longrightarrow & E' & \longrightarrow & S & \longrightarrow & 0
\end{array}
$$

where we note that the middle arrow must be an isomorphism by the Five Lemma. The reader should consult [Ben98, CE99] or the appendix of [ASS06] for details. From this discussion, we see that the quiver of an algebra encodes information about its modules of length at most 2.

To explain the importance of the quiver of an algebra, we need to recall the definition of an admissible ideal. Let Q be a finite quiver. The *arrow ideal* J of the path algebra $\Bbbk Q$ is the ideal with basis the set of paths of length at least 1. An ideal $I \subseteq \Bbbk Q$ is said to be an *admissible ideal* if there exists $n \geq 2$ such that $J^n \subseteq I \subseteq J^2$. The following theorem is due to Gabriel [Gab80] and can be found in [Ben98, Proposition 4.1.7] and [ASS06, Theorem II.3.7], for example.

Theorem 17.1. *Let A be a basic finite dimensional algebra over an algebraically closed field $\Bbbk$. Then $A \cong \Bbbk Q(A)/I$ for some admissible ideal I. Conversely, if I is an admissible ideal of $\Bbbk Q$, then $A = \Bbbk Q/I$ is a basic finite dimensional algebra and $Q(A) \cong Q$.*

Note that if A is a finite dimensional algebra and B is the unique basic algebra that is Morita equivalent to A (i.e., such that A-mod and B-mod are equivalent), then $Q(A) \cong Q(B)$. Representing a basic algebra as a quotient of a path algebra by an admissible ideal is sometimes called finding a *quiver presentation* of the algebra. It follows from Theorem 17.1 that if A is a finite dimensional algebra with $\operatorname{rad}(A)^2 = 0$, then A is determined up to Morita equivalence by $Q(A)$. The next theorem is also due to Gabriel [Gab72]. See [Ben98, Proposition 4.2.4] and [ASS06, Theorem VII.1.7] for a proof.

Theorem 17.2. *Let $\Bbbk$ be an algebraically closed field and Q a finite acyclic quiver. Then $\Bbbk Q$ is hereditary. Moreover, a basic finite dimensional $\Bbbk$-algebra A is hereditary if and only if $Q(A)$ is acyclic and $A \cong \Bbbk Q(A)$.*

In other words, up to Morita equivalence, the basic finite dimensional algebras over an algebraically closed field are the path algebras of acyclic quivers.

17.2 Projective indecomposable modules for $\mathscr{R}$-trivial monoid algebras

In this section we give an explicit description of the projective indecomposable modules for the algebra $\Bbbk M$ of an $\mathscr{R}$-trivial monoid M. The results here are from the paper of Margolis and the author [MS12a], simultaneously generalizing earlier results of Saliola for left regular bands [Sal07] and Denton, Hivert, Schilling, and Thiéry [DHST11] for $\mathscr{J}$-trivial monoids. Fix for this section a field $\Bbbk$ and a finite monoid M.

One defines an equivalence relation $\widetilde{\mathscr{L}}$ on M, following Fountain, Gomes, and Gould [FGG99], by putting $m \widetilde{\mathscr{L}} n$ if $me = m$ if and only if $ne = n$, for all idempotents $e \in E(M)$, that is, $m \widetilde{\mathscr{L}} n$ if and only if they have the same idempotent right identities. The $\widetilde{\mathscr{L}}$-class of an element m will be denoted $\widetilde{L}_m$. One can dually define the $\widetilde{\mathscr{R}}$-relation on a monoid by $m \widetilde{\mathscr{R}} n$ if $em = m$ if and only if $en = n$ for all $e \in E(M)$. We write $\widetilde{R}_m$ for the $\widetilde{\mathscr{R}}$-class of $m \in M$. The reason for the notation $\widetilde{\mathscr{L}}$ is furnished by the following proposition.

Proposition 17.3. *Let $m, n \in M$. Then $m \mathscr{L} n$ implies $m \widetilde{\mathscr{L}} n$. The converse holds if m, n are regular.*

Proof. If $m \mathscr{L} n$, then $n = um$ with $u \in M$. Thus we have that if $e \in E(M)$ with $me = m$, then $ne = ume = um = n$. Dually, we have that $ne = n$ implies $me = m$. Thus $m \widetilde{\mathscr{L}} n$. Next assume that m, n are regular and that $m \widetilde{\mathscr{L}} n$. Then $m = mam$ for some $a \in M$. But $(am)^2 = am$ and so $nam = n$ because $m \widetilde{\mathscr{L}} n$. Therefore $n \in Mm$. A dual argument shows that $m \in Mn$ and so $m \mathscr{L} n$, as required. $\qquad\square$

Thus $\mathscr{L}$ and $\widetilde{\mathscr{L}}$ coincide for regular monoids. It is not in general true that every $\widetilde{\mathscr{L}}$-class of a monoid contains an idempotent. However, this is the case for $\mathscr{R}$-trivial monoids.

Proposition 17.4. *Let M be a finite $\mathscr{R}$-trivial monoid. Then each $\widetilde{\mathscr{L}}$-class of M contains an idempotent (unique up to $\mathscr{L}$-equivalence).*

Proof. The uniqueness statement follows from Proposition 17.3. Let $m \in M$ and let $N = \{n \in M \mid mn = m\}$. Then N is a submonoid of M. Let e be an idempotent of the minimal ideal of N. We claim that $m \widetilde{\mathscr{L}} e$. Indeed, if $f \in E(M)$ and $ef = e$, then trivially $mf = mef = me = m$. Conversely, if $mf = m$, then $f \in N$ and ef is in the minimal ideal of N. By stability (Theorem 1.13), it follows that $eN = efN$. But N is an $\mathscr{R}$-trivial monoid because M is one. Thus $e = ef$. We conclude that $m \widetilde{\mathscr{L}} e$, as required. $\qquad\square$

We now wish to put a $\Bbbk M$-module structure on $\Bbbk \widetilde{L}_m$ for $m \in M$.

Proposition 17.5. *Let $e \in E(M)$ and $X = Me \setminus \widetilde{L}_e$. Then X is a left ideal. Therefore, $\Bbbk Me/\Bbbk X \cong \Bbbk \widetilde{L}_e$ has a natural $\Bbbk M$-module structure given by*

$$m \odot z = \begin{cases} mz, & \text{if } mz \in \widetilde{L}_e \\ 0, & \text{else} \end{cases}$$

for $m \in M$ and $z \in \widetilde{L}_e$.

Proof. First note that because $ee = e$, we have that $me = m$ for all $m \in \widetilde{L}_e$. Thus $\widetilde{L}_e \subseteq Me$. Trivially, if $f \in E(M)$ and $ef = e$, then $mf = m$ for all $m \in Me$. Hence if $x \in Me \setminus \widetilde{L}_e$, then there exists $f \in E(M)$ with $xf = x$ and $ef \neq e$. But then $mxf = mx$ for all $m \in M$ and so $mx \notin \widetilde{L}_e$. We conclude that X is a left ideal. The remainder of the proposition is immediate. $\square$

We drop from now on the notation "$\odot$".

Let M be an $\mathscr{R}$-trivial monoid. In the proof of Corollary 2.7 it was shown that each regular $\mathscr{J}$-class of M is an $\mathscr{L}$-class. In light of this, and Corollary 5.7, the simple $\Bbbk M$-modules are all one-dimensional and in bijection with $\mathscr{L}$-classes of idempotents. To each such $\mathscr{L}$-class L_e (with $e \in E(M)$), we have the corresponding simple S_{L_e} with underlying vector space $\Bbbk$ and with action

$$mc = \begin{cases} c, & \text{if } MeM \subseteq MmM \\ 0, & \text{else} \end{cases}$$

for $m \in M$ and $c \in \Bbbk$. Our goal is to show that $\Bbbk \widetilde{L}_e$ is a projective indecomposable module with simple quotient S_{L_e}.

Proposition 17.6. *Let M be an $\mathscr{R}$-trivial monoid and $e \in E(M)$. Then there is a surjective $\Bbbk M$-module homomorphism $\eta_e \colon \Bbbk \widetilde{L}_e \longrightarrow S_{L_e}$ given by*

$$\eta_e(m) = \begin{cases} 1, & \text{if } m \in L_e \\ 0, & \text{if } m \in I(e) \end{cases}$$

for $m \in \widetilde{L}_e$. Moreover, $\ker \eta_e = \mathrm{rad}(\Bbbk \widetilde{L}_e)$.

Proof. If $n \in M$ with $MeM \subseteq MnM$, then $nm \in L_e = J_e$ for all $m \in L_e$ by Corollary 2.7, as J_e is coprime and hence is the minimal ideal of $M \setminus I(e)$ by Lemma 2.2. Otherwise, if $n \in I(e)$, then we have that $nL_e \subseteq Me \cap I(e) = Me \setminus L_e$. As $\widetilde{L}_e \setminus L_e$ is contained in the left ideal $Me \setminus L_e$, it follows that η_e is a $\Bbbk M$-module homomorphism, which is clearly surjective. The inclusion $\mathrm{rad}(\Bbbk \widetilde{L}_e) \subseteq \ker \eta_e$ holds because S_{L_e} is simple. For the converse, it suffices by Theorem A.5 to prove that $\ker \eta_e \subseteq \mathrm{rad}(\Bbbk M)\Bbbk \widetilde{L}_e$.

Note that $\ker \eta_e$ is spanned by the elements of the form $x - x'$ with $x, x' \in L_e$ and the elements $y \in \widetilde{L}_e \setminus L_e$. First let $y \in \widetilde{L}_e \setminus L_e$. By Corollary 11.6, we have that $y - y^\omega \in \mathrm{rad}(\Bbbk M)$. Observe that $y^\omega \in Me \setminus \widetilde{L}_e$. Indeed, we have that $\widetilde{L}_e \subseteq Me$ and so $My^\omega \subseteq My \subsetneq Me$. By Proposition 17.3 it follows that $y^\omega \notin \widetilde{L}_e$. So we have in the module $\Bbbk\widetilde{L}_e$ that $(y - y^\omega)e = ye = y$ because $y^\omega e = y^\omega \notin \widetilde{L}_e$. We conclude that $y \in \mathrm{rad}(\Bbbk M)\Bbbk\widetilde{L}_e$.

Next suppose that $x, x' \in L_e$. Then $x - x' \in \mathrm{rad}(\Bbbk M)$ by Corollary 11.6, as $x, x' \in E(M)$ and $MxM = Mx'M$. But $(x - x')e = x - x'$ because $x, x' \in Me$. Thus $x - x' \in \mathrm{rad}(\Bbbk M)\Bbbk\widetilde{L}_e$. This establishes that $\ker \eta_e \subseteq \mathrm{rad}(\Bbbk\widetilde{L}_e)$.   □

As a corollary, we may deduce that $\eta_e \colon \Bbbk\widetilde{L}_e \longrightarrow S_{L_e}$ is a projective cover.

Theorem 17.7. *Let M be a finite $\mathscr{R}$-trivial monoid and $\Bbbk$ a field. Let $e_1, \ldots, e_s$ form a complete set of representatives of the $\mathscr{L}$-classes of idempotents of M. Then the modules $\Bbbk\widetilde{L}_{e_1}, \ldots, \Bbbk\widetilde{L}_{e_s}$ form a complete set of representatives of the isomorphism classes of projective indecomposable $\Bbbk M$-modules,*

$$\Bbbk M \cong \bigoplus_{i=1}^{s} \Bbbk\widetilde{L}_{e_i}$$

and $\eta_{e_i} \colon \Bbbk\widetilde{L}_{e_i} \longrightarrow S_{L_{e_i}}$ is a projective cover for $i = 1, \ldots, s$.

Proof. One has $\Bbbk M / \mathrm{rad}(\Bbbk M) \cong \Bbbk\Lambda(M) \cong \Bbbk^{|\Lambda(M)|} = \Bbbk^s$ by Corollaries 11.6 and 9.5. It follows that $\Bbbk M / \mathrm{rad}(\Bbbk M) \cong \bigoplus_{i=1}^{s} S_{L_{e_i}}$ as a $\Bbbk M$-module. Let $V = \bigoplus_{i=1}^{s} \Bbbk\widetilde{L}_{e_i}$. Then $V / \mathrm{rad}(V) \cong \bigoplus_{i=1}^{s} S_{L_{e_i}}$ by Proposition 17.6. Thus there is a projective cover $\pi \colon \Bbbk M \longrightarrow V$ by Theorem A.17. The homomorphism π is an epimorphism. On the other hand, $M = \bigcup_{i=1}^{s} \widetilde{L}_{e_i}$, by Proposition 17.4, and the union is disjoint. Thus we have that

$$|M| = \sum_{i=1}^{s} |\widetilde{L}_{e_i}| = \dim V$$

and so the epimorphism π is an isomorphism by dimension considerations. We conclude that $V \cong \Bbbk M$ and hence each $\Bbbk\widetilde{L}_{e_i}$ is a projective $\Bbbk M$-module. It now follows from Proposition 17.6 that

$$\eta_{e_i} \colon \Bbbk\widetilde{L}_{e_i} \longrightarrow S_{L_{e_i}}$$

is a projective cover and hence $\Bbbk\widetilde{L}_{e_i}$ is a projective indecomposable $\Bbbk M$-module, for $i = 1, \ldots, s$, by Theorem A.17. This completes the proof.   □

A complete set of orthogonal primitive idempotents for the algebra of an $\mathscr{R}$-trivial monoid was constructed in [BBBS11].

The following result will be used in the next two sections.

Proposition 17.8. *Let M be an $\mathscr{R}$-trivial monoid and $\Bbbk$ a field. Let $e, f \in E(M)$. Then $\mathrm{Ext}^1_{\Bbbk M}(S_{L_e}, S_{L_f}) \cong \mathrm{Hom}_{\Bbbk M}(\mathrm{rad}(\Bbbk\widetilde{L}_e), S_{L_f})$.*

Proof. Let $\widetilde{L}_e$ be the $\widetilde{\mathscr{L}}$-class of $e \in E(M)$. Then we have the exact sequence

$$0 \longrightarrow \mathrm{rad}(\Bbbk\widetilde{L}_e) \longrightarrow \Bbbk\widetilde{L}_e \longrightarrow S_{L_e} \longrightarrow 0$$

and hence an exact sequence

$$0 \longrightarrow \mathrm{Hom}_{\Bbbk M}(S_{L_e}, S_{L_f}) \longrightarrow \mathrm{Hom}_{\Bbbk M}(\Bbbk\widetilde{L}_e, S_{L_f})$$

$$\longrightarrow \mathrm{Hom}_{\Bbbk M}(\mathrm{rad}(\Bbbk\widetilde{L}_e), S_{L_f}) \longrightarrow \mathrm{Ext}^1_{\Bbbk M}(S_{L_e}, S_{L_f}) \longrightarrow 0$$

by the projectivity of $\Bbbk\widetilde{L}_e$. The map

$$\mathrm{Hom}_{\Bbbk M}(S_{L_e}, S_{L_f}) \longrightarrow \mathrm{Hom}_{\Bbbk M}(\Bbbk\widetilde{L}_e, S_{L_f})$$

is an isomorphism because $\mathrm{rad}(\Bbbk\widetilde{L}_e)$ is contained in the kernel of any homomorphism from $\Bbbk\widetilde{L}_e$ to the simple module S_{L_f} and hence $\mathrm{Ext}^1_{\Bbbk M}(S_{L_e}, S_{L_f}) \cong \mathrm{Hom}_{\Bbbk M}(\mathrm{rad}(\Bbbk\widetilde{L}_e), S_{L_f})$. □

Let us end this section with an explicit computation of the Cartan matrix of the algebra of an $\mathscr{R}$-trivial monoid from [MS12a], generalizing earlier results of Saliola [Sal07] and Denton *et al.* [DHST11]. The reader should refer to Appendix C for the definition of the Möbius function of a poset.

Theorem 17.9. *Let M be a finite $\mathscr{R}$-trivial monoid and $\Bbbk$ an algebraically closed field. Let $e_1, \ldots, e_s$ form a complete set of idempotent representatives of the regular $\mathscr{J}$-classes of M. Then the Cartan matrix of M is given by*

$$C_{ij} = \sum_{Me_kM \subseteq Me_iM} |e_k M \cap \widetilde{L}_{e_j}| \cdot \mu(Me_kM, Me_iM)$$

where μ is the Möbius function of the lattice $\Lambda(M) = \{MeM \mid e \in E(M)\}$.

Proof. Recall that C_{ij} is $[\Bbbk\widetilde{L}_{e_j} : S_{L_{e_i}}]$ because $\Bbbk\widetilde{L}_{e_j} \longrightarrow S_{L_{e_j}}$ is a projective cover. The character χ of $\Bbbk\widetilde{L}_{e_j}$ is given by

$$\chi(m) = \left| \{x \in \widetilde{L}_{e_j} \mid mx = x\} \right|.$$

Therefore, we have $\chi(e) = |eM \cap \widetilde{L}_{e_j}|$ for $e \in E(M)$. The result now follows from Corollary 7.21. □

Let us specialize this result to the case of a $\mathscr{J}$-trivial monoid M. Since each regular $\mathscr{J}$-class of M, and hence each regular $\mathscr{L}$-class, contains a single idempotent, it follows that each $\widetilde{\mathscr{L}}$-class of M contains a unique idempotent by Proposition 17.3. If $m \in M$, denote by m^- the unique idempotent in $\widetilde{L}_m$. Each $\widetilde{\mathscr{R}}$-class will also contain a unique idempotent by the dual of Proposition 17.3. The unique idempotent in $\widetilde{R}_m$ will be denoted by m^+. One then has that $m^+mm^- = m$ and if $e, f \in E(M)$ with $emf = m$, then $em^+ = m^+$ and $m^-f = m^-$. The following result is from the paper [DHST11].

Theorem 17.10. *Let M be a finite $\mathscr{J}$-trivial monoid and $\Bbbk$ an algebraically closed field. Suppose that $E(M) = \{e_1, \ldots, e_s\}$. Then the Cartan matrix of M is given by*

$$C_{ij} = \left|\{m \in M \mid m^+ = e_i,\ m^- = e_j\}\right|$$

for $1 \leq i, k \leq s$.

Proof. Observe that $m \in e_i M \cap \widetilde{L}_{e_j}$ if and only if $m^- = e_j$ and $Mm^+M \subseteq Me_iM$ because in a $\mathscr{J}$-trivial monoid $MeM \subseteq MfM$ if and only if $fe = e$ for all $e, f \in E(M)$ by the dual of Corollary 2.7. Thus we have that

$$\left|e_i M \cap \widetilde{L}_{e_j}\right| = \sum_{Me_k M \subseteq Me_i M} \left|\{m \in M \mid m^+ = e_k,\ m^- = e_j\}\right|. \tag{17.2}$$

An application of Möbius inversion (Theorem C.6) to (17.2) yields

$$\left|\{m \in M \mid m^+ = e_i,\ m^- = e_j\}\right| = \sum_{Me_k M \subseteq Me_i M} \left|e_k M \cap \widetilde{L}_{e_j}\right| \mu(Me_k M, Me_i M)$$

from which the result follows in conjunction with Theorem 17.9.   □

Since $\mathscr{R}$-trivial monoids have basic algebras over any algebraically closed field, being triangularizable over any field, it is natural to compute the quivers of their algebras. This was accomplished in [MS12a]. We present here some particular cases that are more accessible.

17.3 The quiver of a left regular band algebra

Fix for this section an algebraically closed field $\Bbbk$. Recall that a left regular band is a regular $\mathscr{R}$-trivial monoid and that each element of a left regular band is idempotent. By Proposition 17.3, we have that $\widetilde{L}_m = L_m$, for $m \in M$, and hence the projective cover of S_{L_e} is $\Bbbk L_e$ by Theorem 17.7.

The following theorem is due to Saliola [Sal07]. A topological proof is given in [MSS15b]. We retain the notation of the previous section.

Theorem 17.11. *Let M be a left regular band and $\Bbbk$ an algebraically closed field. Let $Q(\Bbbk M)$ be the quiver of $\Bbbk M$. Then the vertex set of $Q(\Bbbk M)$ is in bijection with the set $\Lambda(M) = \{MmM \mid m \in M\}$. There are no arrows from the simple module corresponding to MmM to the simple module corresponding to MnM unless $MmM \subsetneq MnM$. If $MmM \subsetneq MnM$, then the number of arrows from the simple module corresponding to MmM to the simple module corresponding to MnM is one less than the number of connected components of the graph $\Gamma(MmM, MnM)$ with vertex set $nM \cap L_m$ and adjacency relation $v \sim w$ if there exists $x \in nM \setminus \{n\}$ with $xv = v$ and $xw = w$.*

Proof. We already know that there is a bijection between $\Lambda(M)$ and simple modules via $MmM \mapsto S_{L_m}$. By Proposition 17.8, we have $\operatorname{Ext}^1_{\Bbbk M}(S_{L_m}, S_{L_n}) \cong \operatorname{Hom}_{\Bbbk M}(\operatorname{rad}(\Bbbk L_m), S_{L_n})$ as $\widetilde{L}_m = L_m$ for $m \in M = E(M)$.

Since the mapping $\Bbbk L_m \longrightarrow S_{L_m}$ sends each basis element $x \in L_m$ to 1, its kernel $\operatorname{rad}(\Bbbk L_m)$ is spanned by all differences $x - y$ with $x, y \in L_m$ and has basis $x - m$ with $x \in L_m \setminus \{m\}$. Suppose that it is not true that $MmM \subsetneq MnM$. Let $\varphi \in \operatorname{Hom}_{\Bbbk M}(\operatorname{rad}(\Bbbk L_m), S_{L_n})$. Then $\varphi(x - m) = n\varphi(x - m) = \varphi(nx - nm) = 0$ because if $MmM = MnM$, then $nx = n = nm$ and if $MmM \not\subseteq MnM$, then n annihilates $\Bbbk L_m$ and hence its submodule $\operatorname{rad}(\Bbbk L_m)$. We conclude that $\varphi = 0$ and hence $\dim \operatorname{Hom}_{\Bbbk M}(\operatorname{rad}(\Bbbk L_m), S_{L_n}) = 0$. Thus there are no arrows from the simple module corresponding to MmM to the simple module corresponding to MnM. We remark that we could have used Proposition 16.9 to reduce to the case that MmM and MnM are comparable but not equal.

So we are left with the case $MmM \subsetneq MnM$. Then $L_{nm} = L_m$ and so without loss of generality we may assume that $nm = m$. Fix, for each connected component C of $\Gamma(MmM, MnM)$, a vertex $\overline{C}$. We write C_x for the component of $x \in nM \cap L_m$. Assume that $m = \overline{C_m}$. Let V be the subspace of $\operatorname{rad}(\Bbbk L_m)$ spanned by all differences $\overline{C} - m$ with C a connected component different than C_m. It is easy to see that the elements $\overline{C_x} - m$ with $C_x \neq C_m$ form a basis for V and so $\dim V$ is one less than the number of connected components of $\Gamma(MmM, MnM)$. Recall that $S_{L_n} = \Bbbk$ as a vector space. Hence there is a $\Bbbk$-vector space homomorphism

$$\Psi \colon \operatorname{Hom}_{\Bbbk M}(\operatorname{rad}(\Bbbk L_m), S_{L_n}) \longrightarrow \operatorname{Hom}_{\Bbbk}(V, \Bbbk)$$

given by $\varphi \mapsto \varphi|_V$. We claim that Ψ is an isomorphism.

Suppose that $\Psi(\varphi) = 0$. Let $x \in L_m$. We need to show that $\varphi(x - m) = 0$. Because $\varphi(x - m) = n\varphi(x - m) = \varphi(nx - nm) = \varphi(nx - m)$, we may assume without loss of generality that $x \in nM \cap L_m$. We claim that if $a \sim b$ are adjacent vertices of $\Gamma(MmM, MnM)$, then $\varphi(a - m) = \varphi(b - m)$. Indeed, there exists $c \in nM \setminus \{n\}$ with $ca = a$ and $cb = b$. Therefore, $\varphi(a - b) = \varphi(c(a - b)) = c\varphi(a - b) = 0$ because c annihilates S_{L_n}. It is then clear that $\varphi(a - m) = \varphi(b - m)$. It follows that $\varphi(a - m) = \varphi(b - m)$ whenever a, b are in the same connected component and hence $\varphi(x - m) = \varphi(\overline{C_x} - m) = 0$. This concludes the proof that Ψ is injective.

Next suppose that $\eta \colon V \longrightarrow \Bbbk$ is a linear map. Define a $\Bbbk$-linear map

$$\varphi \colon \operatorname{rad}(\Bbbk L_m) \longrightarrow S_{L_n}$$

by

$$\varphi(x - m) = \eta(\overline{C_{nx}} - m) \tag{17.3}$$

for $x \in L_m \setminus \{m\}$. Note that $\varphi(m - m) = 0 = \eta(\overline{C_{nm}} - m)$ because $nm = m$ and $\overline{C_m} = m$. Thus (17.3) is also valid for $x = m$.

If $MnM \subseteq MzM$, then $nz = n$ and so

$$\varphi(z(x - m)) = \varphi(zx - m + m - zm)$$
$$= \eta(\overline{C_{nzx}} - m) - \eta(\overline{C_{nzm}} - m)$$
$$= \eta(\overline{C_{nx}} - m)$$
$$= \varphi(x - m)$$
$$= z\varphi(x - m)$$

since $nzx = nx$, $nzm = nm = m$, and $\overline{C_m} = m$.

If $MmM \not\subseteq MzM$, then $\varphi(z(x - m)) = \varphi(0) = z\varphi(x - m)$. If $MmM \subseteq MzM \subsetneq MnM$, then

$$\varphi(z(x - m)) = \varphi(zx - m + m - zm) = \eta(\overline{C_{nzx}} - m) - \eta(\overline{C_{nzm}} - m).$$

But $nz \in nM \setminus \{n\}$ implies that $nzx \sim nzm$ and hence $\overline{C_{nzx}} = \overline{C_{nzm}}$. Thus $\varphi(z(x - m)) = 0 = z\varphi(x - m)$. This completes the proof that φ is a $\Bbbk M$-module homomorphism. By construction, $\Psi(\varphi) = \eta$. We conclude that $\mathrm{Hom}_{\Bbbk M}(\mathrm{rad}(\Bbbk L_m), S_{L_n}) \cong \mathrm{Hom}_{\Bbbk}(V, \Bbbk)$. As V has dimension one less than the number of components of $\Gamma(MmM, MnM)$, the result follows.   □

The reader should check that the graph $\Gamma(MmM, MnM)$ depends up to isomorphism only on MmM and MnM and not on the choice of m or n.

17.4 The quiver of a $\mathscr{J}$-trivial monoid algebra

Fix for this section a finite $\mathscr{J}$-trivial monoid M and an algebraically closed field $\Bbbk$. Recall that if $m \in M$, then m^+ and m^- denote the unique idempotents in $\widetilde{R}_m$ and $\widetilde{L}_m$, respectively.

Let us say that $m \in M \setminus E(M)$ is *weakly irreducible* if $m = uv$ with $u^+ = m^+$ and $v^- = m^-$ implies that $u = m^+$ or $v = m^-$. The following theorem is due to Denton, Hivert, Schilling, and Thiéry [DHST11]. Our proof is novel to the text.

Theorem 17.12. *Let M be a $\mathscr{J}$-trivial monoid and $\Bbbk$ an algebraically closed field. Let $Q(\Bbbk M)$ be the quiver of $\Bbbk M$. Then the vertex set of $Q(\Bbbk M)$ is in bijection with $E(M)$. The number of arrows from the simple module corresponding to e to the simple module corresponding to f is the number of weakly irreducible elements $m \in M$ with $m^+ = f$ and $m^- = e$.*

Proof. As M is $\mathscr{J}$-trivial, there is a bijection between $E(M)$ and simple modules via $e \mapsto S_{L_e}$. We have that

$$\mathrm{Ext}^1_{\Bbbk M}(S_{L_e}, S_{L_f}) \cong \mathrm{Hom}_{\Bbbk M}(\mathrm{rad}(\Bbbk \widetilde{L}_e), S_{L_f})$$

by Proposition 17.8.

The unique regular element of $\widetilde{L}_e$ is $\{e\}$ and hence $\mathrm{rad}(\Bbbk\widetilde{L}_e)$ has basis

$$\widetilde{L}_e \setminus \{e\} = \{m \in Me \setminus \{e\} \mid m^- = e\}$$

by Proposition 17.6. Let V be the subspace of $\mathrm{rad}(\Bbbk\widetilde{L}_e)$ with basis the set X of weakly irreducible elements $m \in M$ with $m^- = e$ and $m^+ = f$. We claim that the restriction map

$$\Psi \colon \mathrm{Hom}_{\Bbbk M}(\mathrm{rad}(\Bbbk\widetilde{L}_e), S_{L_f}) \longrightarrow \mathrm{Hom}_{\Bbbk}(V, \Bbbk)$$

is an isomorphism where we recall that $S_{L_f} \cong \Bbbk$ as a $\Bbbk$-vector space.

Suppose that $\Psi(\varphi) = 0$ and $x \in \widetilde{L}_e \setminus \{e\}$. We must show that $\varphi(x) = 0$. Observe that

$$\varphi(x) = f\varphi(x) = \varphi(fx). \tag{17.4}$$

First note that if $fx \notin \widetilde{L}_e$, then $fx = 0$ in $\Bbbk\widetilde{L}_e$ and so $\varphi(x) = 0$ by (17.4). If $fx \in \widetilde{L}_e$, i.e., $(fx)^- = e$, then replacing x by fx and appealing to (17.4), we may assume without loss of generality that $x = fx$. But $fx = x$ implies that $Mx^+M \subseteq MfM$. If the containment is strict, then $\varphi(x) = \varphi(x^+x) = x^+\varphi(x) = 0$. We are thus left with the case $x^+ = f$. If x is not weakly irreducible, then $x = uv$ with $v^- = e$ and $u^+ = f$ and $u \neq f$, $v \neq e$. Then $v \in \mathrm{rad}(\Bbbk\widetilde{L}_e)$ and $\varphi(x) = \varphi(uv) = u\varphi(v) = 0$ because $MuM \subsetneqq MfM$. The remaining case is that $x^+ = f$, $x^- = e$ and x is weakly irreducible. But then $x \in X \subseteq V$ and so $\varphi(x) = 0$ by assumption. It follows that Ψ is injective.

Next assume that $\eta \colon V \longrightarrow \Bbbk$ is a linear map and define a mapping

$$\varphi \colon \mathrm{rad}(\Bbbk\widetilde{L}_e) \longrightarrow S_{L_f},$$

for $x \in \widetilde{L}_e \setminus \{e\}$, by

$$\varphi(x) = \begin{cases} \eta(fx), & \text{if } fx \in X \\ 0, & \text{else.} \end{cases}$$

We check that φ is a $\Bbbk M$-module homomorphism. It clearly restricts on V to η. Let $x \in \widetilde{L}_e \setminus \{e\}$. If $MfM \subseteq MzM$, then $fz = f$ and so $fx = fzx$. It follows that $\varphi(zx) = \varphi(x) = z\varphi(x)$. Next suppose that $MfM \nsubseteq MzM$. Then $MfzM \subsetneqq MfM$. If $(fzx)^+ \neq f$, then $fzx \notin X$ and so $\varphi(zx) = 0 = z\varphi(x)$. So suppose that $(fzx)^+ = f$. If $(fzx)^- \neq e$, then again $fzx \notin X$ and so $\varphi(zx) = 0 = z\varphi(x)$. So assume that $(fzx)^- = e$. Then since $(fz)^+(fzx) = fzx$ we obtain $(fz)^+f = f$. But $f(fz) = fz$ implies that $f(fz)^+ = (fz)^+$ and so $f = (fz)^+$. But since $fz \neq f$, $x^- = e$ and $x \neq e$, this shows that $fzx = (fz)x$ is not weakly irreducible and so $fzx \notin X$. Therefore, $\varphi(zx) = 0 = z\varphi(x)$. This completes the proof that φ is a $\Bbbk M$-module homomorphism. It follows that there are $\dim V = |X|$ arrows from the simple module corresponding to e to the simple module corresponding to f. This completes the proof of the theorem. $\qquad\square$

17.5 Sample quiver computations

In this section, we compute the quivers of several left regular band and $\mathscr{J}$-trivial monoid algebras. Throughout this section, $\Bbbk$ will be an algebraically closed field.

17.5.1 Left regular bands

Let B be a finite set and let $F(B)$ be the free left regular band on B. The reader is referred to Section 14.3.1 for details and notation. In particular, there is an isomorphism of $\Lambda(F(B))$ with $(\mathscr{P}(B), \supseteq)$ sending $F(B)wF(B)$ to $c(w)$ where $c(w)$ is the content of w, that is, the set of letters appearing in w. Denote by L_X the $\mathscr{L}$-class of elements with content X and fix $w_X \in F(B)$ with content X for each $X \subseteq B$.

Lemma 17.13. *Let $X \supsetneq Y$ and consider the graph*

$$\Gamma = \Gamma(F(B)w_X F(B), F(B)w_Y F(B))$$

(as per Theorem 17.11). Then the vertex set of Γ is the set of words of content X having w_Y as a prefix. Two words are connected by an edge if and only if they have the same prefix of length $|w_Y|+1$. Thus each connected component of Γ is a complete graph and there is one connected component for each element of $X \setminus Y$.

Proof. As $w_Y F(B)$ consists of all words with w_Y as a prefix, the description of the vertex set is clear. Suppose that $u, v \in w_Y F(B) \cap L_X$ are words connected by an edge. Then there exists $x \in w_Y F(B) \setminus \{w_Y\}$ with $x \odot u = u$ and $x \odot v = v$. But then x is a common prefix of u and v. Since $x \neq w_Y$, we conclude that $|x| > |w_Y|$ and so u, v have the same prefix of length $|w_Y| + 1$. Conversely, if $u, v \in w_Y F(B) \cap L_X$ have the same prefix x of length $|w_Y|+1$, then $x \odot u = u$ and $x \odot v = v$ and so u and v are adjacent as $x \in w_Y F(A) \setminus \{w_Y\}$. This proves that Γ is as described. It follows that Γ has a connected component for each $a \in X \setminus Y$ consisting of all words with content X and prefix $w_Y a$ and each component is a complete graph. This completes the proof. □

An immediate consequence of Lemma 17.13 and Theorem 17.11 is the following result of K. Brown, which first appeared in [Sal07].

Theorem 17.14. *The quiver $Q(\Bbbk F(B))$ of the algebra of the free left regular band on a finite set B is isomorphic to the quiver with vertex set the power set $\mathscr{P}(B)$ and edges as follows. There are no edges from X to Y unless $X \supsetneq Y$. If $X \supsetneq Y$, then there are $|X| - |Y| - 1$ edges from X to Y.*

K. Brown used Theorem 17.14 and Gabriel's theory of quivers to prove that $\Bbbk F(B)$ is hereditary; the result was first published in [Sal07]. A homological and topological proof appears in [MSS15b].

Next we compute the quiver of the algebra of L^n where we recall that $L = \{1, A, B\}$ with 1 as the identity and $xy = x$ for all $x, y \in \{A, B\}$. See Section 14.3.3 for details. Recall that $\Lambda(L^n)$ is isomorphic to $(\mathscr{P}([n]), \supseteq)$ via the map $L^n x L^n \mapsto c(x)$ where $c(x) = \{i \mid x_i \neq 1\}$ is the support of $x = (x_1, \ldots, x_n)$. For $S \subseteq [n]$, let L_S be the set of elements x with $c(x) = S$ and let e_S be the vector with $x_i = A$ if $i \in S$ and 1, otherwise.

Lemma 17.15. *Let* $S \supsetneq T$. *Then* $\Gamma = \Gamma(L^n e_S L^n, L^n e_T L^n)$ *(defined as per Theorem 17.11) has vertex set all* $x \in L^n$ *with* $c(x) = S$ *and with* $x_i = A$ *for all* $i \in T$. *If* $|S| = |T| + 1$, *then* Γ *consists of two isolated vertices. Otherwise,* Γ *is connected.*

Proof. First note that $e_T L^n$ consists of all elements $x \in L^n$ with $x_i = A$ for $i \in T$. Thus Γ has vertex set as in the statement of the lemma. Suppose first that $|S| = |T| + 1$, say, $S = T \cup \{i\}$. Then the only two vertices of Γ are x, y where x is obtained by replacing the 1 in position i of e_T by A and y is obtained by replacing it with B. Moreover, the only elements of $e_T L^n \setminus \{e_T\}$ that are not in $I(e_S)$ are x, y. As $xy = x$ and $yx = y$, we conclude that there are no edges in Γ in this case.

Next assume that $|S| > |T| + 1$. Notice that all vertices of Γ agree in each coordinate outside of $S \setminus T$. We claim that if two distinct vertices x, y of Γ satisfy $x_j = y_j$ for some $j \in S \setminus T$, then they are adjacent. Suppose that $x_j = y_j = X \in \{A, B\}$. Let z be obtained from e_T by changing coordinate j to X. Then $z \in e_T L^n \setminus \{e_T\}$ and $zx = x$, $zy = y$. Thus x, y are connected by an edge.

Now let x, y be arbitrary vertices. If they are not adjacent, then by the above $x_k \neq y_k$ for all $k \in S \setminus T$. As $|S| > |T| + 1$, we can find distinct $i, j \in S \setminus T$. Let z be the vertex of Γ given by

$$z_k = \begin{cases} x_i, & \text{if } k = i \\ y_k, & \text{else.} \end{cases}$$

Then $z_i = x_i$ implies that z is adjacent to x and $z_j = y_j$ implies that z is adjacent to y. We conclude that Γ is connected. $\square$

From Lemma 17.15 and Theorem 17.11 we obtain the following result (which is a special case of a result of Saliola [Sal09]).

Theorem 17.16. *The quiver* $Q(\Bbbk L^n)$ *is isomorphic to the Hasse diagram of* $(\mathscr{P}([n]), \supseteq)$, *that is, the quiver with vertex set* $\mathscr{P}([n])$ *and with an edge from* S *to* T *if and only if* $S \supseteq T$ *and* $|S| = |T| + 1$.

17.5.2 $\mathscr{J}$-trivial monoids

We next compute the quiver of the algebra of the Catalan monoid. The *Catalan monoid* C_n is the monoid of all mappings $f \colon [n] \longrightarrow [n]$ such that:

(i) $i \leq f(i)$ for all $i \in [n]$;

(ii) $i \leq j$ implies $f(i) \leq f(j)$.

In other words, C_n is the monoid of all order-preserving and nondecreasing maps on $[n]$. The reason for the name is that it is well known that the cardinality of C_n is the nth Catalan number $\frac{1}{n+1}\binom{2n}{n}$ (cf. [Sta99, Exercise 6.19(s)]).

Proposition 17.17. *The Catalan monoid C_n is $\mathscr{J}$-trivial.*

Proof. We claim that $C_n f C_n \subseteq C_n g C_n$ implies that g is pointwise below f. From the claim it is immediate that C_n is $\mathscr{J}$-trivial. Indeed, if $f = hgk$ and $i \in [n]$, then $i \leq k(i)$ and $g(i) \leq h(g(i))$ imply that $f(i) = h(g(k(i))) \geq h(g(i)) \geq g(i)$ as h, g are order-preserving. □

A deep theorem of Simon [Sim75] implies that if M is a finite $\mathscr{J}$-trivial monoid, then there exists $n \geq 1$ and a submonoid N of C_n such that M is a homomorphic image of N. See [Pin86] for details. The Catalan monoid has been studied by several authors, under different names, in the context of representation theory [HT09, DHST11, Gre12, GM14].

In order to further understand C_n, we must compute its set of idempotents. Let $P_n = \{S \subseteq [n] \mid n \in S\}$ be the collection of subsets of $[n]$ containing n. Notice that $f(n) = n$ for all $f \in C_n$ because $n \leq f(n) \in [n]$. Therefore, the image of each element of C_n belongs to P_n. Recall that the *rank* of a mapping is the cardinality of its image.

Proposition 17.18. *For each $S \in P_n$, there is a unique idempotent $e_S \in C_n$ with image S. Moreover, one has*

$$e_S(i) = \min\{s \in S \mid i \leq s\}$$

for $S \in P_n$.

Proof. First suppose that $e, f \in E(C_n)$ have the same image set S. As e fixes S, we conclude that $ef = f$ and similarly $fe = e$. Hence $e = f$ by $\mathscr{J}$-triviality. Thus there is at most one idempotent with image set S. Let us verify that e_S is an idempotent of C_n with image set S.

Trivially, $e_S([n]) \subseteq S$. Moreover, if $s \in S$, then $e_S(s) = s$ by definition. Thus e_S is an idempotent mapping with image S. It remains to prove that $e_S \in C_n$. If $i \in [n]$, then $i \leq e_S(i)$ from the definition of e_S. If $j \leq i$, then $j \leq i \leq e_S(i) \in S$. Thus $e_S(j) \leq e_S(i)$ by definition of e_S. This completes the proof. □

If M is a $\mathscr{J}$-trivial monoid, then, for idempotents $e, f \in E(M)$, one has that $e \leq f$ if and only if $MeM \subseteq MfM$ by Corollary 2.7 and its dual.

Proposition 17.19. *Let $S, T \in P_n$. Then $C_n e_S C_n \subseteq C_n e_T C_n$ if and only if $S \subseteq T$. This occurs if and only if $e_S e_T = e_S = e_T e_S$.*

Proof. As discussed before the proposition, we have $C_n e_S C_n \subseteq C_n e_T C_n$ if and only if $e_S \leq e_T$ because C_n is $\mathcal{J}$-trivial. So suppose that $C_n e_S C_n \subseteq C_n e_T C_n$. Then $e_T e_S = e_S$ and so the image S of e_S is contained in the image T of e_T. Conversely, if $S \subseteq T$, then e_T fixes the image S of e_S and so $e_T e_S = e_S$, whence $C_n e_S C_n \subseteq C_n e_T C_n$. This completes the proof. $\square$

Our next goal is to compute f^+ and f^- for $f \in C_n$. It will turn out that f is, in fact, uniquely determined by the pair (f^+, f^-). If $S, T \in P_n$, let us write $S \preceq T$ if $|S| = |T|$ and if

$$S = \{i_1 < \ldots < i_k = n\} \quad \text{and} \quad T = \{j_1 < \ldots < j_k = n\}, \qquad (17.5)$$

with $i_r \leq j_r$ for all $1 \leq r \leq k$. It is routine to verify that this is a partial order on P_n.

Proposition 17.20. *Let $f \in C_n$. Let T be the image of f and let*

$$S = \{s \in [n] \mid s = \max f^{-1}(f(s))\}.$$

Then $f^- = e_S$ and $f^+ = e_T$. Moreover, $S \preceq T$ and if S and T are as in (17.5), then $f(x) = j_r$ if and only if $i_{r-1} < x \leq i_r$ where we take $i_0 = 0$. Hence f is uniquely determined by f^- and f^+. Conversely, if $S \preceq T$ and are as in (17.5), then f, defined as above, belongs to C_n and satisfies $f^- = e_S$ and $f^+ = e_T$.

Proof. Clearly, $e_X f = f$ if and only if X contains the range of f and so $e_T = f^+$ by Proposition 17.19. Suppose that $f e_X = f$. For $s \in S$, we have that $s \leq e_X(s)$. But also $f(e_X(s)) = f(s)$ and so by definition of S we must have $e_X(s) \leq s$. Thus e_X fixes S, that is, $S \subseteq X$. It follows that $f e_X = f$ implies that $e_S \leq e_X$ by Proposition 17.19. Thus to prove that $f^- = e_S$, it remains to show that $f e_S = f$.

If $i \in [n]$ and $s = \max f^{-1}(f(i))$, then $s \in S$ and $i \leq s$. Thus $i \leq e_S(i) \leq s$ by Proposition 17.18. Therefore, $f(s) = f(i) \leq f(e_S(i)) \leq f(s)$ and so $f(i) = f(e_S(i))$. This proves that $f = f e_S$.

From the definition of S, it easily follows that $f|_S$ is injective with image T. Thus $|S| = |T|$. Let S and T be as in (17.5). Then since f is order-preserving and nondecreasing, we have that $f(x) = j_r$ if and only if $i_{r-1} < x \leq i_r$, where we take $i_0 = 0$, and $i_r \leq j_r$. In particular, we have $S \preceq T$.

For the converse, if $S \preceq T$ are as in (17.5) and we define f as above, then f is trivially order-preserving. If $i_{r-1} < x \leq i_r$, then $f(x) = j_r \geq i_r \geq x$ and so f is nondecreasing. By construction and the first part of the proposition, $f^+ = e_T$ and $f^- = e_S$. $\square$

As a consequence of Proposition 17.20 and Theorem 17.10, we may deduce that the Cartan matrix of $\Bbbk C_n$ is the zeta function of the poset $(P_n, \succeq)$ (see Appendix C for the definition of the zeta function of a poset).

Corollary 17.21. *Let $\Bbbk$ be an algebraically closed field. The Cartan matrix of $\Bbbk C_n$ is the zeta function of the poset $(P_n, \succeq)$ (viewed as a $P_n \times P_n$-matrix).*

We now aim to characterize the weakly irreducible elements of C_n. First we need a lemma.

Lemma 17.22. *Suppose that $g, h \in C_n$ with $g^- = h^+$. Then $(gh)^+ = g^+$ and $(gh)^- = h^-$.*

Proof. Let us put $g^- = e_U = h^+$, $g^+ = e_T$ and $h^- = e_S$. Then $h(S) = U$, $g(U) = T = g([n])$, and $|S| = |U| = |T|$ by Proposition 17.20. Therefore, $gh(S) = T = g([n])$. We deduce that T is the image of gh and hence $(gh)^+ = e_T$ by Proposition 17.20. It follows that gh has rank $|T|$. Trivially, $ghe_S = gh$ and so if $(gh)^- = e_R$, then $R \subseteq S$ by Proposition 17.19. But $|R|$ is the rank of gh, which is $|T| = |S|$ and so $R = S$. Thus $(gh)^- = e_S$. This completes the proof. $\qquad\square$

If $S, T \subseteq [n]$, then their symmetric difference is $S \triangle T = (S \setminus T) \cup (T \setminus S)$.

Proposition 17.23. *One has that $f \in C_n$ is weakly irreducible if and only if $f^- = e_S$ and $f^+ = e_T$ where T covers S in the partial order $\preceq$, that is, $S \triangle T = \{i, i+1\}$ with $i \in S$ and $i+1 \in T$.*

Proof. From the definition of $\preceq$, it is straightforward to verify that T covers S if and only if $|S| = |T|$ and $S \triangle T = \{i, i+1\}$ with $i \in S$ and $i+1 \in T$. It follows from Proposition 17.20 that $f \notin E(C_n)$ if and only if $S \neq T$ and that in this case $S \prec T$. So assume that $f \notin E(C_n)$.

Suppose first that $f = gh$ with $g^+ = e_T$, $h^- = e_S$, $g \neq e_T$, and $h \neq e_S$. Then $h^+ = e_U$ with $S \prec U$ and $g^- = e_V$ with $V \prec T$. Notice that $|U| = |S| = |T| = |V|$ is the rank of f, g, h and $U = h([n])$. Therefore, we must have that $g|_U$ is injective. Since g is order-preserving and nondecreasing, we deduce using Proposition 17.20 that $U \preceq V$ and hence $S \prec U \preceq V \prec T$. Thus T does not cover S.

Conversely, if T does not cover S, then we can find U with $S \prec U \prec T$. By Proposition 17.20, we can find g, h with $g^- = U$, $g^+ = T$, $h^- = S$, and $h^+ = U$. Moreover, note that $g, h \notin E(C_n)$. Then Lemma 17.22 yields that $(gh)^+ = e_T$ and $(gh)^- = e_S$. Therefore, $gh = f$ by Proposition 17.20 and hence f is not weakly irreducible. $\qquad\square$

Putting together Propositions 17.20 and 17.23 with Theorem 17.12, we obtain the following result, which was proved in Denton et al. [DHST11] in a different formulation using 0-Hecke monoids.

Theorem 17.24. *The quiver $Q(\Bbbk C_n)$ of the algebra of the Catalan monoid C_n is isomorphic to the Hasse diagram of $(P_n, \preceq)$. The latter has vertex set the subsets of $[n]$ containing n and there is an edge from S to T if and only if their symmetric difference is $S \triangle T = \{i, i+1\}$ with $i \in S$ and $i+1 \in T$.*

In fact, the algebra $\Bbbk C_n$ is isomorphic to the algebra of the category associated with the poset P_n (cf. Example 8.2) or, equivalently, isomorphic to the opposite of the incidence algebra of this poset (see Appendix C for the definition). This was first proved by Hivert and Thiéry [HT09]. An alternative proof was given in Grensing [Gre12]. Our proof here is new to the best of our knowledge.

Theorem 17.25. *Let C_n denote the Catalan monoid of degree n and $\Bbbk$ a field. Then $\Bbbk C_n$ is isomorphic to the category algebra of the poset P_n. Equivalently, $\Bbbk C_n$ is isomorphic to the incidence algebra of $(P_n, \succeq)$ over $\Bbbk$.*

Proof. Denote also by P_n the category associated with the poset $(P_n, \preceq)$ as per Example 8.2. Let $A = \Bbbk C_n$ and $B = \Bbbk P_n$. Note that $\dim B = |C_n| = \dim A$ because there is a bijection between the arrow set of P_n and C_n taking (S, T) with $S \preceq T$ to the unique element $g_{S,T} \in C_n$ with $g_{S,T}^- = e_S$ and $g_{S,T}^+ = e_T$ as per Proposition 17.20. The strategy for the proof is as follows. We shall define an injective homomorphism $\rho \colon B^{op} \longrightarrow \operatorname{End}_A(A) \cong A^{op}$ (the isomorphism is by Proposition A.20). Then since $\dim A = \dim B$, we will be able to conclude that ρ is an isomorphism and hence $A \cong B$. To implement this scheme, we shall use that $A \cong \bigoplus_{S \in P_n} \Bbbk \widetilde{L}_{e_S}$ by Theorem 17.7.

So let (S, T) be an arrow of P. Then we can define a $\Bbbk$-linear map $\rho_{S,T} \colon \Bbbk \widetilde{L}_{e_T} \longrightarrow \Bbbk \widetilde{L}_{e_S}$ on the basis by $\rho_{S,T}(f) = f g_{S,T}$. Note that $(f g_{S,T})^- = g_{S,T}^- = e_S$ by Lemma 17.22 because $f^- = e_T = g_{S,T}^+$. In fact, $\rho_{S,T}$ is a left A-module homomorphism. Indeed, if $f \in \widetilde{L}_{e_T}$ and $h \in C_n$ with $hf \in \widetilde{L}_{e_T}$, then $h(f g_{S,T}) \in \widetilde{L}_{e_S}$ by Lemma 17.22 and $\rho_{S,T}(hf) = (hf)g_{S,T} = h(f g_{S,T}) = h\rho_{S,T}(f)$. On the other hand, if $hf \notin \widetilde{L}_{e_T}$, then since $hfe_T = hf$, we must have $(hf)^- = e_U$ with $U \subsetneq T$ by Proposition 17.19. Therefore, hf has rank $|U| < |T|$. It follows that $hf g_{S,T}$ has rank at most $|U|$, which is less than $|S|$. We conclude that $h(f g_{S,T}) \notin \widetilde{L}_{e_S}$ and so $h(f g_{S,T}) = 0$ in $\Bbbk \widetilde{L}_{e_S}$. Thus $\rho_{S,T}(hf) = \rho_{S,T}(0) = 0 = h(f g_{S,T}) = h\rho_{S,T}(f)$. This proves that $\rho_{S,T}$ is an A-module homomorphism.

Using the vector space decomposition

$$\operatorname{End}_A \left(\bigoplus_{S \in P_n} \Bbbk \widetilde{L}_{e_S} \right) = \bigoplus_{S, T \in P_n} \operatorname{Hom}_A(\Bbbk \widetilde{L}_{e_T}, \Bbbk \widetilde{L}_{e_S}) \tag{17.6}$$

we can view $\rho_{S,T}$ as an element of $\operatorname{End}_A \left(\bigoplus_{S \in P_n} \Bbbk \widetilde{L}_{e_S} \right)$. We check that

$$\rho \colon B^{op} \longrightarrow \operatorname{End}_A \left(\bigoplus_{S \in P_n} \Bbbk \widetilde{L}_{e_S} \right)$$

given by $\rho((S, T)) = \rho_{S,T}$, for $S \preceq T$, is a homomorphism. It is immediate that if $T = U$, then $\rho_{S,T} \rho_{U,V} = \rho_{S,V}$ because $g_{T,V} g_{S,T} = g_{S,V}$ by a combination of Proposition 17.20 and Lemma 17.22. On the other hand, if $T \neq U$, then

$\rho_{S,T}\rho_{U,V} = 0$ because $\rho_{U,V}$ has image contained in the summand $\Bbbk\widetilde{L}_{e_U}$, which is annihilated by $\rho_{S,T}$ whenever $T \neq U$. Thus ρ is a homomorphism.

To see that ρ is injective, note that it takes each basis element (S,T) of B^{op} with $S \preceq T$ to a distinct summand $\mathrm{Hom}_A(\Bbbk\widetilde{L}_{e_T}, \Bbbk\widetilde{L}_{e_S})$ of (17.6). Thus ρ takes the basis of B^{op} to a linearly independent set and hence is injective. The theorem now follows via the argument given in the first paragraph of the proof. $\square$

As the quiver of the category algebra of a poset P is always the Hasse diagram of P (cf. [ASS06]), this leads to another proof of Theorem 17.24.

17.6 Exercises

17.1. Compute the quiver Q of $\Bbbk M$ where $M = \{1, x_1, \ldots, x_n\}$ with 1 the identity and $x_i x_j = x_i$ for $i = 1, \ldots, n$ and prove that $\Bbbk M \cong \Bbbk Q$.

17.2. Prove Theorem 17.9 over $\mathbb{C}$ using Theorem 7.31.

17.3. Let M be a finite $\mathscr{L}$-trivial monoid and $\Bbbk$ a field. Let $e \in E(M)$ and let S_{J_e} be the simple $\Bbbk M$-module with apex e. Let V be the $\Bbbk$-vector space of mappings $f : eM \longrightarrow \Bbbk$ vanishing on $eM \setminus \widetilde{R}_e$ with $\Bbbk M$-module structure given by $(mf)(x) = f(xm)$ for $x \in eM$ and $m \in M$. Prove that V is an injective indecomposable module with simple socle isomorphic to S_{J_e} consisting of those mappings $f \in V$ vanishing on $\widetilde{R}_e \setminus R_e$ and constant on R_e.

17.4. Prove that $\Bbbk L^n \cong I(\mathscr{P}([n]), \Bbbk)$ (the incidence algebra of $\mathscr{P}([n])$ defined in Appendix C) where $L = \{1, A, B\}$ with $xy = x$ for $x, y \in \{A, B\}$.

17.5. Prove that the quiver of $\Bbbk \Sigma_n$ is isomorphic to the Hasse diagram of the lattice Π_n.

17.6. Compute the quiver of $\mathbb{C}M$ for M the monoid in Exercise 16.10.

17.7. Let M be a finite $\mathscr{J}$-trivial monoid whose idempotents are central (i.e., commute with each element of M). Prove that each edge of the quiver of $\Bbbk M$ is a loop for any algebraically closed field $\Bbbk$.

17.8. Let M be a finite $\mathscr{J}$-trivial monoid with commuting idempotents. Construct a complete set of orthogonal primitive idempotents for M. (Hint: use that $E(M)$ is a lattice.)

17.9. Consider the monoid HD_n given by the presentation

$$HD_n = \langle a, b \mid a^2 = a, b^2 = b, w_n(a,b) = w_n(b,a) \rangle$$

for $n \geq 1$ where $w_n(x,y) = xyxy\cdots$ is the alternating word in x, y starting with x of length n.

(a) Prove that HD_n is a finite $\mathscr{J}$-trivial monoid with $2n$ elements.
(b) Prove that $E(HD_n) = \{1, a, b, w_n(a, b)\}$.
(c) Let $\Bbbk$ be an algebraically closed field. Compute the quiver of $\Bbbk HD_n$.

17.10. Let G be a finite group. Let $\mathscr{P}_1(G) = \{X \subseteq G \mid 1 \in X\}$ equipped with the product $AB = \{ab \mid a \in A, b \in B\}$. Prove that $\mathscr{P}_1(G)$ is a $\mathscr{J}$-trivial monoid, $E(\mathscr{P}_1(G))$ is the set of subgroups of G and that the natural partial order on $E(\mathscr{P}_1(G))$ is reverse inclusion.

17.11. Compute the quiver of $\mathbb{C}\mathscr{P}_1(\mathbb{Z}_6)$. See Exercise 17.10 for the definition.

17.12. Compute the quiver of $\mathbb{C}\mathscr{P}_1(S_3)$. See Exercise 17.10 for the definition.

17.13. Let A be a finite dimensional algebra over an algebraically closed field $\Bbbk$. Let e, f be primitive idempotents of A and put $S = Ae/\mathrm{rad}(A)e$ and $S' = Af/\mathrm{rad}(A)f$. Prove that $\mathrm{Ext}^1_A(S, S') \cong f[\mathrm{rad}(A)/\mathrm{rad}^2(A)]e$.

17.14. Compute the quiver of the algebra $\Bbbk M$ for M the left regular band $\{1, a_i, b_i \mid 1 \leq i \leq n\}$ with 1 the identity and product given by

$$x_i y_j = \begin{cases} y_j, & \text{if } i < j \\ x_i, & \text{else} \end{cases}$$

for $x, y \in \{a, b\}$.

17.15. Let M be the left regular band $\{1, a, b, x, y, z, z'\}$ with multiplication table

	1	a	b	x	y	z	z'
1	1	a	b	x	y	z	z'
a	a	a	a	x	x	z	z'
b	b	b	b	x	x	z	z'
x	x	x	x	x	x	z'	z'
y	y	y	y	y	y	z'	z'
z	z	z	z	z'	z'	z	z'
z'	z'	z'	z'	z'	z'	z'	z'

(a) Compute the quiver of $\Bbbk M$.
(b) Prove that $\Bbbk L_x \otimes \Bbbk L_z \cong S_{L_a} \oplus S_{L_1}$, which is not projective, and hence the tensor product of projective modules need not be projective.

18

Further Developments

This chapter highlights some further developments in the representation theory of finite monoids whose detailed treatment is beyond the scope of this text. No proofs are presented.

18.1 Monoids of Lie type

There is a well-developed theory of linear algebraic monoids, due principally to Putcha and Renner [Put88, Ren05]. It is a beautiful interplay between the theory of algebraic groups and semigroup theory. Putcha introduced the finite analogues of linear algebraic monoids, called *finite monoids of Lie type*, see [Put89, Put95] or [Ren05, Chapter 10]. We do not attempt to give the definition here, but the example to keep in mind is the monoid $M_n(\mathbb{F}_q)$ of $n \times n$ matrices over the field $\mathbb{F}_q$ of q elements.

An exciting theorem, due to Okniński and Putcha [OP91], asserts the semisimplicity of the complex algebra of a finite monoid of Lie type.

Theorem 18.1. *Let M be a finite monoid of Lie type. Then $\mathbb{C}M$ is semisimple. In particular, $\mathbb{C}M_n(\mathbb{F}_q)$ is semisimple.*

This result was improved later by Putcha [Put99] who proved that $\Bbbk M$ is semisimple so long as the characteristic of $\Bbbk$ does not divide the order of the group of units of M. Independently, Kovács [Kov92] gave the elementary (but ingenious) direct approach to the study of the monoid algebra of $M_n(\mathbb{F}_q)$ that we developed in Section 5.6.

The modular representation theory of finite monoids of Lie type has also been studied. Motivated by problems in stable homotopy theory, Harris and Kuhn proved the following delightful result [HK88].

© Springer International Publishing Switzerland 2016
B. Steinberg, *Representation Theory of Finite Monoids*,
Universitext, DOI 10.1007/978-3-319-43932-7_18

Theorem 18.2. *Let $q = p^m$ with p prime and $m \geq 1$. Let ρ be an irreducible representation of $M_n(\mathbb{F}_q)$ over the algebraic closure of $\mathbb{F}_p$. Then the restriction of ρ to the general linear group $GL_n(\mathbb{F}_q)$ is also irreducible.*

This theorem was extended by Putcha and Renner to arbitrary finite monoids of Lie type [PR93], cf. [Ren05, Theorem 10.10].

18.2 The representation theory of the full transformation monoid

In Section 5.3 we presented Putcha's results on the irreducible representations of the full transformation monoid. In this section, we discuss further aspects of the representation theory of T_n.

Recall that a finite dimensional algebra A has finite representation type if there are only finitely many isomorphism classes of finite dimensional indecomposable A-modules. Let T_n denote the monoid of all self-mappings on an n-element set. Ponizovskiĭ proved that $\mathbb{C}T_n$ has finite representation type for $n \leq 3$ and conjectured that this was true for all n [Pon87]. Putcha disproved Ponizovskiĭ's conjecture by showing that $\mathbb{C}T_n$ does not have finite representation type for $n \geq 5$ [Put98]. He also computed the quiver of $\mathbb{C}T_4$. Ringel [Rin00] computed a quiver presentation for $\mathbb{C}T_4$ and proved that it is of finite representation type and has global dimension 3. (It is easy to check that $\mathbb{C}T_n$ also has global dimension $n - 1$ for $n = 1, 2, 3$.) We summarize this discussion in the following theorem.

Theorem 18.3. *The algebra $\mathbb{C}T_n$ of the full transformation monoid of degree n has finite representation type if and only if $n \leq 4$.*

It is an open question to compute the quiver of $\mathbb{C}T_n$ in full generality. The author [Ste16b] has recently proved that the quiver of $\mathbb{C}T_n$ is acyclic, for all $n \geq 1$, and has computed the global dimension of $\mathbb{C}T_n$.

Theorem 18.4 (Steinberg). *The global dimension of $\mathbb{C}T_n$ is $n - 1$ for all $n \geq 1$.*

The author and V. Mazorchuk have computed the characteristic tilting module and the Ringel dual of $\mathbb{C}T_n$ with respect to its natural structure of a quasi-hereditary algebra in an appendix to [Ste16b].

18.3 The representation theory of left regular bands

The representation theory of left regular bands was put to good effect in the analysis of Markov chains [BD98, BHR99, Bro00, Bro04, AD10, CG12, Sal12] and in the representation theory of Coxeter groups and Solomon's descent algebra [Bro00, Bro04, AM06, Sch06, Sal08, Sal10]. It was therefore natural to delve more deeply into their representation theory. This has turned out to be a treasure trove of fascinating interconnections between semigroup theory, combinatorics, and topology.

The initial steps were taken by Saliola [Sal07], who computed a complete set of orthogonal primitive idempotents, the projective indecomposable modules, the Cartan matrix, and the quiver for a left regular band algebra. For hyperplane face monoids [BHR99], Saliola computed a quiver presentation, calculated all Ext-spaces between simple modules, and proved that the monoid algebra is the Koszul dual of the incidence algebra of the intersection lattice of the arrangement [Sal09]. Key to his approach were resolutions of the simple modules obtained by Brown and Diaconis [BD98] using the cellular chain complexes of associated zonotopes. (Zonotopes are images of hypercubes under affine maps; they are dual in a sense that can be made precise to hyperplane arrangements [Zie95].) Saliola proved that these resolutions are, in fact, the minimal projective resolutions. The reader is referred to [Ben98, ASS06] for the definition of a minimal projective resolution.

A more detailed analysis of left regular band algebras was then undertaken by the author together with Margolis and Saliola. The first set of results can be found in [MSS15b]; a self-contained monograph describing further results by the same authors is [MSS15a]. The remainder of this section will require some familiarity with algebraic topology.

Fix a field $\Bbbk$ and a left regular band M. We continue to denote the lattice of principal left ideals by $\Lambda(M) = \{Mm \mid m \in M\}$. Let $\sigma \colon M \longrightarrow \Lambda(M)$ be the natural homomorphism given by $\sigma(m) = Mm$. Inspired by hyperplane theory and the theory of oriented matroids [BLVS$^+$99], for $X \in \Lambda(M)$, the submonoid

$$M_{\geq X} = \{m \in M \mid \sigma(m) \geq X\}$$

is called the *contraction* of M to X. There is a natural surjective homomorphism $\rho_X \colon \Bbbk M \longrightarrow \Bbbk M_{\geq X}$ given on $m \in M$ by

$$\rho_X(m) = \begin{cases} m, & \text{if } m \in M_{\geq X} \\ 0, & \text{else.} \end{cases}$$

Moreover, the simple $\Bbbk M$-module S_X associated with X is just the inflation of the trivial $\Bbbk M_{\geq X}$-module via ρ_X.

If P is a finite poset, then the *order complex* $\Delta(P)$ is the simplicial complex with vertex set P and whose simplices are the chains (or totally ordered subsets) of P. Let us view $M = E(M)$ as a poset via the natural order. Because $m \leq n$ if and only if $mM \subseteq nM$ by Proposition 2.9, the action of M on itself by left multiplication is order-preserving. Therefore, M acts by simplicial maps on $\Delta(M)$. Moreover, $\Delta(M)$ is a contractible simplicial complex because 1 is a maximum element of M and hence a cone point of $\Delta(M)$. It follows that the augmented simplicial chain complex for $\Delta(M)$ is a resolution of the trivial module by $\Bbbk M$-modules. More generally, the augmented simplicial chain complex of $\Delta(M_{\geq X})$ is a resolution of the simple module S_X by $\Bbbk M$-modules via inflation along ρ_X. In [MSS15a], we prove that this is a projective resolution.

Theorem 18.5. *Let M be a left regular band and $\Bbbk$ a field. Let $X \in \Lambda(M)$. Then the augmented simplicial chain complex of $\Delta(M_{\geq X})$ yields a finite projective resolution of the simple module S_X associated with X.*

Let us say that M is a *CW left regular band* if $M_{\geq X}$ is the face poset of a regular CW complex (a CW complex whose attaching maps are homeomorphisms) for all $X \in \Lambda(M)$. This regular CW complex is unique up to cellular isomorphism (cf. [Bjö84, BLVS$^+$99]). Note that M is the contraction of M to the minimum element of $\Lambda(M)$ and so M itself is the face poset of a regular CW complex. Examples of CW left regular bands include hyperplane face monoids [BHR99], oriented matroids [BLVS$^+$99], and complex hyperplane face monoids [BZ92, Bjö08]. In this case, one has that $M_{\geq X}$ acts by cellular maps on the corresponding regular CW complex (which must, in fact, be homeomorphic to a closed ball). In our joint work with Margolis and Saliola [MSS15a], the following theorem is proved. It generalizes Saliola's results for hyperplane face monoids [Sal09].

Theorem 18.6. *Let M be a CW left regular band and $\Bbbk$ a field.*

(i) The augmented cellular chain complex of the regular CW complex associated with $M_{\geq X}$ provides the minimal projective resolution of the simple module S_X corresponding to $X \in \Lambda(M)$.

(ii) The quiver Q of $\Bbbk M$ is the Hasse diagram of $\Lambda(M)$.

(iii) $\Bbbk M \cong \Bbbk Q/I$ where I is the (admissible) ideal generated by the sum of all paths in Q of length 2.

(iv) $\Bbbk M$ is a (graded) Koszul algebra with Koszul dual the incidence algebra of $\Lambda(M)$.

(v) The global dimension of $\Bbbk M$ is the dimension of the regular CW complex whose face poset is M.

Finding a quiver presentation for the algebra of an arbitrary left regular band is an outstanding open question.

The author, with Margolis and Saliola, computed the Ext-spaces between simple modules for left regular band algebras in [MSS15b]. As a consequence, we were able to compute the global dimension of the algebras of all the examples appearing in the literature, as well as new ones. The main result of [MSS15b] is as follows.

Theorem 18.7. *Let M be a left regular band and $\Bbbk$ a field. Let $X, Y \in \Lambda(M)$. Fix e_Y with $Me_Y = Y$. Then*

$$\mathrm{Ext}^n_{\Bbbk M}(S_X, S_Y) = \begin{cases} \Bbbk, & \text{if } X = Y,\ n = 0 \\ \widetilde{H}^{n-1}(\Delta(e_Y M_{\geq X} \setminus \{e_Y\}); \Bbbk), & \text{if } Y > X,\ n \geq 1 \\ 0, & \text{else} \end{cases}$$

where $\widetilde{H}^q(K; \Bbbk)$ denotes the reduced cohomology in dimension q of the simplicial complex K with coefficients in $\Bbbk$.

As a consequence, we obtain an alternative description of the quiver of a left regular band.

Corollary 18.8. *Let M be a left regular band and $\Bbbk$ a field. Then the quiver of $\Bbbk M$ can be identified with the quiver with vertex set $\Lambda(M)$ and arrows described as follows. There are no arrows $X \longrightarrow Y$ unless $X < Y$, in which case the number of arrows $X \longrightarrow Y$ is one fewer than the number of connected components of $\Delta(e_Y M_{\geq X} \setminus \{e_Y\})$ where $Y = Me_Y$.*

The reader is invited to prove directly that Theorem 17.11 and Corollary 18.8 give isomorphic quivers.

The proof of Theorem 18.7 in [MSS15b] goes through classifying spaces of small categories and Quillen's celebrated Theorem A [Qui73]. In the monograph [MSS15a], we give a more direct proof using Theorem 18.5.

Let us consider a sample application. It is well known, and fairly easy to show, that if the Hasse diagram of a poset P is a forest, then each connected component of $\Delta(P)$ is contractible (a proof can be found in [MSS15b], for example). If M is a left regular band whose Hasse diagram is a tree rooted at 1, then the Hasse diagram of $e_Y M_{\geq X} \setminus \{e_Y\}$ will be a forest for all $X < Y$ in $\Lambda(M)$. Theorem 18.7 then implies that $\Bbbk M$ has global dimension at most 1, that is, $\Bbbk M$ is hereditary. For instance, if $F(B)$ is the free left regular band on a set B, then the natural partial order is the opposite of the prefix order on repetition-free words. Therefore, the Hasse diagram of $F(B)$ is a tree rooted at 1. We can thus deduce K. Brown's result that the algebra $\Bbbk F(B)$ is hereditary from Theorem 18.7.

In [MSS15b] we proved that finite dimensional path algebras are left regular band algebras.

Theorem 18.9. *Let Q be a finite acyclic quiver and $\Bbbk$ a field. Then the path algebra $\Bbbk Q$ is isomorphic to the algebra of a left regular band.*

In light of Gabriel's theorem [Gab72], it follows that each hereditary finite dimensional algebra over an algebraically closed field is Morita equivalent to the algebra of a left regular band. In other words, the representation theory of left regular bands is at least as rich as the representation theory of quivers!

Another interesting open question is to classify left regular band algebras of finite representation type. Path algebras of finite representation type were famously classified by Gabriel [Gab73] (cf. [Ben98, ASS06]).

18.4 The Burnside problem for linear monoids

In this section, we lift the restriction that all monoids are finite. A monoid M is said to be *periodic* if, for all $m \in M$, there exist $c, d > 0$ such that $m^c = m^{c+d}$. Of course, finite monoids are periodic. The Burnside problem concerns hypotheses that guarantee that a finitely generated periodic monoid must be finite. A classical theorem of Schur [CR88, Theorem 36.2] states that any finitely generated periodic group of matrices over a field is finite. This was extended to monoids by McNaughton and Zalcstein [MZ75].

Theorem 18.10 (McNaughton-Zalcstein). *Let $M \leq M_n(\Bbbk)$ be a finitely generated periodic monoid of matrices. Then M is finite.*

A number of proofs of Theorem 18.10 have since appeared, cf. [Str83, FGG97, Okn98, Ste12b]. The key difference between the group case and the monoid case lies in the process of passing to the completely reducible case because monoid homomorphisms do not have a kernel. The completely reducible case was handled for groups by Burnside [CR88, Theorem 36.1] and applies *mutatis mutandis* to monoids.

One elegant approach is based on an induction on the number of composition factors for $\Bbbk^n$ as a module over the subalgebra A spanned by $M \leq M_n(\Bbbk)$, where the base case of a simple module is handled by Burnside's theorem. For the inductive step, one puts M in a 2×2-block upper triangular form such that the diagonal block monoids are finite by induction. Then one applies to the projection to the block diagonal the following highly nontrivial theorem of Brown [Bro71] (see also [RS09, Theorem 4.2.4]). A semigroup S is said to be *locally finite* if each finitely generated subsemigroup of S is finite.

Theorem 18.11 (Brown). *Let $\varphi \colon S \longrightarrow T$ be a semigroup homomorphism such that T is locally finite and $\varphi^{-1}(e)$ is locally finite for each $e \in E(T)$. Then S is locally finite.*

The reader should note that the special case of Brown's theorem where S and T are groups is trivial.

Appendix A

Finite Dimensional Algebras

This appendix reviews the necessary background material from the theory of finite dimensional algebras. Standard texts covering most of this subject matter are [CR88, Lam91, Ben98, ASS06]. Readers familiar with this material are urged to skim this chapter or skip it entirely. Very few proofs will be given here, as the results can be found in the references. Let us remark that, unlike the case of group algebras, monoid algebras are seldom semisimple, even over the complex numbers. This forces us to use more of the theory of finite dimensional algebras than would be encountered in a first course on the ordinary representation theory of finite groups. Some of the more elaborate tools, like projective covers, are not used except for in the final chapters of the text. Most of the text uses nothing beyond Wedderburn theory.

Let us fix for the rest of this appendix a field $\Bbbk$ and a finite dimensional $\Bbbk$-algebra A.

A.1 Semisimple modules and algebras

An A-module S is *simple* if $S \neq 0$ and the only submodules of S are 0 and S. Equivalently, a nonzero module S is simple if $Av = S$ for all nonzero vectors $v \in S$. Notice that $S = Av$, for $v \neq 0$, implies that S is a quotient of A. Thus every simple A-module is finite dimensional. Schur's lemma asserts that there are very few homomorphisms between simple modules. See [CR88, Lemma 27.3].

Lemma A.1 (Schur). *If S, S' are simple A-modules, then every nonzero homomorphism $\varphi \colon S \longrightarrow S'$ is an isomorphism. In particular, $\mathrm{End}_A(S)$ is a finite dimensional division algebra over $\Bbbk$. If $\Bbbk$ is algebraically closed, then $\mathrm{End}_A(S) = \Bbbk \cdot 1_S \cong \Bbbk$.*

© Springer International Publishing Switzerland 2016
B. Steinberg, *Representation Theory of Finite Monoids*,
Universitext, DOI 10.1007/978-3-319-43932-7

An A-module M is *semisimple* if $M = \bigoplus_{\alpha \in F} S_\alpha$ for some family of simple submodules $\{S_\alpha \mid \alpha \in F\}$. The following proposition is the content of [CR88, Theorem 15.3].

Proposition A.2. *Let M be an A-module. Then the following are equivalent.*

(i) M is semisimple.

(ii) $M = \sum_{\alpha \in F} S_\alpha$ with $S_\alpha \leq M$ simple for all $\alpha \in F$.

(iii) For each submodule $N \leq M$, there is a submodule $N' \leq M$ such that $M = N \oplus N'$.

It follows that every A-module V has a unique maximal semisimple submodule $\mathrm{soc}(V)$, called the *socle* of V. Indeed, $\mathrm{soc}(V)$ is the sum of all the simple submodules of V; it is semisimple by Proposition A.2.

Proposition A.3. *The subcategory of semisimple A-modules is closed under taking submodules, quotient modules, and direct sums.*

Proof. Closure under direct sum is clear from the definition. Closure under quotients is immediate from Proposition A.2(ii) and the fact that a quotient of a simple module is either 0 or simple. Closure under submodules follows from closure under quotients because each submodule is a direct summand by Proposition A.2(iii) and hence a quotient. □

A *maximal submodule* of a module V is a proper submodule which is maximal with respect to the inclusion ordering on the set of proper submodules of V. If V is an A-module, we define $\mathrm{rad}(V)$ to be the intersection of all maximal submodules of V. The next proposition is essentially [ASS06, Corollary I.3.8].

Proposition A.4. *Let V be a finite dimensional A-module.*

(i) V is semisimple if and only if $\mathrm{rad}(V) = 0$.

(ii) $V/\mathrm{rad}(V)$ is semisimple.

(iii) If $W \leq V$, then V/W is semisimple if and only if $\mathrm{rad}(V) \subseteq W$.

We can view A, itself, as a finite dimensional A-module called the *regular module*. Maximal submodules of A are just maximal left ideals and so $\mathrm{rad}(A)$ is the intersection of all maximal left ideals of A. However, $\mathrm{rad}(A)$ enjoys a number of additional properties. Recall that an ideal I of an algebra A is *nilpotent* if $I^n = 0$ for some $n \geq 1$.

The following theorem can be extracted from [ASS06, Corollary 1.4], [Ben98, Proposition 1.2.5 and Theorem 1.2.7], and [CR88, Theorem 25.24] with a little effort.

Theorem A.5. *Let A be a finite dimensional $\Bbbk$-algebra.*

(i) $\mathrm{rad}(A)$ is the intersection of all maximal right ideals of A.

(ii) $\mathrm{rad}(A)$ is a two-sided ideal.

(iii) $\mathrm{rad}(A)$ is the intersection of the annihilators of the simple A-modules.

(iv) If V is a finite dimensional A-module, then $\mathrm{rad}(V) = \mathrm{rad}(A) \cdot V$.

(v) An A-module M is semisimple if and only if $\mathrm{rad}(A) \cdot M = 0$.
(vi) $\mathrm{rad}(A)$ *is nilpotent.*
(vii) $\mathrm{rad}(A)$ *is the largest nilpotent ideal of A.*

One also has Nakayama's lemma (cf. [Ben98, Lemma 1.2.3]).

Lemma A.6 (Nakayama). *Let V be a finite dimensional A-module. Then one has that* $\mathrm{rad}(A) \cdot V = V$ *implies* $V = 0$.

A finite dimensional $\Bbbk$-algebra A is *semisimple* if the regular module A is a semisimple module. Wedderburn famously characterized semisimple algebras. See [Ben98, Theorem 1.3.4 and Theorem 1.3.5].

Theorem A.7 (Wedderburn). *The following are equivalent for a finite dimensional $\Bbbk$-algebra A.*

(i) A is semisimple.
(ii) Each A-module is semisimple.
(iii) $\mathrm{rad}(A) = 0$.
(iv) $A \cong \prod_{i=1}^{r} M_{n_i}(D_i)$ *where the D_i are finite dimensional division algebras over $\Bbbk$.*

Moreover, if A is semisimple, then A has finitely many simple modules $S_1, \ldots, S_r$ up to isomorphism and in (iv) (after reordering) one has $D_i \cong \mathrm{End}_A(S_i)^{op}$ and $n_i = \dim S_i / \dim D_i$. Furthermore, there is an A-module isomorphism

$$A \cong \bigoplus_{i=1}^{r} n_i \cdot S_i.$$

If, in addition, $\Bbbk$ is algebraically closed, then $D_i = \Bbbk$ for all $i = 1, \ldots, r$.

The class of semisimple algebras is closed under quotients.

Proposition A.8. *Quotients of semisimple finite dimensional algebras are semisimple.*

Proof. If A is semisimple and I is an ideal, then A/I is a direct sum of simple submodules as an A-module and hence as an A/I-module.   □

Factoring a finite dimensional algebra by its radical results in a semisimple algebra. More precisely, we have the following, which combines elements of [ASS06, Corollary 1.4] and [CR88, Theorem 25.24].

Theorem A.9. *Let A be a finite dimensional $\Bbbk$-algebra.*

(i) $A/\mathrm{rad}(A)$ is semisimple.
(ii) $A/\mathrm{rad}(A)$-mod can be identified with the full subcategory of A-mod whose objects are the semisimple A-modules.
(iii) If I is an ideal of A with A/I semisimple, then $\mathrm{rad}(A) \subseteq I$.

In particular, the simple A-modules are the inflations of simple $A/\operatorname{rad}(A)$-modules and hence A has only finitely many isomorphism classes of simple A-modules.

A useful corollary of Theorem A.9, which is exploited in Chapter 11, is the following.

Corollary A.10. *Suppose that A is a finite dimensional $\Bbbk$-algebra and I is a nilpotent ideal such that A/I is semisimple. Then $I = \operatorname{rad}(A)$.*

Proof. The inclusion $I \subseteq \operatorname{rad}(A)$ follows from Theorem A.5 and the inclusion $\operatorname{rad}(A) \subseteq I$ follows from Theorem A.9.   □

Another theorem of Wedderburn, which is employed elsewhere in the text, gives a criterion for nilpotency of an ideal. An element $a \in A$ is *nilpotent* if $a^n = 0$ for some $n \geq 1$. Clearly, each element of a nilpotent ideal is nilpotent. The following strong converse is due to Wedderburn, cf. [CR88, Theorem 27.27] (where we note that the hypothesis that the field is algebraically closed is unnecessary since having a basis of nilpotent elements is preserved under extension of scalars).

Theorem A.11. *Let A be a finite dimensional $\Bbbk$-algebra. Then an ideal I of A is nilpotent if and only if it is spanned by a set of nilpotent elements.*

For one of the exercises, we shall need the following proposition (cf. [Lam91, Proposition 11.7]). Recall that an A-module V is *faithful* if its annihilator is 0, that is, $aV = 0$ implies $a = 0$.

Proposition A.12. *Let A be a finite dimensional $\Bbbk$-algebra with a faithful simple module. Then A is simple, i.e., has no proper nonzero ideals.*

Next we discuss the Jordan-Hölder theorem for finite dimensional modules. For a proof, see [CR88, Theorem 13.7]. If V is a finite dimensional A-module, then a *composition series* for V is an unrefinable chain of submodules

$$0 = V_0 \subsetneq V_1 \subsetneq \cdots \subsetneq V_n = V.$$

We call n the *length* of the composition series and the simple modules V_i/V_{i-1} the *composition factors*.

Theorem A.13 (Jordan-Hölder). *Let V be a finite dimensional A-module and let*

$$0 = V_0 \subsetneq \cdots \subsetneq V_n = V$$
$$0 = W_0 \subsetneq \cdots \subsetneq W_m = V$$

be composition series for V. Then $m = n$ and there exists $\sigma \in S_n$ such that $V_i/V_{i-1} \cong W_{\sigma(i)}/W_{\sigma(i)-1}$ for $i = 1, \ldots, n$.

It follows from the Jordan-Hölder theorem that we can unambiguously define the *length* of V to be the length of a composition series for V and we can define, for a simple A-module S, its *multiplicity* as a composition factor in V to be the number $[V : S]$ of composition factors in some composition series for V that are isomorphic to S.

The following theorem was originally proved by Frobenius and Schur in the context of complex group algebras and can be found as [CR88, Theorem 27.8].

Theorem A.14. *Let $\Bbbk$ be an algebraically closed field and A a finite dimensional $\Bbbk$-algebra. Suppose that $S_1, \ldots, S_r$ form a complete set of representatives of the isomorphism classes of simple A-modules. Fix a basis for each S_i and let $\varphi^{(k)} : A \longrightarrow M_{n_k}(\Bbbk)$ be the $\Bbbk$-algebra homomorphism given by sending $a \in A$ to the matrix*

$$\varphi^{(k)}(a) = \left(\varphi_{ij}^{(k)}(a) \right)$$

of the operator $v \mapsto av$. Then the linear functionals

$$\varphi_{ij}^{(k)} : A \longrightarrow \Bbbk$$

with $1 \leq i, j \leq n_k$ and $1 \leq k \leq r$ are linearly independent over $\Bbbk$.

A.2 Indecomposable modules

A nonzero A-module M is *indecomposable* if $M = M' \oplus M''$ implies $M' = 0$ or $M'' = 0$. Every simple module is indecomposable, but the converse only holds when A is semisimple. The Krull-Schmidt theorem [CR88, Theorem 14.5] asserts that each finite dimensional module admits an essentially unique decomposition into a direct sum of indecomposable modules.

Theorem A.15 (Krull-Schmidt). *If A is a finite dimensional $\Bbbk$-algebra and V is a finite dimensional A-module, then $V = \bigoplus_{i=1}^{s} M_i$ with the M_i indecomposable submodules. Moreover, if $V \cong \bigoplus_{i=1}^{r} N_i$ with the N_i indecomposable, then $r = s$ and there is $\sigma \in S_r$ such that $M_i \cong N_{\sigma(i)}$ for $i = 1, \ldots, r$.*

A finite dimensional algebra is said to be of *finite representation type* if it has only finitely many isomorphism classes of finite dimensional indecomposable modules.

Applying the Krull-Schmidt theorem to the regular module A, we have that $A = \bigoplus_{i=1}^{s} P_i$ where the P_i are indecomposable and, moreover, projective. The following theorem combines aspects of [CR88, Theorem 54.11, Corollary 54.13 and Corollary 54.14] and the discussion on Page 14 of [Ben98].

Theorem A.16. *Let A be a finite dimensional $\Bbbk$-algebra and suppose that*

$$A = \bigoplus_{i=1}^{s} P_i$$

is a decomposition of A into indecomposable submodules.

(i) Every projective indecomposable module is isomorphic to P_i for some $i = 1, \ldots, s$.
(ii) $P_i / \operatorname{rad}(P_i)$ is simple.
(iii) $P_i \cong P_j$ if and only if $P_i / \operatorname{rad}(P_i) \cong P_j / \operatorname{rad}(P_j)$.
(iv) One has that

$$A/\operatorname{rad}(A) = \bigoplus_{i=1}^{s} P_i/\operatorname{rad}(P_i).$$

In particular, each simple A-module is isomorphic to one the form $P/\operatorname{rad}(P)$ for some projective indecomposable (unique up to isomorphism) and the multiplicity of $P/\operatorname{rad}(P)$ as a direct summand in $A/\operatorname{rad}(A)$ coincides with the multiplicity of P as a direct summand in A.

Projective indecomposable modules are also called *principal indecomposable modules* in the literature.

If V is a finite dimensional A-module and P a finite dimensional projective A-module, then an epimorphism $\varphi \colon P \longrightarrow V$ is called a *projective cover* if $W + \ker \varphi = P$ implies $W = P$ for W a submodule of P. Using Nakayama's lemma, one can show that this is equivalent to $\ker \varphi \subseteq \operatorname{rad}(P)$. The following is the content of [ASS06, Theorem I.5.8] (which assumes throughout that the ground field is algebraically closed, but does not need it for this result).

Theorem A.17. *Let $S_1, \ldots, S_r$ form a complete set of representatives of the isomorphism classes of simple A-modules and let $P_1, \ldots, P_r$ form a complete set of representatives of the isomorphism classes of projective indecomposable A-modules, ordered so that $P_i / \operatorname{rad}(P_i) \cong S_i$.*

(i) The canonical epimorphism $\eta \colon P_i \longrightarrow P_i / \operatorname{rad}(P_i)$ is a projective cover.
(ii) If V is a finite dimensional A-module with

$$V/\operatorname{rad}(V) \cong \bigoplus_{i=1}^{r} n_i \cdot S_i,$$

then there is a projective cover

$$\psi \colon \bigoplus_{i=1}^{r} n_i \cdot P_i \longrightarrow V.$$

(iii) If $\varphi \colon P \longrightarrow V$ and $\psi \colon Q \longrightarrow V$ are projective covers. Then there is an isomorphism $\tau \colon P \longrightarrow Q$ such that

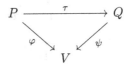

commutes.

In other words, each finite dimensional module has a projective cover and the corresponding projective module is unique up to isomorphism. It is not the case that the isomorphism τ in Theorem A.17(iii) is unique.

A.3 Idempotents

The set $E(A)$ of idempotents of a finite dimensional algebra A plays a crucial role in the theory. Two idempotents $e, f \in E(A)$ are *orthogonal* if $ef = 0 = fe$.

Proposition A.18. *Let $e, f \in E(A)$ be orthogonal idempotents. Then $e + f$ is an idempotent and $A(e + f) = Ae \oplus Af$.*

Proof. We compute $(e + f)^2 = e^2 + ef + fe + f^2 = e + f$ since e, f are orthogonal idempotents. As $a(e+f) = ae+af$, $e(e+f) = e$ and $f(e+f) = f$, trivially $A(e + f) = Ae + Af$. If $a \in Ae \cap Af$, then $ae = a = af$ and so $a = af = aef = 0$. Thus $A(e + f) = Ae \oplus Af$. □

If $e \in E(A)$, then $(1 - e)^2 = 1 - 2e + e^2 = 1 - e$ and so $1 - e \in E(A)$. Trivially, e and $1 - e$ are orthogonal and so Proposition A.18 admits the following corollary.

Corollary A.19. *If $e \in E(A)$, then $A = Ae \oplus A(1 - e)$ and hence Ae is a projective module.*

A collection $\{e_1, \ldots, e_n\}$ of pairwise orthogonal idempotents of A is called a *complete set of orthogonal idempotents* if $1 = e_1 + \cdots + e_n$. In this case $A \cong Ae_1 \oplus \cdots \oplus Ae_n$ by repeated application of Proposition A.18.

Note that if A is finite dimensional and $e \in E(A)$, then $\mathrm{rad}(Ae) = \mathrm{rad}(A)Ae = \mathrm{rad}(A)e$. Proposition 1.8 has the following analogue for algebras [Ben98, Lemma 1.3.3].

Proposition A.20. *Let $e \in E(A)$ and let M be an A-module.*

(i) $\mathrm{Hom}_A(Ae, M) \cong eM$ via $\varphi \mapsto \varphi(e)$.
(ii) $\mathrm{End}_A(Ae) \cong (eAe)^{op}$.

In particular, $\mathrm{End}_A(A) \cong A^{op}$.

If L is a direct summand in the regular module A, then there is a projection $\pi \colon A \longrightarrow L$. If $\iota \colon L \longrightarrow A$ is the inclusion, then $\iota\pi \in \mathrm{End}_A(A)$ is an idempotent and so by Proposition A.20 there exists $e \in E(A)$ with $\iota\pi(1) = e$. Therefore, $L = \iota\pi(A) = A\iota\pi(1) = Ae$ and we have proved the following.

Proposition A.21. *Every direct summand of the regular module A is of the form Ae with $e \in E(A)$.*

A nonzero idempotent $e \in E(A)$ is said to be *primitive* if $e = e_1 + e_2$ with e_1, e_2 orthogonal implies that $e_1 = 0$ or $e_2 = 0$.

Proposition A.22. *Let $e \in E(A) \setminus \{0\}$. Then the following are equivalent.*

(i) e is primitive.
(ii) Ae is indecomposable.
(iii) $E(eAe) = \{0, e\}$.

Proof. Suppose that e is primitive and let $f \in E(eAe)$. Then $f, e - f$ are orthogonal idempotents and $e = f + (e - f)$. Thus $f = 0$ or $e - f = 0$ by primitivity of e. This shows that (i) implies (iii). Assume that (iii) holds. If $Ae = V \oplus W$, then there is an idempotent $\pi \in \text{End}_A(Ae)$ with $\pi(Ae) = V$. But $\text{End}(Ae) \cong (eAe)^{op}$ by Proposition A.20 and so $\pi = 0$ or $\pi = 1_{Ae}$. Thus $V = 0$ or $W = 0$ and so Ae is indecomposable. This proves that (iii) implies (ii). Assume that (ii) holds. If $e = e_1 + e_2$ with e_1, e_2 orthogonal idempotents, then $Ae = Ae_1 \oplus Ae_2$ by Proposition A.18. Thus $Ae_1 = 0$ or $Ae_2 = 0$, that is, we have $e_1 = 0$ or $e_2 = 0$. This establishes that e is primitive and completes the proof. □

We say that a complete set of orthogonal idempotents $\{e_1, \ldots, e_s\}$ is a *complete set of orthogonal primitive idempotents* if each e_i is primitive. The next theorem is essentially [Ben98, Corollary 1.7.4] and the discussion following it.

Theorem A.23. *Let A be a finite dimensional $\Bbbk$-algebra.*

(i) $A = \bigoplus_{i=1}^s Ae_i$ with $e_i \in E(A)$ is a direct sum decomposition into projective indecomposable modules if and only if $e_1, \ldots, e_s$ form a complete set of orthogonal primitive idempotents.
(ii) If $\{e_1, \ldots, e_s\}$ is a complete set of orthogonal primitive idempotents for A, then $\{e_1 + \text{rad}(A), \ldots, e_s + \text{rad}(A)\}$ is a complete set of orthogonal primitive idempotents for $A/\text{rad}(A)$.
(iii) Every complete set of orthogonal primitive idempotents for $A/\text{rad}(A)$ is of the form $\{e_1 + \text{rad}(A), \ldots, e_s + \text{rad}(A)\}$ with $\{e_1, \ldots, e_s\}$ a complete set of orthogonal primitive idempotents for A.

It follows that the projective indecomposable A-modules are of the form Ae and the simple A-modules are of the form $Ae/\text{rad}(A)e$ with e a primitive idempotent.

Another proposition concerning idempotents that is used in the text is the following, which combines [CR88, Theorem 54.12] and [CR88, Theorem 54.16].

Proposition A.24. *Let A be a finite dimensional algebra over a field $\Bbbk$ and let e be a primitive idempotent of A. Let $S = Ae/\text{rad}(A)e$ be the corresponding simple module. If V is a finite dimensional A-module, then $\text{Hom}_A(Ae, V) \cong eV \neq 0$ if and only if S is a composition factor of V. Moreover, if $\Bbbk$ is algebraically closed, then $\dim eV = [V : S]$.*

The *center* $Z(A)$ of A consists of those elements of A that commute with all elements of A. A *central idempotent* is an idempotent e of $Z(A)$. Notice that $eAe = Ae = AeA$ in this case. If $e_1, \ldots, e_n$ form a complete set of orthogonal central idempotents, then the direct sum decomposition $A = Ae_1 \oplus \cdots \oplus Ae_n$ is actually a direct product decomposition $A = Ae_1 \times \cdots \times Ae_n$ of $\Bbbk$-algebras. It is known that $(E(Z(A)), \leq)$ is a finite boolean algebra (where as usual, idempotents are ordered by $e \leq f$ if $ef = e = fe$). The atoms of $E(Z(A))$ are called the *central primitive idempotents*. Equivalently, a central primitive idempotent is a nonzero central idempotent e such that $e = e_1 + e_2$ with e_1, e_2 central and orthogonal implies that $e_1 = 0$ or $e_2 = 0$ or, equivalently, such that $E(Z(Ae)) = \{0, e\}$. The central primitive idempotents form a complete set of orthogonal idempotents. If e is a central primitive idempotent, then Ae is an *indecomposable algebra*, i.e., cannot be expressed as a direct product of two nonzero algebras, and is called a *block* of A. The reader is referred to [Ben98, Section 1.8] for details.

If e is a central idempotent of A and V is a simple A-module, then eV is a submodule and hence $eV = 0$ or $eV = V$. Consequently, if $e_1, \ldots, e_n$ form a complete set of orthogonal central idempotents of A and V is a simple A-module, then there is a unique $i \in \{1, \ldots, n\}$ with $e_iV = V$ and $e_jV = 0$ for $j \neq i$. The A-module structure on V then comes from the Ae_i-module structure on V via inflation along the projection $A \longrightarrow Ae_i$. A similar remark holds for indecomposable modules. Thus simple modules, and more generally indecomposable modules, belong to a block.

A.4 Duality and Morita equivalence

Let A be a finite dimensional $\Bbbk$-algebra. Using the vector space dual, we can turn left A-modules into right A-modules, or equivalently left A^{op}-modules. More precisely, if V is a left A-module, then $D(V) = \operatorname{Hom}_{\Bbbk}(V, \Bbbk)$ is a right A-module where $(\varphi a)(v) = \varphi(av)$ for $a \in A$, $v \in V$ and $\varphi \in \operatorname{Hom}_{\Bbbk}(V, \Bbbk)$. The contravariant functor $D \colon A\text{-mod} \longrightarrow A^{op}\text{-mod}$ is called the *standard duality*. We write D_A if the algebra A is not clear from context. The standard duality enjoys the following properties; see [ASS06, Section I.2.9].

Theorem A.25. *Let A be a finite dimensional $\Bbbk$-algebra.*

(i) $D_{A^{op}} \circ D_A \cong 1_{A\text{-mod}}$.
(ii) D is a contravariant equivalence of categories.
(iii) D sends projective modules to injective modules.
(iv) D send injective modules to projective modules.
(v) D sends simple modules to simple modules.
(vi) D preserves indecomposability.

We recall here that a functor $F \colon C \longrightarrow D$ is an *equivalence* of categories if there is a functor $G \colon D \longrightarrow C$, called a *quasi-inverse*, such that the functors $G \circ F$ and $F \circ G$ are naturally isomorphic to the identity functor on C and D, respectively. Equivalently, one has that F is an equivalence if and only if it is fully faithful and essentially surjective. A functor $F \colon C \longrightarrow D$ is *fully faithful* if, for c, c' objects of C, one has that $F \colon \mathrm{Hom}(c, c') \longrightarrow \mathrm{Hom}(F(c), F(c'))$ is a bijection and F is *essentially surjective* if each object of D is isomorphic to an object of the form $F(c)$ with c an object of C. The reader is referred to [Mac98] for details.

As a consequence we obtain a description of the injective indecomposable modules (cf. [ASS06, Corollary I.5.17]).

Corollary A.26. *Let A be a finite dimensional $\Bbbk$-algebra and let $e_1, \ldots, e_s$ form a complete set of orthogonal primitive idempotents.*

  (i) The injective indecomposable A-modules are precisely those of the form $D(e_i A)$ for $i = 1, \ldots, s$ (up to isomorphism).
  (ii) $A e_i / \mathrm{rad}(A) e_i \cong \mathrm{soc}(D(e_i A))$ for $i = 1, \ldots, s$.
  (iii) $D(e_i A) \cong D(e_j A)$ if and only if $A e_i \cong A e_j$, if and only if $A e_i / \mathrm{rad}(A) e_i \cong A e_j / \mathrm{rad}(A) e_j$.

Two finite dimensional $\Bbbk$-algebras A, B are said to be *Morita equivalent* if the categories A-mod and B-mod are equivalent.

Example A.27. If A is a finite dimensional $\Bbbk$-algebra and $n \geq 1$, then A is Morita equivalent to $M_n(A)$. The equivalence sends an A-module V to V^n where

$$(a_{ij})(v_1, \ldots, v_n) = \left(\sum_{j=1}^{n} a_{1j} v_j, \ldots, \sum_{j=1}^{n} a_{nj} v_j \right)$$

gives the $M_n(A)$-module structure.

Notice that if A is semisimple and B is Morita equivalent to A, then B is also semisimple. Indeed, A is semisimple if and only if each finite dimensional A-module is injective by Proposition A.2. But an equivalence of categories sends injective modules to injective modules and so we conclude that each finite dimensional B-module is injective. Therefore, B is semisimple.

Chapter 4 proves a special case of the following well-known theorem, cf. [Ben98, Theorem 2.2.6].

Theorem A.28. *Let A and B be finite dimensional $\Bbbk$-algebras. Then A is Morita equivalent to B if and only if there is a finite dimensional projective A-module P such that $\mathrm{End}_A(P) \cong B^{op}$ and each projective indecomposable A-module is isomorphic to a direct summand in P.*

Any equivalence of categories $F \colon A$-mod $\longrightarrow B$-mod, for finite dimensional $\Bbbk$-algebras A and B, is of the form $V \mapsto U \otimes_A V$ with U a finite dimensional B-A-bimodule and thus $F \colon \mathrm{Hom}_A(V, W) \longrightarrow \mathrm{Hom}_B(F(V), F(W))$ is always a $\Bbbk$-vector space isomorphism. See [Ben98, Section 2.2] for details.

Appendix B

Group Representation Theory

This appendix surveys those aspects of the representation theory of finite groups that are used throughout the text. The final section provides a brief overview of the representation theory of the symmetric group in characteristic zero. Monoid representation theory very much builds on group representation theory and so one should have a solid foundation in the latter subject before attempting to master the former. Good references for the representation theory of finite groups are [Isa76, Ser77, CR88]. See also [Ste12a]. We mostly omit proofs here, although we occasionally do provide a sketch or complete proof for convenience of the reader.

B.1 Group algebras

If G is a group and $\Bbbk$ a field, then the *group algebra* $\Bbbk G$ is the $\Bbbk$-algebra of formal sums $\sum_{g \in G} c_g g$ with $g \in G$, $c_g \in \Bbbk$ and with only finitely many $c_g \neq 0$. The multiplication is defined by

$$\left(\sum_{g \in G} c_g g \right) \cdot \left(\sum_{g \in G} d_g g \right) = \sum_{g,h \in G} c_g d_h gh.$$

Note that $\Bbbk G$ is finite dimensional precisely when G is finite.

A *matrix representation* of G over $\Bbbk$ is a homomorphism $\rho \colon G \longrightarrow GL_n(\Bbbk)$ for some $n \geq 0$ where $GL_n(\Bbbk)$ is the general linear group of degree n over $\Bbbk$. We say that matrix representations $\rho, \psi \colon G \longrightarrow GL_n(\Bbbk)$ are *equivalent* if there is a matrix $T \in GL_n(\Bbbk)$ such that $T^{-1}\psi(g)T = \rho(g)$ for all $g \in G$.

If V is a finite dimensional $\Bbbk G$-module, then choosing a basis for V gives rise to a matrix representation of G (which we say is *afforded* by V) and different bases yield equivalent representations. In fact, equivalence classes of matrix representations are in bijection with isomorphism classes

B. Steinberg, *Representation Theory of Finite Monoids*,
Universitext, DOI 10.1007/978-3-319-43932-7

of $\Bbbk G$-modules. Representations afforded by simple modules are termed *irreducible* and those afforded by semisimple modules are called *completely reducible*.

The first fundamental theorem about group algebras is Maschke's characterization of semisimplicity (cf. [CR88, Theorem 15.6]).

Theorem B.1 (Maschke). *Let G be a finite group and $\Bbbk$ a field. Then $\Bbbk G$ is semisimple if and only if either the characteristic of $\Bbbk$ does not divide the order of G.*

Let $\Bbbk$ be an algebraically closed field of characteristic zero. Then G is abelian if and only if $\Bbbk G$ is commutative. Since $\Bbbk G$ is semisimple and $\Bbbk$ is algebraically closed, we deduce from Wedderburn's theorem that this will be the case if and only if each simple $\Bbbk G$-module is one-dimensional. We have thus proved the following result.

Proposition B.2. *Let $\Bbbk$ be an algebraically closed field of characteristic zero and let G be a finite group. Then G is abelian if and only if $\Bbbk G \cong \Bbbk^G$, which is equivalent to each simple $\Bbbk G$-module being one-dimensional.*

B.2 Group character theory

Let us now specialize to the case of the field of complex numbers, although much of what we say holds over any algebraically closed field of characteristic zero. Fix a finite group G for the section.

If $\rho \colon G \longrightarrow GL_n(\mathbb{C})$ is a representation, then the *character* of ρ is the mapping $\chi_\rho \colon G \longrightarrow \mathbb{C}$ defined by $\chi_\rho(g) = \mathrm{Tr}(\rho(g))$ where $\mathrm{Tr}(A)$ is the trace of a matrix A. It is clear that equivalent representations have the same character and hence if V is a finite dimensional $\mathbb{C}G$-module, then we can define the *character* χ_V of V by putting $\chi_V = \chi_\rho$ for any representation ρ afforded by V. We sometimes say that V *affords* the character χ. Note that $\chi_V(1) = \dim V$. A character is called *irreducible* if it is the character of a simple module. One useful property of characters is the following (a proof of which can be found on page 221 of [CR88]).

Proposition B.3. *Let χ be a character of a finite group G. Then $\chi(g^{-1}) = \overline{\chi(g)}$ for $g \in G$.*

Isomorphic modules are easily seen to have the same character. In fact, the character determines the module up to isomorphism. Let us enter into a bit of detail. A mapping $f \colon G \longrightarrow \mathbb{C}$ is called a *class function* if it is constant on conjugacy classes. The set $\mathrm{Cl}(G)$ of class functions on G is a subspace of $\mathbb{C}^G$ whose dimension is the number of conjugacy classes of G and, in fact, it is a subalgebra where $\mathbb{C}^G$ is made a $\mathbb{C}$-algebra via pointwise operations. We can define an inner product on $\mathbb{C}^G$ by putting

$$\langle f, h \rangle_G = \frac{1}{|G|} \sum_{g \in G} f(g)\overline{h(g)}.$$

If G is understood, we just write $\langle f, h \rangle$. Clearly, each character is a class function because the trace is constant on similarity classes of matrices. Our statement of the first orthogonality relations combines [Lam91, Theorem 8.16] and [Lam91, Corollary 8.17].

Theorem B.4 (First orthogonality relations). *Let $S_1, \ldots, S_r$ form a complete set of representatives of the isomorphism classes of simple $\mathbb{C}G$-modules. Then the irreducible characters $\chi_{S_1}, \ldots, \chi_{S_r}$ form an orthonormal basis for the space $\mathrm{Cl}(G)$ of class functions.*

As an immediate consequence, one can compute the number of isomorphism classes of simple $\mathbb{C}G$-modules.

Corollary B.5. *Let G be a finite group. The number of isomorphism classes of simple $\mathbb{C}G$-modules coincides with the number of conjugacy classes of G.*

Retaining the notation of Theorem B.4, if V is a $\mathbb{C}G$-module with

$$V \cong \bigoplus_{i=1}^{r} m_i \cdot S_i,$$

then $\chi_V = m_1 \chi_{S_1} + \cdots + m_r \chi_{S_r}$ and so we deduce from Theorem B.4 that $m_i = \langle \chi_V, \chi_{S_i} \rangle$. Therefore, V is determined up to isomorphism by its character. We summarize this discussion in the following corollary.

Corollary B.6. *Let $S_1, \ldots, S_r$ form a complete set of representatives of the isomorphism classes of simple $\mathbb{C}G$-modules and let V be a finite dimensional $\mathbb{C}G$-module. Then*

$$V \cong \bigoplus_{i=1}^{r} m_i \cdot S_i$$

where $m_i = \langle \chi_V, \chi_{S_i} \rangle$. Thus two finite dimensional $\mathbb{C}G$-modules V and W are isomorphic if and only if $\chi_V = \chi_W$.

It is for this reason that group representation theory over $\mathbb{C}$ reduces to character theory.

Theorem B.4 and Corollary B.6 lead to the following criterion for simplicity.

Corollary B.7. *Let V be a finite dimensional $\mathbb{C}G$-module. Then V is simple if and only if $\langle \chi_V, \chi_V \rangle = 1$.*

Proof. Necessity is clear from Theorem B.4. For sufficiency, let $S_1, \ldots, S_r$ form a complete set of non-isomorphic simple $\mathbb{C}G$-module and observe that if

$$V \cong \bigoplus_{i=1}^{r} m_i \cdot S_i,$$

then $\chi_V = m_1 \chi_{S_1} + \cdots + m_r \chi_{S_r}$. By Theorem B.4, we then have that $1 = \langle \chi_V, \chi_V \rangle = m_1^2 + \cdots + m_r^2$ and so there exists i such that $m_i = 1$ and $m_j = 0$ for $j \neq i$. We conclude that $V \cong S_i$ is simple. $\qquad\square$

The next theorem summarizes some further properties of group representations that we shall not require in the sequel, but which give some flavor of the subject. Proofs can be found in [CR88, (27.21), Theorem 33.7 and Theorem 33.8].

Theorem B.8. *Let G be a finite group and let $d_1, \ldots, d_r$ be the dimensions of a complete set of representatives of the isomorphism classes of simple $\mathbb{C}G$-modules.*

(i) $d_1^2 + \cdots + d_r^2 = |G|$.
(ii) d_i divides $|G|$ for $i = 1, \ldots, r$.
(iii) The number of one-dimensional representations of G is $|G/[G,G]|$.
(iv) If $\chi_1, \ldots, \chi_r$ are the irreducible characters of G, then the

$$e_i = \frac{\chi(1)}{|G|} \sum_{g \in G} \overline{\chi(g)} \cdot g,$$

for $i = 1, \ldots, r$, form a complete set of orthogonal central primitive idempotents of $\mathbb{C}G$.

Let $\chi_1, \ldots, \chi_r$ be the distinct irreducible characters of G and let $C_1, \ldots, C_r$ be the conjugacy classes of G. The *character table* of G is the $r \times r$ matrix $X(G)$ given by $X(G)_{ij} = \chi_i(C_j)$ where $\chi_i(C_j)$ is the value that χ_i takes on the conjugacy class C_j. Sometime we index the rows of $X(G)$ by the irreducible characters and the columns by the conjugacy classes.

The next theorem is a reformulation of what is often called the second orthogonality relations. Recall that if $A = (a_{ij}) \in M_r(\mathbb{C})$, then $A^* = (\overline{a_{ji}})$ is the conjugate-transpose.

Theorem B.9 (Second orthogonality relations). *Let G be a finite group with irreducible characters $\chi_1, \ldots, \chi_r$ and conjugacy classes $C_1, \ldots, C_r$. Put $z_i = |C_i|/|G|$, for $i = 1, \ldots, r$, and let Z be the $r \times r$ diagonal matrix with $Z_{ii} = z_i$. Then $X(G)ZX(G)^* = I$ and hence $X(G)$ is invertible with inverse $ZX(G)^*$.*

Proof. Let $A = X(G)ZX(G)^*$. Then we compute that

$$A_{ij} = \frac{1}{|G|} \sum_{k=1}^{r} \chi_i(C_k) \cdot |C_k| \cdot \overline{\chi_j(C_k)} = \frac{1}{|G|} \sum_{g \in G} \chi_i(g)\overline{\chi_j(g)} = \langle \chi_i, \chi_j \rangle = I_{ij}$$

where the last equality uses the first orthogonality relations. $\qquad\square$

B.3 Permutation modules

Let G be a finite group and Ω a finite G-set. Let $\Bbbk$ be a field. Then $\Bbbk\Omega$ is a $\Bbbk G$-module by extending linearly the action of G on Ω, i.e.,

$$g \cdot \sum_{\alpha \in \Omega} c_\alpha \alpha = \sum_{\alpha \in \Omega} c_\alpha g\alpha$$

for $g \in G$. One usually calls $\Bbbk\Omega$ a *permutation module*. Recall that Ω is a *transitive G-set* if G has a single orbit on Ω. In other words, G acts transitively on Ω if, for all $\alpha, \beta \in \Omega$, there exists $g \in G$ with $g\alpha = \beta$. If G acts transitively on Ω and if $H \leq G$ is the stabilizer of $\omega \in \Omega$, then $|\Omega| = [G : H]$. Set $\lfloor \Omega \rfloor = \sum_{\omega \in \Omega} \omega$. This vector will play an important role.

If V is a $\Bbbk G$-module, then one puts

$$V^G = \{v \in V \mid gv = v, \forall g \in G\}.$$

So V^G is the subspace of fixed vectors. Observe that V^G is the submodule of $\mathrm{soc}(V)$ spanned by all copies of the trivial $\Bbbk G$-module and is isomorphic to the direct sum of $\dim V^G$ copies of the trivial module. In particular, if $\Bbbk = \mathbb{C}$ and $\dim V < \infty$, then $\dim V^G = \langle \chi_V, \chi_1 \rangle$ where χ_1 is the character of the trivial module.

Proposition B.10. *Let G be a finite group, Ω a transitive G-set, and $\Bbbk$ a field of characteristic 0. Then $(\Bbbk\Omega)^G = \Bbbk\lfloor\Omega\rfloor$. Moreover, if*

$$U = \frac{1}{|G|} \sum_{g \in G} g,$$

then U is an idempotent with $U \cdot \Bbbk\Omega = (\Bbbk\Omega)^G$ and $U\beta = \frac{1}{|\Omega|}\lfloor\Omega\rfloor$ for $\beta \in \Omega$.

Proof. Put $V = \Bbbk\Omega$. It is clear that $g\lfloor\Omega\rfloor = \lfloor\Omega\rfloor$ for all $g \in G$ because G permutes Ω. Suppose that $v = \sum_{\alpha \in \Omega} c_\alpha \alpha$ belongs to V^G. Fix $\omega \in \Omega$. If $\alpha \in \Omega$ and $g \in G$ with $g\omega = \alpha$, then the coefficient of α in gv is c_ω. As $gv = v$ for all $g \in G$, we conclude that $v = c_\omega \lfloor\Omega\rfloor$. This establishes that $V^G = \Bbbk\lfloor\Omega\rfloor$.

If $h \in G$, then since left multiplication by h permutes G, we have $hU = U$. It follows easily that $U^2 = U$ and that $U \cdot V \subseteq V^G$. But if $v \in V^G$, then

$$Uv = \frac{1}{|G|} \sum_{g \in G} gv = \frac{1}{|G|} \sum_{g \in G} v = v$$

and so $U \cdot V = V^G$ (note that this argument works for any $\Bbbk G$-module V).

Let $\beta \in \Omega$ and let $H \leq G$ be the stabilizer of β. Fix $g \in G$ with $g\beta = \alpha$. Then, for $g' \in G$, one has that $g'\beta = \alpha$ if and only if $g' \in gH$. Therefore, we may compute

$$U\beta = \frac{1}{|G|} \sum_{g \in G} g\beta = \frac{|H|}{|G|} \sum_{\alpha \in \Omega} \alpha = \frac{1}{|\Omega|}\lfloor\Omega\rfloor$$

as $[G : H] = |\Omega|$. This completes the proof. $\qquad\square$

For a G-set Ω and $g \in G$, put $\mathrm{Fix}(g) = \{\alpha \in \Omega \mid g\alpha = \alpha\}$. Note that

$$\chi_{\mathbb{C}\Omega}(g) = |\mathrm{Fix}(g)| \tag{B.1}$$

by considering the matrix representation afforded by $\mathbb{C}\Omega$ with respect to the basis Ω of $\mathbb{C}\Omega$. In particular, taking $\Omega = G$ with the regular action of G yields

$$\chi_{\mathbb{C}G}(g) = \begin{cases} |G|, & \text{if } g = 1 \\ 0, & \text{else,} \end{cases} \tag{B.2}$$

a fact used elsewhere in the text.

An immediate corollary of Proposition B.10 is the following result.

Corollary B.11. *Let Ω be a finite G-set for a finite group G and suppose that G has s orbits on Ω. Let χ_1 be the trivial character of G. Then*

$$s = \langle \chi_{\mathbb{C}\Omega}, \chi_1 \rangle = \frac{1}{|G|} \sum_{g \in G} |\mathrm{Fix}(g)|,$$

i.e., the average number of fixed points of an element of G is the number of orbits of G on Ω.

Proof. If $\Omega_1, \ldots, \Omega_s$ are the orbits of G, then $\mathbb{C}\Omega = \bigoplus_{i=1}^{s} \mathbb{C}\Omega_i$. Therefore, we have that

$$\langle \chi_{\mathbb{C}\Omega}, \chi_1 \rangle = \sum_{i=1}^{s} \langle \chi_{\mathbb{C}\Omega_i}, \chi_1 \rangle = s$$

where the last equality uses that $\langle \chi_{\Omega_i}, \chi_1 \rangle = \dim(\mathbb{C}\Omega_i)^G = 1$ by Proposition B.10. Taking into account (B.1), the corollary follows. $\qquad\square$

Corollary B.11 is sometimes called the Cauchy-Frobenius-Burnside lemma.

If Ω is a G-set, let us define the *augmentation submodule* $\mathrm{Aug}(\Bbbk\Omega)$ to be the submodule of $\Bbbk\Omega$ consisting of those $v = \sum_{\alpha \in \Omega} c_\alpha \alpha$ with $\sum_{\alpha \in \Omega} c_\alpha = 0$. Note that $\Bbbk\Omega = \mathrm{Aug}(\Bbbk\Omega) \oplus \Bbbk\lfloor\Omega\rfloor$ if $\Bbbk$ is of characteristic zero.

If Ω is a G-set, then Ω^2 is a G-set via $g(\alpha, \beta) = (g\alpha, g\beta)$. The diagonal

$$\Delta = \{(\alpha, \alpha) \in \Omega^2 \mid \alpha \in \Omega\}$$

is obviously G-invariant and hence so is its complement $\Omega^2 \setminus \Delta$. One says that G acts 2-*transitively* or *doubly transitively* on Ω if G is transitive on $\Omega^2 \setminus \Delta$.

Proposition B.12. *Let G be a finite group and Ω a transitive G-set. Then G is 2-transitive if and only if $\mathrm{Aug}(\mathbb{C}\Omega)$ is simple.*

Proof. Let θ be the character afforded by $\mathbb{C}\Omega$ and let χ_1 be the trivial character of G. It is then clear from (B.1) that $\chi_{\mathbb{C}\Omega^2} = \theta^2$. Also $\mathbb{C}\Omega \cong \mathbb{C}\Delta$ since Ω and Δ are isomorphic G-sets. Let $\Upsilon = \Omega^2 \setminus \Delta$. The direct sum decomposition

Fig. B.1. The Young diagram of $\lambda = (3, 3, 2, 1)$

$\mathbb{C}\Omega^2 = \mathbb{C}\Upsilon \oplus \mathbb{C}\Delta$ shows that $\theta^2 = \chi_{\mathbb{C}\Upsilon} + \theta$. Also note that if μ is the character afforded by $\mathrm{Aug}(\mathbb{C}\Omega)$, then $\mu = \theta - \chi_1$ because $\mathbb{C}\Omega = \mathrm{Aug}(\mathbb{C}\Omega) \oplus \mathbb{C}\lfloor\Omega\rfloor$.

We observe that

$$\langle \theta^2, \chi_1 \rangle = \frac{1}{|G|} \sum_{g \in G} \theta^2(g) = \langle \theta, \theta \rangle$$

since θ is integer-valued. Therefore, we compute

$$\langle \mu, \mu \rangle = \langle \theta, \theta \rangle - 2\langle \theta, \chi_1 \rangle + \langle \chi_1, \chi_1 \rangle = \langle \theta^2, \chi_1 \rangle - \langle \theta, \chi_1 \rangle = \langle \theta^2 - \theta, \chi_1 \rangle$$
$$= \langle \chi_{\mathbb{C}\Upsilon}, \chi_1 \rangle$$

using that $\langle \theta, \chi_1 \rangle = 1 = \langle \chi_1, \chi_1 \rangle$ by Proposition B.10.

It follows from Corollary B.11 that $\langle \mu, \mu \rangle$ is the number of orbits of G on Υ and hence $\mathrm{Aug}(\mathbb{C}\Omega)$ is simple if and only if G is doubly transitive by Corollary B.7. $\qquad\qquad\square$

B.4 The representation theory of the symmetric group

We review here some elements of the representation theory of the symmetric group S_n. Standard references for this material include [Jam78, JK81, CR88, FH91, Mac95, Sta99, Sag01, CSST10, Ste12a]. No proofs are provided. We assume here that $\Bbbk$ is a field of characteristic 0. Put $[n] = \{1, \ldots, n\}$ for $n \geq 0$.

A *partition* of n is a nonincreasing sequence $\lambda = (\lambda_1, \ldots, \lambda_m)$ of positive integers with $\lambda_1 + \cdots + \lambda_m = n$. The λ_i are called the *parts* of λ. The set of partitions of n is denoted $\mathcal{P}_n$. We allow an empty partition () of 0. We shall write i^m as short hand for m consecutive occurrences of i. For example $(3^2, 2, 1^4)$ is short hand for $(3, 3, 2, 1, 1, 1, 1)$.

It is convenient to represent partitions by Young diagrams. If $\lambda = (\lambda_1, \ldots, \lambda_m)$ is a partition of n, then the *Young diagram* of λ consists of n boxes placed into m rows where the ith row has λ_i boxes. For example, the Young diagram of $\lambda = (3, 3, 2, 1)$ is depicted in Figure B.1. Conversely, any diagram consisting of n boxes arranged into rows such that the number of boxes in each row is nonincreasing (going from top to bottom) is the Young diagram of a unique partition of n.

The *cycle type* of $f \in S_n$ is the partition $\lambda(f) = (\lambda_1, \ldots, \lambda_m)$ where m is the number of orbits of f and the λ_i are the sizes of the orbits of f (with multiplicities) listed in nonincreasing order. For example, if $f \in S_{10}$ has cycle decomposition $(1\ 2\ 3)(4\ 5)(6\ 7\ 9)$, then $\lambda(f) = (3^2, 2, 1^2)$. It is well known

$$\begin{array}{|c|c|c|}\hline 1 & 2 & 3 \\\hline 4 & 5 \\\cline{1-2} 6 \\\cline{1-1}\end{array} \qquad \begin{array}{|c|c|c|}\hline 1 & 3 & 4 \\\hline 2 & 5 \\\cline{1-2} 6 \\\cline{1-1}\end{array} \qquad \begin{array}{|c|c|c|}\hline 1 & 2 & 4 \\\hline 3 & 6 \\\cline{1-2} 5 \\\cline{1-1}\end{array}$$

Fig. B.2. Some standard Young tableaux of shape $(3, 2, 1)$

that two permutations are conjugate if and only if they have the same cycle type. The conjugacy class of permutations of cycle type λ will be denoted C_λ.

The simple $\Bbbk S_n$-modules are indexed by partitions λ of n. There is a canonical labeling and the simple module associated with a partition λ is denoted by S_λ and is called the *Specht module* associated with λ. We recall the construction for completeness.

If λ is a partition of n, then a *Young tableau* of shape λ, or λ-*tableau*, is an array t of integers obtained by placing $1, \ldots, n$ into the boxes of the Young diagram for λ. There are clearly $n!$ tableaux of shape λ. A *standard Young tableau* of shape λ is a tableau whose entries are increasing along each row and column. For example, if $\lambda = (3, 2, 1)$, then some standard tableaux are as in Figure B.2.

To construct S_λ, let t_λ be the λ-tableau with the numbers $1, \ldots, n$ arranged in order, from left to right, starting at the top left entry; in other words, j is in the jth box of t_λ. For example, $t_{(3,2,1)}$ is the first tableau in Figure B.2. If t is a λ-tableau, then R_t denotes the subgroup of S_n preserving the rows of t and C_t the subgroup preserving the columns of t. Consider the elements

$$a_\lambda = \sum_{f \in R_{t_\lambda}} f \quad \text{and} \quad b_\lambda = \sum_{f \in C_{t_\lambda}} \operatorname{sgn}(f) f$$

of $\Bbbk S_n$ where

$$\operatorname{sgn}(f) = \begin{cases} 1, & \text{if } f \text{ is even} \\ -1, & \text{if } f \text{ is odd.} \end{cases}$$

The *Young symmetrizer* associated with λ is $c_\lambda = a_\lambda b_\lambda$ and the Specht module is $S_\lambda = \Bbbk S_n c_\lambda$.

We remark that $S_{(n)}$ is the trivial $\Bbbk S_n$-module and $S_{(1^n)}$ is the one-dimensional sign representation $\operatorname{sgn} \colon S_n \longrightarrow \Bbbk \setminus \{0\}$ sending a permutation to its sign. Also, as $c_\lambda \in \mathbb{Q} S_n$, it follows that S_λ is defined over $\mathbb{Q}$, that is, there is a basis of S_λ so that each permutation is sent to a matrix with rational entries. The dimension of S_λ is known to be the number f_λ of standard Young tableaux of shape λ; see [FH91, Page 57].

$$\begin{array}{|c|c|c|}\hline 1 & 2 & 3 \\\hline 4 & 5 \\\cline{1-2}\end{array} \quad \sim \quad \begin{array}{|c|c|c|}\hline 3 & 1 & 2 \\\hline 5 & 4 \\\cline{1-2}\end{array}$$

Fig. B.3. Two equivalent tableaux of shape $\lambda = (3, 2)$

Let us also provide an alternative, and more explicit, description of S_λ. The group S_n acts freely and transitively on the set of λ-tableaux by applying $f \in S_n$ to the entries of the boxes of a tableau; the result of applying $f \in S_n$ to a tableau t is denoted ft. For example, if $f = (1\ 3\ 2)$ and

$$t = \begin{array}{|c|c|}\hline 1 & 3 \\\hline 4 & 2 \\\hline\end{array},$$

then

$$ft = \begin{array}{|c|c|}\hline 3 & 2 \\\hline 4 & 1 \\\hline\end{array}.$$

Let us define an equivalence relation $\sim$ on the set of λ-tableaux by putting $t_1 \sim t_2$ if they have the same entries in each row. Two equivalent Young tableaux are shown in Figure B.3. A $\sim$-equivalence class of λ-tableaux is called a λ-*tabloid* or a *tabloid of shape* λ. We write $[t]$ for the tabloid of a tableau t. The set of all tabloids of shape λ is denoted T^λ.

We put $T_\lambda = [t_\lambda]$. So, for example, $T_{(3,2,1)}$ is the tabloid corresponding to the first tableau in Figure B.2. If $t_1 \sim t_2$ and $f \in S_n$, then $ft_1 \sim ft_2$ and hence there is a well-defined action of S_n on T^λ given by putting $f[t] = [ft]$ for $f \in S_n$ and t a λ-tableau.

The action of S_n on the set of λ-tabloids is transitive as it was already transitive on λ-tableaux. Clearly, R_t is the stabilizer of $[t]$. Therefore, if $\lambda = (\lambda_1, \ldots, \lambda_\ell)$, then the stabilizer of T_λ is

$$R_{t_\lambda} = S_{\{1,\ldots,\lambda_1\}} \times S_{\{\lambda_1+1,\ldots,\lambda_1+\lambda_2\}} \times \cdots \times S_{\{\lambda_1+\cdots+\lambda_{\ell-1}+1,\ldots,n\}}.$$

Thus $|T^\lambda| = [S_n : R_{t_\lambda}] = n!/\lambda_1! \cdots \lambda_\ell!$. The subgroup R_{t_λ} is called the *Young subgroup* associated with the partition λ and the associated permutation module $\Bbbk T^\lambda$ is denoted M_λ.

For example, if $\lambda = (n-1, 1)$, then two λ-tableaux are equivalent if and only if they have the same entry in the second row and hence $M_{(n-1,1)}$ is isomorphic to the natural $\Bbbk S_n$-module $\Bbbk^n$. On the other hand, if $\lambda = (n)$, then there is only one λ-tabloid and so $M_{(n)}$ is the trivial module.

It turns out that M_λ contains S_λ as a distinguished simple submodule. For a λ-tableau t, the element

$$\varepsilon_t = \sum_{g \in C_t} \mathrm{sgn}(g) g[t]$$

of M_λ is called the *polytabloid* associated with t. The action of S_n on λ-tableaux is compatible with the definition of polytabloids in the following sense: if $f \in S_n$ and t is a λ-tableau, then $f\varepsilon_t = \varepsilon_{ft}$. The key step in proving this is the claim that $C_{ft} = fC_t f^{-1}$. Indeed, if X_i is the set of entries of column i of t, then $f(X_i)$ is the set of entries of column i of ft. Since g stabilizes X_i if and only if fgf^{-1} stabilizes $f(X_i)$, the claim follows. Thus we have that

$$f\varepsilon_t = \sum_{g \in C_t} \mathrm{sgn}(g) fgf^{-1} f[t]$$

$$= \sum_{h \in C_{ft}} \mathrm{sgn}(f^{-1}hf) h[ft] = \varepsilon_{ft}$$

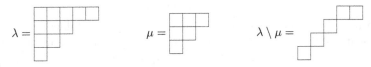

$$\lambda = \qquad \mu = \qquad \lambda \setminus \mu =$$

Fig. B.4. A horizontal strip

where we have made the substitution $h = fgf^{-1}$. One then has that

$$S_\lambda \cong \Bbbk S_n \varepsilon_{t_\lambda} = \Bbbk\{\varepsilon_t \mid t \in T^\lambda\} \le M_\lambda.$$

Moreover, the polytabloids ε_t with t a standard Young tableau form a basis for S_λ.

Suppose that $r \le n$ and that V is a $\Bbbk S_r$-module and W is a $\Bbbk S_{n-r}$-module. Their outer product $V \boxtimes W$ is the $\Bbbk S_n$-module defined as follows. We view $S_r \times S_{n-r}$ as a subgroup of S_n by identifying S_r with the permutations fixing $\{r+1, \ldots, n\}$ pointwise and S_{n-r} with the permutations fixing $[r]$ pointwise. Note that $V \otimes W$ is a $\Bbbk[S_r \times S_{n-r}]$-module via the action $(f, g)(v \otimes w) = fv \otimes gw$ for $f \in S_r$, $g \in S_{n-r}$, $v \in V$, and $w \in W$. The *outer product* of V and W is then the induced representation

$$V \boxtimes W = \Bbbk S_n \otimes_{\Bbbk[S_r \times S_{n-r}]} (V \otimes W).$$

Outer products with the trivial module are easily decomposed thanks to what is sometimes known as *Pieri's rule*.

If λ is a partition of n and μ is a partition of $r \le n$, then we write $\mu \subseteq \lambda$ if the Young diagram of μ is contained in the Young diagram of λ. We say that the skew diagram $\lambda \setminus \mu$ is a *horizontal strip* if no two boxes of $\lambda \setminus \mu$ are in the same column. For example, if $\lambda = (5, 3, 2, 1)$ and $\mu = (3, 2, 1)$, then $\mu \subseteq \lambda$ and $\lambda \setminus \mu$ is a horizontal strip; see Figure B.4. The empty diagram is considered to be a horizontal strip. The following theorem is [JK81, Corollary 2.8.3].

Theorem B.13 (Pieri's rule). *Let μ be a partition of r with $0 \leq r \leq n$. Then the decomposition*

$$S_\mu \boxtimes S_{(n-r)} = \bigoplus_{\substack{\mu \subseteq \lambda \\ \lambda \backslash \mu \text{ is a horizontal strip}}} S_\lambda$$

holds.

Notice that the decomposition in Theorem B.13 is multiplicity-free, i.e., has no isomorphic summands.

B.5 Exercises

B.1. Let G be a group and let V, W be finite dimensional $\mathbb{C}G$-modules. Then $\mathrm{Hom}_\mathbb{C}(V, W)$ is a $\mathbb{C}G$-module via $(g\varphi)(v) = g\varphi(g^{-1}v)$.

(a) Prove that $\mathrm{Hom}_\mathbb{C}(V, W)^G = \mathrm{Hom}_{\mathbb{C}G}(V, W)$.
(b) Prove that the character θ of $\mathrm{Hom}_\mathbb{C}(V, W)$ is given by $\theta(g) = \chi(g)\overline{\psi(g)}$ where ψ is the character of V and χ is the character of W.
(c) Use the previous parts and that $U = \frac{1}{|G|} \sum_{g \in G} g$ is the projector to the fixed subspace of a $\mathbb{C}G$-module to prove the first orthogonality relations.

B.2. A representation $\rho \colon G \longrightarrow GL_n(\mathbb{C})$ of a group G is *unitary* if $\rho(g^{-1}) = \rho(g)^*$ for all $g \in G$. Prove that every representation of a finite group over $\mathbb{C}$ is equivalent to a unitary representation.

B.3. Suppose that G is a finite group and $\Bbbk$ is a field of characteristic p dividing $|G|$. Prove that the trivial $\Bbbk G$-module is not projective. (Hint: if it were, there would be a primitive idempotent e such that $\Bbbk Ge \cong \Bbbk$; show that $e = c \cdot \sum_{g \in G} g$ with $c \in \Bbbk$ and $|G| \cdot c = 1$.)

B.4. Let G be a finite group and $H \leq G$ a subgroup. Let V be a $\Bbbk H$-module and W a $\Bbbk G$-module. Prove that $\mathrm{Hom}_{\Bbbk G}(\Bbbk G \otimes_{\Bbbk H} V, W) \cong \mathrm{Hom}_{\Bbbk H}(V, W)$ where we view W as a $\Bbbk H$-module via restriction of scalars. This is known as *Frobenius reciprocity*.

Appendix C

Incidence Algebras and Möbius Inversion

The purpose of this appendix is to familiarize the reader with an important combinatorial technique that is used throughout the text. The theory of incidence algebras and Möbius inversion for posets was developed by Rota [Rot64] and can be considered as part of the origins of algebraic combinatorics. It provides a highly conceptual generalization of the principle of inclusion-exclusion. A thorough introduction, including techniques for computing the Möbius function of a poset and connections with algebraic topology, can be found in Stanley's classic text [Sta97, Chapter 3]. Throughout this appendix, P will denote a finite poset.

C.1 The incidence algebra of a poset

Let $\Bbbk$ be a field. The *incidence algebra* $I(P, \Bbbk)$ of a finite poset P over $\Bbbk$ is the set of all mappings $f \colon P \times P \longrightarrow \Bbbk$ such that $f(p, q) = 0$ if $p \not\leq q$. One can profitably think of f as a $P \times P$-upper triangular matrix. Matrix multiplication then translates into the following associative multiplication on $I(P, \Bbbk)$:

$$f * g(p, q) = \sum_{p \leq r \leq q} f(p, r)g(r, q). \tag{C.1}$$

The proof of the following proposition is left to the reader as an exercise.

Proposition C.1. *The incidence algebra $I(P, \Bbbk)$ of a finite poset P is a finite dimensional $\Bbbk$-algebra with respect to pointwise addition of functions and the product (C.1). The multiplicative identity is given by the Kronecker function*

$$\delta(p, q) = \begin{cases} 1, & \text{if } p = q \\ 0, & \text{else} \end{cases}$$

for $p, q \in P$.

© Springer International Publishing Switzerland 2016
B. Steinberg, *Representation Theory of Finite Monoids*,
Universitext, DOI 10.1007/978-3-319-43932-7

The *zeta function* of P is the mapping $\zeta \colon P \times P \longrightarrow \Bbbk$ given by

$$\zeta(p, q) = \begin{cases} 1, & \text{if } p \leq q \\ 0, & \text{else.} \end{cases}$$

It is easy to see that if k is the length of the longest chain in P, then one has that $(\delta - \zeta)^{k+1} = 0$, and so $\zeta = \delta - (\delta - \zeta)$ is a unit with inverse

$$\mu = \sum_{n=0}^{k} (\delta - \zeta)^n.$$

The mapping μ is called the *Möbius function* of P. We shall use the notation ζ_P and μ_P if the poset P is not clear from context.

A recursive formula for μ can also be given, which is essentially the classical adjoint formula for the inverse of a matrix restricted to the special case of an upper triangular matrix. See [Sta97, Chapter 3, Section 7] for details.

Proposition C.2. *Let P be a finite poset. Then*

$$\mu(p, q) = \begin{cases} 1, & \text{if } p = q \\ -\sum_{p \leq r < q} \mu(p, r), & \text{if } p < q \\ 0, & \text{else} \end{cases}$$

for all $p, q \in P$.

Example C.3. If $P = \{0, 1\}$ with $0 < 1$, then $\mu(0, 0) = 1 = \mu(1, 1)$ and $\mu(0, 1) = -1$.

The Möbius function is compatible with the formation of direct products. Recall that if P, Q are posets, then $P \times Q$ is a poset with $(p, q) \leq (p', q')$ if and only if $p \leq p'$ and $q \leq q'$. The following can be found in [Sta97, Proposition 3.8.2].

Proposition C.4. *Let P and Q be finite posets. Then $\mu_{P \times Q}((p, q), (p', q')) = \mu_P(p, p') \mu_Q(q, q')$.*

Since the power set of an n-element set is isomorphic to $\{0, 1\}^n$, Proposition C.4 and Example C.3 combine to yield the following result.

Corollary C.5. *Let $\mathscr{P}(X)$ be the power set of a finite set X. Then one has $\mu(Y, Z) = (-1)^{|Z| - |Y|}$ for $Y \subseteq Z$.*

The Möbius function of a poset is a fundamental invariant in poset theory, which also has connections with algebraic topology. In this text, our primary interest is due to Rota's Möbius inversion theorem. We shall state it in the following form.

Theorem C.6. *Let P be a finite poset and V a $\Bbbk$-vector space. Suppose that $f\colon P \longrightarrow V$ is a function and define $g\colon P \longrightarrow V$ by*

$$g(p) = \sum_{q \le p} f(q).$$

Then we have that

$$f(p) = \sum_{q \le p} g(q)\mu(q,p)$$

for all $p \in P$.

Proof. Let W be the $\Bbbk$-vector space of all mappings $h\colon P \longrightarrow V$ (with pointwise operations). We make W into a right $I(P,\Bbbk)$-module by putting

$$h * a(p) = \sum_{q \le p} h(q)a(q,p)$$

for $h \in W$, $a \in I(P,\Bbbk)$ and $p \in P$. Then we compute

$$f * \zeta(p) = \sum_{q \le p} f(q) = g(p),$$

and so $f = f * \zeta * \mu = g * \mu$. But, for $p \in P$, we have

$$g * \mu(p) = \sum_{q \le p} g(q)\mu(q,p).$$

The theorem follows.   □

C.2 Exercises

C.1. Compute the Möbius function of the poset $\{1,\dots,n\}$ with the usual ordering.

C.2. Prove Proposition C.2.

C.3. Prove Proposition C.4.

C.4. Prove Corollary C.5.

C.5. Let P and Q be finite posets and suppose that $[p,p'] \cong [q,q']$. Prove that $\mu_P(p,p') = \mu_Q(q,q')$.

C.6. This exercise requires some familiarity with algebraic topology. Let P be a finite poset. If $p \le q$, let $\Delta(p,q)$ be the simplicial complex with vertex set $(p,q) = \{x \in P \mid p < x < q\}$ and with simplices the chains in (p,q). Prove that $\mu(p,q) = \chi(\Delta(p,q)) - 1$ where $\chi(K)$ denotes the Euler characteristic of a simplicial complex K.

References

[ABC13]   J. Araújo, W. Bentz, P.J. Cameron, Groups synchronizing a transformation of non-uniform kernel. Theor. Comput. Sci. **498**, 1–9 (2013)

[AC14]   J. Araújo, P.J. Cameron, Primitive groups synchronize non-uniform maps of extreme ranks. J. Comb. Theory Ser. B **106**, 98–114 (2014)

[ACS15]   J. Araújo, P.J. Cameron, B. Steinberg, Between primitive and 2-transitive: synchronization and its friends. ArXiv e-prints, November 2015

[AD10]   C.A. Athanasiadis, P. Diaconis, Functions of random walks on hyperplane arrangements. Adv. Appl. Math. **45**(3), 410–437 (2010)

[AKS14a]   A. Ayyer, S. Klee, A. Schilling, Combinatorial Markov chains on linear extensions. J. Algebraic Comb. **39**(4), 853–881 (2014)

[AKS14b]   A. Ayyer, S. Klee, A. Schilling, Markov chains for promotion operators, in *Algebraic Monoids, Group Embeddings, and Algebraic Combinatorics*, ed. by M. Can, Z. Li, B. Steinberg, Q. Wang. Fields Institute Communications, vol. 71 (Springer, New York, 2014), pp. 285–304

[Alm94]   J. Almeida, *Finite Semigroups and Universal Algebra*. Series in Algebra, vol. 3 (World Scientific, River Edge, NJ, 1994). Translated from the 1992 Portuguese original and revised by the author

[AM06]   M. Aguiar, S. Mahajan, *Coxeter Groups and Hopf Algebras*. Fields Institute Monographs, vol. 23 (American Mathematical Society, Providence, RI, 2006). With a foreword by N. Bergeron

[AMSV09]   J. Almeida, S. Margolis, B. Steinberg, M. Volkov, Representation theory of finite semigroups, semigroup radicals and formal language theory. Trans. Am. Math. Soc. **361**(3), 1429–1461 (2009)

[AO08]   M. Aguiar, R.C. Orellana, The Hopf algebra of uniform block permutations. J. Algebraic Comb. **28**(1), 115–138 (2008)

[APT92]   M. Auslander, M.I. Platzeck, G. Todorov, Homological theory of idempotent ideals. Trans. Am. Math. Soc. **332**(2), 667–692 (1992)

© Springer International Publishing Switzerland 2016   295
B. Steinberg, *Representation Theory of Finite Monoids*,
Universitext, DOI 10.1007/978-3-319-43932-7

296   References

[AR67]   W.W. Adams, M.A. Rieffel, Adjoint functors and derived functors with an application to the cohomology of semigroups. J. Algebra **7**, 25–34 (1967)

[ARS97]   M. Auslander, I. Reiten, S.O. Smalø, *Representation Theory of Artin Algebras*. Cambridge Studies in Advanced Mathematics, vol. 36 (Cambridge University Press, Cambridge, 1997). Corrected reprint of the 1995 original

[AS06]   F. Arnold, B. Steinberg, Synchronizing groups and automata. Theor. Comput. Sci. **359**(1–3), 101–110 (2006)

[AS09]   J. Almeida, B. Steinberg, Matrix mortality and the Černý-Pin conjecture, in *Developments in Language Theory*. Lecture Notes in Comput. Sci., vol. 5583 (Springer, Berlin, 2009), pp. 67–80

[ASS06]   I. Assem, D. Simson, A. Skowroński, *Elements of the Representation Theory of Associative Algebras. Vol. 1: Techniques of Representation Theory*. London Mathematical Society Student Texts, vol. 65 (Cambridge University Press, Cambridge, 2006)

[ASST15a]   A. Ayyer, A. Schilling, B. Steinberg, N.M. Thiéry, Directed nonabelian sandpile models on trees. Commun. Math. Phys. **335**(3), 1065–1098 (2015)

[ASST15b]   A. Ayyer, A. Schilling, B. Steinberg, N.M. Thiéry, Markov chains, $\mathscr{R}$-trivial monoids and representation theory. Int. J. Algebra Comput. **25**(1–2), 169–231 (2015)

[BBBS11]   C. Berg, N. Bergeron, S. Bhargava, F. Saliola, Primitive orthogonal idempotents for R-trivial monoids. J. Algebra **348**, 446–461 (2011)

[BBD82]   A.A. Beĭlinson, J. Bernstein, P. Deligne, Faisceaux pervers, in *Analysis and Topology on Singular Spaces, I (Luminy, 1981)*. Astérisque, vol. 100 (Soc. Math. France, Paris, 1982), pp. 5–171

[BBD99]   L.J. Billera, K.S. Brown, P. Diaconis, Random walks and plane arrangements in three dimensions. Am. Math. Mon. **106**(6), 502–524 (1999)

[BD92]   D. Bayer, P. Diaconis, Trailing the dovetail shuffle to its lair. Ann. Appl. Probab. **2**(2), 294–313 (1992)

[BD98]   K.S. Brown, P. Diaconis, Random walks and hyperplane arrangements. Ann. Probab. **26**(4), 1813–1854 (1998)

[BD12]   R. Boltje, S. Danz, Twisted split category algebras as quasi-hereditary algebras. Arch. Math. (Basel) **99**(6), 589–600 (2012)

[BD15]   R. Boltje, S. Danz, Quasi-hereditary structure of twisted split category algebras revisited. J. Algebra **440**, 317–353 (2015)

[Ben98]   D.J. Benson, *Representations and Cohomology: I, Basic Representation Theory of Finite Groups and Associative Algebras*. Cambridge Studies in Advanced Mathematics, vol. 30, 2nd edn. (Cambridge University Press, Cambridge, 1998)

[BHR99]   P. Bidigare, P. Hanlon, D. Rockmore, A combinatorial description of the spectrum for the Tsetlin library and its generalization to hyperplane arrangements. Duke Math. J. **99**(1), 135–174 (1999)

[Bjö84]   A. Björner, Posets, regular CW complexes and Bruhat order. Eur. J. Comb. **5**(1), 7–16 (1984)

[Bjö08]   A. Björner, Random walks, arrangements, cell complexes, greedoids, and self-organizing libraries, in *Building Bridges*. Bolyai Soc. Math. Stud., vol. 19 (Springer, Berlin, 2008), pp. 165–203

[Bjö09]   A. Björner, Note: random-to-front shuffles on trees. Electron. Commun. Probab. **14**, 36–41 (2009)

[BLVS⁺99]   A. Björner, M. Las Vergnas, B. Sturmfels, N. White, G.M. Ziegler, *Oriented Matroids*. Encyclopedia of Mathematics and Its Applications, vol. 46, 2nd edn. (Cambridge University Press, Cambridge, 1999)

[BPR10]   J. Berstel, D. Perrin, C. Reutenauer, *Codes and Automata*. Encyclopedia of Mathematics and Its Applications, vol. 129 (Cambridge University Press, Cambridge, 2010)

[BR90]   J. Berstel, C. Reutenauer, Zeta functions of formal languages. Trans. Am. Math. Soc. **321**(2), 533–546 (1990)

[BR11]   J. Berstel, C. Reutenauer, *Noncommutative Rational Series with Applications*. Encyclopedia of Mathematics and Its Applications, vol. 137 (Cambridge University Press, Cambridge, 2011)

[Bra64]   R. Brauer, A note on theorems of Burnside and Blichfeldt. Proc. Am. Math. Soc. **15**, 31–34 (1964)

[Bro71]   T.C. Brown, An interesting combinatorial method in the theory of locally finite semigroups. Pac. J. Math. **36**, 285–289 (1971)

[Bro00]   K.S. Brown, Semigroups, rings, and Markov chains. J. Theor. Probab. **13**(3), 871–938 (2000)

[Bro04]   K.S. Brown, Semigroup and ring theoretical methods in probability, in *Representations of Finite Dimensional Algebras and Related Topics in Lie Theory and Geometry*. Fields Inst. Commun., vol. 40 (American Mathematical Society, Providence, RI, 2004), pp. 3–26

[Bur55]   W. Burnside, *Theory of Groups of Finite Order*, 2nd edn. (Dover, New York, 1955)

[BZ92]   A. Björner, G.M. Ziegler, Combinatorial stratification of complex arrangements. J. Am. Math. Soc. **5**(1), 105–149 (1992)

[Č64]   J. Černý. A remark on homogeneous experiments with finite automata. Mat.-Fyz. Časopis Sloven. Akad. Vied **14**, 208–216 (1964)

[Car86]   R.W. Carter, Representation theory of the 0-Hecke algebra. J. Algebra **104**(1), 89–103 (1986)

[CE99]   H. Cartan, S. Eilenberg, *Homological Algebra*. Princeton Landmarks in Mathematics (Princeton University Press, Princeton, NJ, 1999). With an appendix by D.A. Buchsbaum, Reprint of the 1956 original

[CEFN14]   T. Church, J.S. Ellenberg, B. Farb, R. Nagpal, FI-modules over Noetherian rings. Geom. Topol. **18**(5), 2951–2984 (2014)

[CG12]   F. Chung, R. Graham, Edge flipping in graphs. Adv. Appl. Math. **48**(1), 37–63 (2012)

[Cli41]   A.H. Clifford, Semigroups admitting relative inverses. Ann. Math. (2) **42**, 1037–1049 (1941)

[Cli42]   A.H. Clifford, Matrix representations of completely simple semigroups. Am. J. Math. **64**, 327–342 (1942)

[CP61]   A.H. Clifford, G.B. Preston, *The Algebraic Theory of Semigroups. Volume I*. Mathematical Surveys, No. 7 (American Mathematical Society, Providence, RI, 1961)

[CP67]   A.H. Clifford, G.B. Preston, *The Algebraic Theory of Semigroups. Volume II*. Mathematical Surveys, No. 7 (American Mathematical Society, Providence, RI, 1967)

[CPS88]   E. Cline, B. Parshall, L. Scott, Finite-dimensional algebras and highest weight categories. J. Reine Angew. Math. **391**, 85–99 (1988)

[CPS96] E. Cline, B. Parshall, L. Scott, Stratifying endomorphism algebras. Mem. Am. Math. Soc. **124**(591), viii+119 (1996)

[CR88] C.W. Curtis, I. Reiner, *Representation Theory of Finite Groups and Associative Algebras*. Wiley Classics Library (Wiley, New York, 1988). Reprint of the 1962 original, A Wiley-Interscience Publication

[CSST08] T. Ceccherini-Silberstein, F. Scarabotti, F. Tolli, *Harmonic Analysis on Finite Groups: Representation Theory, Gelfand Pairs and Markov Chains*. Cambridge Studies in Advanced Mathematics, vol. 108 (Cambridge University Press, Cambridge, 2008)

[CSST10] T. Ceccherini-Silberstein, F. Scarabotti, F. Tolli, *Representation Theory of the Symmetric Groups: The Okounkov-Vershik Approach, Character Formulas, and Partition Algebras*. Cambridge Studies in Advanced Mathematics, vol. 121 (Cambridge University Press, Cambridge, 2010)

[DE15] S. Danz, K. Erdmann, Crossed products as twisted category algebras. Algebr. Represent. Theory **18**(2), 281–296 (2015)

[Den11] T. Denton, A combinatorial formula for orthogonal idempotents in the 0-Hecke algebra of the symmetric group. Electron. J. Comb. **18**(1), Research Paper 28, 20 pp. (2011) (electronic)

[DF95] R.P. Dobrow, J.A. Fill, On the Markov chain for the move-to-root rule for binary search trees. Ann. Appl. Probab. **5**(1), 1–19 (1995)

[DHP03] M. Dieng, T. Halverson, V. Poladian, Character formulas for q-rook monoid algebras. J. Algebraic Comb. **17**(2), 99–123 (2003)

[DHST11] T. Denton, F. Hivert, A. Schilling, N. Thiéry, On the representation theory of finite $\mathcal{J}$-trivial monoids. Sém. Lothar. Comb. **64**, Art. B64d, 34 pp. (2011) (electronic)

[Dia88] P. Diaconis, *Group Representations in Probability and Statistics*. Institute of Mathematical Statistics Lecture Notes—Monograph Series, vol. 11 (Institute of Mathematical Statistics, Hayward, CA, 1988)

[Dia98] P. Diaconis, From shuffling cards to walking around the building: an introduction to modern Markov chain theory, in *Proceedings of the International Congress of Mathematicians, Vol. I (Berlin, 1998)*, number Extra vol. I (1998), pp. 187–204

[DK94] Y.A. Drozd, V.V. Kirichenko, *Finite-dimensional Algebras* (Springer, Berlin, 1994). Translated from the 1980 Russian original and with an appendix by V. Dlab

[DR89] V. Dlab, C.M. Ringel, Quasi-hereditary algebras. Ill. J. Math. **33**(2), 280–291 (1989)

[DS81] P. Diaconis, M. Shahshahani, Generating a random permutation with random transpositions. Z. Wahrsch. Verwandte Geb. **57**(2), 159–179 (1981)

[Dub98] L. Dubuc, Sur les automates circulaires et la conjecture de Černý. RAIRO Inform. Théor. Appl. **32**(1–3), 21–34 (1998)

[ECH+92] D.B.A. Epstein, J.W. Cannon, D.F. Holt, S.V.F. Levy, M.S. Paterson, W.P. Thurston, *Word Processing in Groups* (Jones and Bartlett Publishers, Boston, MA, 1992)

[Eil74] S. Eilenberg, *Automata, Languages, and Machines. Volume A*. Pure and Applied Mathematics, vol. 58 (Academic, New York, 1974)

[Eil76] S. Eilenberg, *Automata, Languages, and Machines. Volume B* (Academic, New York, 1976). Pure and Applied Mathematics, vol. 59. With two chapters ("Depth decomposition theorem" and "Complexity of semigroups and morphisms") by B. Tilson.

[Fay05] M. Fayers, 0-Hecke algebras of finite Coxeter groups. J. Pure Appl. Algebra **199**(1–3), 27–41 (2005)

[FGG97] A. Freedman, R.N. Gupta, R.M. Guralnick, Shirshov's theorem and representations of semigroups. Pac. J. Math. **181**(3), 159–176 (1997). Olga Taussky-Todd: in memoriam

[FGG99] J. Fountain, G.M.S. Gomes, V. Gould, Enlargements, semiabundancy and unipotent monoids. Commun. Algebra **27**(2), 595–614 (1999)

[FH91] W. Fulton, J. Harris, *Representation Theory. A First Course*. Readings in Mathematics, Graduate Texts in Mathematics, vol. 129 (Springer, New York, 1991)

[FH96] J.A. Fill, L. Holst, On the distribution of search cost for the move-to-front rule. *Random Structures Algorithms* **8**(3), 179–186 (1996)

[Fil96] J.A. Fill, An exact formula for the move-to-front rule for self-organizing lists. J. Theor. Probab. **9**(1), 113–160 (1996)

[Gab72] P. Gabriel, Unzerlegbare Darstellungen. I. Manuscripta Math. **6**, 71–103 (1972). Correction, ibid. **6**, 309 (1972)

[Gab73] P. Gabriel, Indecomposable representations. II, in *Symposia Mathematica, Vol. XI (Convegno di Algebra Commutativa, INDAM, Rome, 1971)* (Academic Press, London, 1973), pp. 81–104

[Gab80] P. Gabriel, Auslander-Reiten sequences and representation-finite algebras, in *Representation Theory, I (Proc. Workshop, Carleton Univ., Ottawa, Ont., 1979)*. Lecture Notes in Math., vol. 831 (Springer, Berlin, 1980), pp. 1–71

[GL96] J.J. Graham, G.I. Lehrer, Cellular algebras. Invent. Math. **123**(1), 1–34 (1996)

[GM09] O. Ganyushkin, V. Mazorchuk, *Classical Finite Transformation Semigroups, An Introduction*. Algebra and Applications, Number 9 (Springer, Berlin, 2009)

[GM14] A.-L. Grensing, V. Mazorchuk, Categorification of the Catalan monoid. Semigroup Forum **89**(1), 155–168 (2014)

[GMS09] O. Ganyushkin, V. Mazorchuk, B. Steinberg, On the irreducible representations of a finite semigroup. Proc. Am. Math. Soc. **137**(11), 3585–3592 (2009)

[GNS00] R.I. Grigorchuk, V.V. Nekrashevich, V.I. Sushchanskiĭ, Automata, dynamical systems, and groups. Tr. Mat. Inst. Steklova **231**(Din. Sist., Avtom. i Beskon. Gruppy), 134–214 (2000)

[Gol70] D.M. Goldschmidt, A group theoretic proof of the $p^a q^b$ theorem for odd primes. Math. Z. **113**, 373–375 (1970)

[GR97] P. Gabriel, A.V. Roiter, *Representations of Finite-Dimensional Algebras* (Springer, Berlin, 1997). Translated from the Russian, With a chapter by B. Keller, Reprint of the 1992 English translation

[Gre51] J.A. Green, On the structure of semigroups. Ann. of Math. (2) **54**, 163–172 (1951)

[Gre55] J.A. Green, The characters of the finite general linear groups. Trans. Am. Math. Soc. **80**, 402–447 (1955)

[Gre80] J.A. Green, *Polynomial Representations of* GL_n. Lecture Notes in Mathematics, vol. 830 (Springer, Berlin, 1980)

[Gre12] A.-L. Grensing, Monoid algebras of projection functors. J. Algebra **369**, 16–41 (2012)

[Gro02] C. Grood, A Specht module analog for the rook monoid. Electron. J. Comb. **9**(1), Research Paper 2, 10 pp. (2002) (electronic)

[Gro06] C. Grood, The rook partition algebra. J. Comb. Theory Ser. A **113**(2), 325–351 (2006)

[Hal04] T. Halverson, Representations of the q-rook monoid. J. Algebra **273**(1), 227–251 (2004)

[Hen72] W.J. Hendricks, The stationary distribution of an interesting Markov chain. J. Appl. Probab. **9**, 231–233 (1972)

[Hig92] P.M. Higgins, *Techniques of Semigroup Theory*. Oxford Science Publications (The Clarendon Press Oxford University Press, New York, 1992). With a foreword by G.B. Preston

[HK88] J.C. Harris, N.J. Kuhn, Stable decompositions of classifying spaces of finite abelian p-groups. Math. Proc. Camb. Philos. Soc. **103**(3), 427–449 (1988)

[How95] J.M. Howie, *Fundamentals of Semigroup Theory*. Oxford Science Publications, London Mathematical Society Monographs. New Series, vol. 12 (The Clarendon Press Oxford University Press, New York, 1995)

[HR01] T. Halverson, A. Ram, q-rook monoid algebras, Hecke algebras, and Schur-Weyl duality. Zap. Nauchn. Sem. S.-Peterburg. Otdel. Mat. Inst. Steklov. (POMI) **283**(Teor. Predst. Din. Sist. Komb. i Algoritm. Metody. 6), 224–250, 262–263 (2001)

[HS97] P.J. Hilton, U. Stammbach, *A Course in Homological Algebra*. Graduate Texts in Mathematics, vol. 4, 2nd edn. (Springer, New York, 1997)

[HS15] W. Hajji, B. Steinberg, A parametrization of the irreducible representations of a compact inverse semigroup. Commun. Algebra **43**(12), 5261–5281 (2015)

[HSH13] W. Hajji, B. Steinberg, D. Handelman, On finite-dimensional representations of compact inverse semigroups. Semigroup Forum **87**(3), 497–508 (2013)

[Hsi09] S.K. Hsiao, A semigroup approach to wreath-product extensions of Solomon's descent algebras. Electron. J. Comb. **16**(1), Research Paper 21, 9 (2009)

[HST13] F. Hivert, A. Schilling, N. Thiéry, The biHecke monoid of a finite Coxeter group and its representations. Algebra Number Theory **7**(3), 595–671 (2013)

[HT09] F. Hivert, N.M. Thiéry, The Hecke group algebra of a Coxeter group and its representation theory. J. Algebra **321**(8), 2230–2258 (2009)

[HZ57] E. Hewitt, H.S. Zuckerman, The irreducible representations of a semigroup related to the symmetric group. Ill. J. Math. **1**, 188–213 (1957)

[Isa76] I.M. Isaacs, *Character Theory of Finite Groups*. Pure and Applied Mathematics, No. 69 (Academic Press [Harcourt Brace Jovanovich Publishers], New York, London, 1976)

[Jam78] G.D. James, *The Representation Theory of the Symmetric Groups*. Lecture Notes in Mathematics, vol. 682 (Springer, Berlin, 1978)

[JK81]   G. James, A. Kerber, *The Representation Theory of the Symmetric Group*. Encyclopedia of Mathematics and Its Applications, vol. 16 (Addison-Wesley, Reading, MA, 1981). With a foreword by P.M. Cohn, With an introduction by G. de B. Robinson

[Kar03]   J. Kari, Synchronizing finite automata on Eulerian digraphs. Theor. Comput. Sci. **295**(1-3), 223-232 (2003). Mathematical foundations of computer science (Mariánské Lázně, 2001)

[KK94]   P. Krasoń, N.J. Kuhn, On embedding polynomial functors in symmetric powers. J. Algebra **163**(1), 281-294 (1994)

[KM09]   G. Kudryavtseva, V. Mazorchuk, On three approaches to conjugacy in semigroups. Semigroup Forum **78**(1), 14-20 (2009)

[Kov92]   L.G. Kovács, Semigroup algebras of the full matrix semigroup over a finite field. Proc. Am. Math. Soc. **116**(4), 911-919 (1992)

[KR68]   K. Krohn, J. Rhodes, Complexity of finite semigroups. Ann. Math. (2) **88**, 128-160 (1968)

[KRT68]   K. Krohn, J. Rhodes, B. Tilson, *Algebraic Theory of Machines, Languages, and Semigroups*, ed. by M.A. Arbib. With a major contribution by K. Krohn, J.L. Rhodes, Chaps. 1, 5-9 (Academic Press, New York, 1968)

[Kuh94a] N.J. Kuhn, Generic representations of the finite general linear groups and the Steenrod algebra. I. Am. J. Math. **116**(2), 327-360 (1994)

[Kuh94b] N.J. Kuhn, Generic representations of the finite general linear groups and the Steenrod algebra. II. *K*-Theory **8**(4), 395-428 (1994)

[Kuh15]   N.J. Kuhn, Generic representation theory of finite fields in nondescribing characteristic. Adv. Math. **272**, 598-610 (2015)

[Lal79]   G. Lallement, *Semigroups and Combinatorial Applications*. Pure and Applied Mathematics, A Wiley-Interscience Publication (Wiley, New York, Chichester, Brisbane, 1979)

[Lam91]   T.Y. Lam, *A First Course in Noncommutative Rings*. Graduate Texts in Mathematics, vol. 131 (Springer, New York, 1991)

[Law98]   M.V. Lawson, *Inverse Semigroups*. The Theory of Partial Symmetries (World Scientific, River Edge, NJ, 1998)

[Li11]   L. Li, A characterization of finite EI categories with hereditary category algebras. J. Algebra **345**, 213-241 (2011)

[Li14]   L. Li, On the representation types of category algebras of finite EI categories. J. Algebra **402**, 178-218 (2014)

[Lin14]   M. Linckelmann, A version of Alperin's weight conjecture for finite category algebras. J. Algebra **398**, 386-395 (2014)

[Lip96]   S. Lipscomb, *Symmetric Inverse Semigroups*. Mathematical Surveys and Monographs, vol. 46 (American Mathematical Society, Providence, RI, 1996)

[LM95]   D. Lind, B. Marcus, *An Introduction to Symbolic Dynamics and Coding* (Cambridge University Press, Cambridge, 1995)

[Lot97]   M. Lothaire, *Combinatorics on Words*. Cambridge Mathematical Library (Cambridge University Press, Cambridge, 1997). With a foreword by R. Lyndon and a preface by D. Perrin, Corrected reprint of the 1983 original, with a new preface by Perrin

[Lot02]   M. Lothaire, *Algebraic Combinatorics on Words*. Encyclopedia of Mathematics and Its Applications, vol. 90 (Cambridge University Press, Cambridge, 2002)

[LP69]   G. Lallement, M. Petrich, Irreducible matrix representations of finite semigroups. Trans. Am. Math. Soc. **139**, 393–412 (1969)

[LPW09] D.A. Levin, Y. Peres, E.L. Wilmer, *Markov Chains and Mixing Times* (American Mathematical Society, Providence, RI, 2009). With a chapter by J.G. Propp, D.B. Wilson

[LS12]   M. Linckelmann, M. Stolorz, On simple modules over twisted finite category algebras. Proc. Am. Math. Soc. **140**(11), 3725–3737 (2012)

[Lüc89] W. Lück, *Transformation Groups and Algebraic K-theory.* Mathematica Gottingensis, Lecture Notes in Mathematics, vol. 1408 (Springer, Berlin, 1989)

[Mac95] I.G. Macdonald, *Symmetric Functions and Hall Polynomials.* Oxford Science Publications, Oxford Mathematical Monographs, 2nd edn. (The Clarendon Press, Oxford University Press, New York, 1995). With contributions by A. Zelevinsky

[Mac98] S. Mac Lane. *Categories for the Working Mathematician.* Graduate Texts in Mathematics, vol. 5, 2nd edn. (Springer, New York, 1998)

[Mal10] M.E. Malandro, Fast Fourier transforms for finite inverse semigroups. J. Algebra **324**(2), 282–312 (2010)

[Mal13] M.E. Malandro, Inverse semigroup spectral analysis for partially ranked data. Appl. Comput. Harmon. Anal. **35**(1), 16–38 (2013)

[Man71] A. Manning, Axiom A diffeomorphisms have rational zeta functions. Bull. Lond. Math. Soc. **3**, 215–220 (1971)

[McA71] D.B. McAlister, Representations of semigroups by linear transformations. I, II. Semigroup Forum **2**(3), 189–263 (1971); ibid. **2**(4), 283–320 (1971)

[McA72] D.B. McAlister, Characters of finite semigroups. J. Algebra **22**, 183–200, (1972)

[Mik10] Mikola, Why aren't representations of monoids studied so much? Math-Overflow (2010), http://mathoverflow.net/q/37115 (version: 30 Aug 2010)

[Mit72] B. Mitchell, Rings with several objects. Adv. Math. **8**, 1–161 (1972)

[MP55]   W.D. Munn, R. Penrose, A note on inverse semigroups. Proc. Camb. Philos. Soc. **51**, 396–399 (1955)

[MQS15] A.M. Masuda, L. Quoos, B. Steinberg, Character theory of monoids over an arbitrary field. J. Algebra **431**, 107–126 (2015)

[MR95]   R. Mantaci, C. Reutenauer, A generalization of Solomon's algebra for hyperoctahedral groups and other wreath products. Commun. Algebra **23**(1), 27–56 (1995)

[MR10]   M. Malandro, D. Rockmore, Fast Fourier transforms for the rook monoid. Trans. Am. Math. Soc. **362**(2), 1009–1045 (2010)

[MS11]   S. Margolis, B. Steinberg, The quiver of an algebra associated to the Mantaci-Reutenauer descent algebra and the homology of regular semigroups. Algebr. Represent. Theory **14**(1), 131–159 (2011)

[MS12a] S. Margolis, B. Steinberg, Quivers of monoids with basic algebras. Compos. Math. **148**(5), 1516–1560 (2012)

[MS12b] V. Mazorchuk, B. Steinberg, Double Catalan monoids. J. Algebraic Comb. **36**(3), 333–354 (2012)

[MS12c] V. Mazorchuk, B. Steinberg, Effective dimension of finite semigroups. J. Pure Appl. Algebra **216**(12), 2737–2753 (2012)

[MSS15a] S. Margolis, F. Saliola, B. Steinberg, Cell complexes, poset topology and the representation theory of algebras arising in algebraic combinatorics and discrete geometry. ArXiv e-prints, August 2015

[MSS15b] S. Margolis, F. Saliola, B. Steinberg, Combinatorial topology and the global dimension of algebras arising in combinatorics. J. Eur. Math. Soc. **17**(12), 3037–3080 (2015)

[Mun55] W.D. Munn, On semigroup algebras. Proc. Camb. Philos. Soc. **51**, 1–15 (1955)

[Mun57a] W.D. Munn, The characters of the symmetric inverse semigroup. Proc. Camb. Philos. Soc. **53**, 13–18 (1957)

[Mun57b] W.D. Munn, Matrix representations of semigroups. Proc. Camb. Philos. Soc. **53**, 5–12 (1957)

[Mun60] W.D. Munn, Irreducible matrix representations of semigroups. Q. J. Math. Oxf. Ser. (2) **11**, 295–309 (1960)

[Mun78] W.D. Munn, Semiunitary representations of inverse semigroups. J. Lond. Math. Soc. (2) **18**(1), 75–80 (1978)

[MZ75] R. McNaughton, Y. Zalcstein, The Burnside problem for semigroups. J. Algebra **34**, 292–299 (1975)

[Nek05] V. Nekrashevych, *Self-Similar Groups*. Mathematical Surveys and Monographs, vol. 117 (American Mathematical Society, Providence, RI, 2005)

[Neu09] P.M. Neumann, Primitive permutation groups and their section-regular partitions. Mich. Math. J. **58**(1), 309–322 (2009)

[Nic71] W.R. Nico, Homological dimension in semigroup algebras. J. Algebra **18**, 404–413 (1971)

[Nic72] W.R. Nico, An improved upper bound for global dimension of semigroup algebras. Proc. Am. Math. Soc. **35**, 34–36 (1972)

[Nor79] P.N. Norton, 0-Hecke algebras. J. Aust. Math. Soc. Ser. A **27**(3), 337–357 (1979)

[Okn91] J. Okniński, *Semigroup Algebras*. Monographs and Textbooks in Pure and Applied Mathematics, vol. 138 (Marcel Dekker, New York, 1991)

[Okn98] J. Okniński, *Semigroups of Matrices*. Series in Algebra, vol. 6 (World Scientific, River Edge, NJ, 1998)

[OP91] J. Okniński, M.S. Putcha, Complex representations of matrix semigroups. Trans. Am. Math. Soc. **323**(2), 563–581 (1991)

[Pag06] R. Paget, Representation theory of q-rook monoid algebras. J. Algebraic Comb. **24**(3), 239–252 (2006)

[Pas14] D.S. Passman, Elementary bialgebra properties of group rings and enveloping rings: an introduction to Hopf algebras. Commun. Algebra **42**(5), 2222–2253 (2014)

[Pat99] A.L.T. Paterson, *Groupoids, Inverse Semigroups, and Their Operator Algebras*. Progress in Mathematics, vol. 170 (Birkhäuser, Boston, MA, 1999)

[Per13] D. Perrin, Completely reducible sets. Int. J. Algebra Comput. **23**(4), 915–941 (2013)

[Pet63] M. Petrich, The maximal semilattice decomposition of a semigroup. Bull. Am. Math. Soc. **69**, 342–344 (1963)

[Pet64] M. Petrich, The maximal semilattice decomposition of a semigroup. Math. Z. **85**, 68–82 (1964)

[Pet84] M. Petrich, *Inverse Semigroups*. Pure and Applied Mathematics (Wiley, New York, 1984)

[Pha91] R.M. Phatarfod, On the matrix occurring in a linear search problem. J. Appl. Probab. **28**(2), 336–346 (1991)

[Pin78] J.-E. Pin, Sur un cas particulier de la conjecture de Cerny, in *Automata, Languages and Programming (Fifth Internat. Colloq., Udine, 1978)*. Lecture Notes in Comput. Sci., vol. 62 (Springer, Berlin, 1978), pp. 345–352

[Pin81] J.-E. Pin, Le problème de la synchronisation et la conjecture de Černý, in *Noncommutative Structures in Algebra and Geometric Combinatorics (Naples, 1978). Quad. "Ricerca Sci."*, vol. 109 (CNR, Rome, 1981), pp. 37–48

[Pin86] J.-E. Pin, *Varieties of Formal Languages*. Foundations of Computer Science (Plenum Publishing Corp., New York, 1986). With a preface by M.-P. Schützenberger, Translated from the French by A. Howie

[Pin97] J.-E. Pin, Syntactic semigroups, in *Handbook of Formal Languages*, vol. 1 (Springer, Berlin, 1997), pp. 679–746

[Pon58] I.S. Ponizovskiĭ, On irreducible matrix representations of finite semigroups. Usp. Mat. Nauk **13**(6(84)), 139–144 (1958)

[Pon87] I.S. Ponizovskiĭ, Some examples of semigroup algebras of finite representation type. Zap. Nauchn. Sem. Leningrad. Otdel. Mat. Inst. Steklov. (LOMI) **160**(Anal. Teor. Chisel i Teor. Funktsii. 8), 229–238, 302 (1987)

[Pon93] J.S. Ponizovskiĭ, Semigroup algebras of finite representation type. Semigroup Forum **46**(1), 1–6 (1993)

[PQ95] D.S. Passman, D. Quinn, Burnside's theorem for Hopf algebras. Proc. Am. Math. Soc. **123**(2), 327–333 (1995)

[PR93] M.S. Putcha, L.E. Renner, The canonical compactification of a finite group of Lie type. Trans. Am. Math. Soc. **337**(1), 305–319 (1993)

[Put88] M.S. Putcha, *Linear Algebraic Monoids*. London Mathematical Society Lecture Note Series, vol. 133 (Cambridge University Press, Cambridge, 1988)

[Put89] M.S. Putcha, Monoids on groups with BN-pairs. J. Algebra **120**(1), 139–169 (1989)

[Put94] M.S. Putcha, Classification of monoids of Lie type. J. Algebra **163**(3), 636–662 (1994)

[Put95] M.S. Putcha, Monoids of Lie type, in *Semigroups, Formal Languages and Groups (York, 1993)*, NATO Adv. Sci. Inst. Ser. C Math. Phys. Sci., vol. 466 (Kluwer Academic, Dordrecht, 1995), pp. 353–367

[Put96] M.S. Putcha, Complex representations of finite monoids. Proc. Lond. Math. Soc. (3) **73**(3), 623–641 (1996)

[Put98] M.S. Putcha, Complex representations of finite monoids. II. Highest weight categories and quivers. J. Algebra **205**(1), 53–76 (1998)

[Put99] M.S. Putcha, Hecke algebras and semisimplicity of monoid algebras. J. Algebra **218**(2), 488–508 (1999)

[Qui73] D. Quillen, Higher algebraic K-theory. I, in *Algebraic K-theory, I: Higher K-Theories (Proc. Conf., Battelle Memorial Inst., Seattle, Wash., 1972)*, Lecture Notes in Math., vol. 341 (Springer, Berlin, 1973), pp. 85–147

[Ren05] L.E. Renner, *Linear Algebraic Monoids*. Invariant Theory and Algebraic Transformation Groups V, Encyclopaedia of Mathematical Sciences, vol. 134 (Springer, Berlin, 2005)

[Rho69a] J. Rhodes, Algebraic theory of finite semigroups. Structure numbers and structure theorems for finite semigroups, in *Semigroups (Proc. Sympos., Wayne State Univ., Detroit, Mich., 1968)*, ed. by K. Folley (Academic Press, New York, 1969), pp. 125–162

[Rho69b] J. Rhodes, Characters and complexity of finite semigroups. J Comb. Theory **6**, 67–85 (1969)

[Rie67] M.A. Rieffel, Burnside's theorem for representations of Hopf algebras. J. Algebra **6**, 123–130 (1967)

[Rin00] C.M. Ringel, The representation type of the full transformation semigroup T_4. Semigroup Forum **61**(3), 429–434 (2000)

[Rot64] G.-C. Rota, On the foundations of combinatorial theory. I. Theory of Möbius functions. Z. Wahrscheinlichkeitstheorie Verw. Gebiete **2**, 340–368 (1964)

[RS09] J. Rhodes, B. Steinberg, *The q-theory of Finite Semigroups*. Springer Monographs in Mathematics (Springer, New York, 2009)

[Ruk78] A.V. Rukolaĭne, The center of the semigroup algebra of a finite inverse semigroup over the field of complex numbers. Zap. Naučn. Sem. Leningrad. Otdel. Mat. Inst. Steklov. (LOMI) **75**, 154–158, 198 (1978). Rings and linear groups

[Ruk80] A.V. Rukolaĭne, Semigroup algebras of finite inverse semigroups over arbitrary fields. Zap. Nauchn. Sem. Leningrad. Otdel. Mat. Inst. Steklov. (LOMI) **103**, 117–123, 159 (1980). Modules and linear groups

[RZ91] J. Rhodes, Y. Zalcstein, Elementary representation and character theory of finite semigroups and its application, in *Monoids and Semigroups with Applications (Berkeley, CA, 1989)* (World Scientific, River Edge, NJ, 1991), pp. 334–367

[Sag01] B.E. Sagan, *The Symmetric Group*. Representations, Combinatorial Algorithms, and Symmetric Functions, Graduate Texts in Mathematics, vol. 203, 2nd edn. (Springer, New York, 2001)

[Sal07] F.V. Saliola, The quiver of the semigroup algebra of a left regular band. Int. J. Algebra Comput. **17**(8), 1593–1610 (2007)

[Sal08] F.V. Saliola, On the quiver of the descent algebra. J. Algebra **320**(11), 3866–3894 (2008)

[Sal09] F.V. Saliola, The face semigroup algebra of a hyperplane arrangement. Can. J. Math. **61**(4), 904–929 (2009)

[Sal10] F.V. Saliola, The Loewy length of the descent algebra of type D. Algebr. Represent. Theory **13**(2), 243–254 (2010)

[Sal12] F. Saliola, Eigenvectors for a random walk on a left-regular band. Adv. Appl. Math. **48**(2), 306–311 (2012)

[Sch58] M.-P. Schützenberger, Sur la représentation monomiale des demi-groupes. C. R. Acad. Sci. Paris **246**, 865–867 (1958)

[Sch06] M. Schocker, The module structure of the Solomon-Tits algebra of the symmetric group. J. Algebra **301**(2), 554–586 (2006)

[Sch08] M. Schocker, Radical of weakly ordered semigroup algebras. J. Algebraic Comb. **28**(1), 231–234 (2008). With a foreword by N. Bergeron

[Ser77]   J.-P. Serre, *Linear Representations of Finite Groups*. Graduate Texts in Mathematics, vol. 42 (Springer, New York, 1977). Translated from the second French edition by L.L. Scott

[Sim75]   I. Simon, Piecewise testable events, in *Automata Theory and Formal Languages (Second GI Conf., Kaiserslautern, 1975)*. Lecture Notes in Comput. Sci., vol. 33 (Springer, Berlin, 1975), pp. 214–222

[Sol67]   L. Solomon, The Burnside algebra of a finite group. J. Comb. Theory **2**, 603–615 (1967)

[Sol02]   L. Solomon, Representations of the rook monoid. J. Algebra **256**(2), 309–342 (2002)

[Sta97]   R.P. Stanley, *Enumerative Combinatorics. Volume 1*. Cambridge Studies in Advanced Mathematics, vol. 49 (Cambridge University Press, Cambridge, 1997). With a foreword by G.-C. Rota, Corrected reprint of the 1986 original

[Sta99]   R.P. Stanley, *Enumerative Combinatorics. Volume 2*. Cambridge Studies in Advanced Mathematics, vol. 62 (Cambridge University Press, Cambridge, 1999). With a foreword by G.-C. Rota and appendix 1 by S. Fomin

[Ste62]   R. Steinberg, Complete sets of representations of algebras. Proc. Am. Math. Soc. **13**, 746–747 (1962)

[Ste06]   B. Steinberg, Möbius functions and semigroup representation theory. J. Comb. Theory Ser. A **113**(5), 866–881 (2006)

[Ste08]   B. Steinberg, Möbius functions and semigroup representation theory. II. Character formulas and multiplicities. Adv. Math. **217**(4), 1521–1557 (2008)

[Ste10a] B. Steinberg, A simple proof of Brown's diagonalizability theorem (October 2010), http://arxiv.org/abs/1010.0716

[Ste10b] B. Steinberg, A theory of transformation monoids: combinatorics and representation theory. Electron. J. Comb. **17**(1), Research Paper 164, 56 pp. (2010) (electronic)

[Ste10c] B. Steinberg, Černý's conjecture and group representation theory. J. Algebr. Comb. **31**(1), 83–109 (2010)

[Ste10d] B. Steinberg, A groupoid approach to discrete inverse semigroup algebras. Adv. Math. **223**(2), 689–727 (2010)

[Ste11]   B. Steinberg, The averaging trick and the Černý conjecture. Int. J. Found. Comput. Sci. **22**(7), 1697–1706 (2011)

[Ste12a] B. Steinberg, *Representation Theory of Finite Groups*. An Introductory Approach, Universitext (Springer, New York, 2012)

[Ste12b] B. Steinberg, Yet another solution to the Burnside problem for matrix semigroups. Can. Math. Bull. **55**(1), 188–192 (2012)

[Ste15]   I. Stein, Algebras of Ehresmann semigroups and categories. ArXiv e-prints, December 2015

[Ste16a] I. Stein, The representation theory of the monoid of all partial functions on a set and related monoids as EI-category algebras. J. Algebra **450**, 549–569 (2016)

[Ste16b] B. Steinberg, The global dimension of the full transformation monoid (with an Appendix by V. Mazorchuk and B. Steinberg). Algebr. Represent. Theory **19**(3), 731–747 (2016)

[Str83] H. Straubing, The Burnside problem for semigroups of matrices, in *Combinatorics on Words (Waterloo, Ont., 1982)* (Academic Press, Toronto, ON, 1983), pp. 279–295

[Str94] H. Straubing, *Finite Automata, Formal Logic, and Circuit Complexity*. Progress in Theoretical Computer Science (Birkhäuser, Boston, MA, 1994)

[SY11] A. Skowroński, K. Yamagata, *Frobenius Algebras. I*. Basic Representation Theory, EMS Textbooks in Mathematics (European Mathematical Society (EMS), Zürich, 2011)

[tD87] T. tom Dieck, *Transformation Groups*. de Gruyter Studies in Mathematics, vol. 8 (Walter de Gruyter, Berlin, 1987)

[Ter99] A. Terras, *Fourier Analysis on Finite Groups and Applications*. London Mathematical Society Student Texts, vol. 43 (Cambridge University Press, Cambridge, 1999)

[Thi12] N. Thiéry, Cartan invariant matrices for finite monoids, in *DMTCS Proceedings, 24th International Conference on Formal Power Series and Algebraic Combinatorics (FPSAC 2012)*. DMTCS, 2012, pp. 887–898

[Til69] B.R. Tilson, Appendix to "Algebraic theory of finite semigroups". On the p-length of p-solvable semigroups: preliminary results, in *Semigroups (Proc. Sympos., Wayne State Univ., Detroit, Mich., 1968)*, ed. by K. Folley (Academic Press, New York, 1969), pp. 163–208

[Vol08] M.V. Volkov, Synchronizing automata and the Černý conjecture, in *Language and Automata Theory and Applications Second International Conference, LATA 2008, Tarragona, Spain, March 13–19, 2008*, ed. by C. Martín-Vide, F. Otto, H. Fernau. Lecture Notes in Computer Science, vol. 5196 (Springer, Berlin, Heidelberg, 2008), pp. 11–27

[Web07] P. Webb, An introduction to the representations and cohomology of categories, in *Group Representation Theory* (EPFL Press, Lausanne, 2007), pp. 149–173

[Web08] P. Webb, Standard stratifications of EI categories and Alperin's weight conjecture. J. Algebra **320**(12), 4073–4091 (2008)

[Wei73] B. Weiss, Subshifts of finite type and sofic systems. Monatsh. Math. **77**, 462–474 (1973)

[Wei94] C.A. Weibel, *An Introduction to Homological Algebra*. Cambridge Studies in Advanced Mathematics, vol. 38 (Cambridge University Press, Cambridge, 1994)

[Zie95] G.M. Ziegler, *Lectures on Polytopes*. Graduate Texts in Mathematics, vol. 152 (Springer, New York, 1995)

Index of Notation

$[m]_\equiv$	Congruence or equivalence class of m with respect to $\equiv$, page 4
$[r]$	The set $\{1,\dots,r\}$ for a natural number r, page 61
$[V:S]$	Multiplicity of the simple module S as a composition factor of V, page 273
$[V]$	Isomorphism class of the module V, page 55
$N_e(V)$	Largest submodule of V annihilated by e, page 42
$\mathrm{Aug}(\Bbbk\Omega)$	Augmentation submodule of the transformation module $\Bbbk\Omega$, page 194
$\mathrm{Aug}(\Bbbk M)$	Augmentation ideal of $\Bbbk M$, page 157
$\Bbbk C$	Category algebra of the category C over $\Bbbk$, page 126
$\Bbbk M$	Monoid algebra of the monoid M over $\Bbbk$, page 54
$\Bbbk Q$	Path algebra of the quiver Q, page 127
$\boxtimes$	Outer product of symmetric group modules, page 288
χ_V	Character of the module V, page 107
$\mathrm{Cl}(M)$	Ring of class functions on M, page 103
Coind	Coinduction functor, page 42
$\lfloor X \rfloor$	Sum of the elements of the set $X \subseteq \Omega$ in $\Bbbk\Omega$, page 192
d	Domain map for a category, page 125
$\Delta(P)$	Order complex of the poset P, page 265
gl. dim A	Global dimension of the algebra A, page 235
ht	Height of an idempotent, page 239
$T_e(V)$	Trace of the projective module associated with e in V, page 42
$\mathrm{im}^\infty T$	Eventual range of T, page 79
Ind	Induction functor, page 42
$\mathrm{Irr}_\Bbbk(M)$	Set of isomorphism classes of simple $\Bbbk M$-modules, page 55
$\mathscr{J}$	Green's relation $\mathscr{J}$, page 7
$\ker\varphi$	The kernel of φ, page 4
$\ker^\infty T$	Maximal subspace on which T acts nilpotently, page 81
$\mathcal{R}_\Bbbk(M)$	Representation ring of M over $\Bbbk$, page 97
$\Lambda(M)$	Lattice of ideals generated by coprime idempotents, page 19

A^*	Free monoid on A, page 177
A^e	Enveloping algebra of the algebra A, page 119
A^{op}	Opposite algebra of the algebra A, page 41
$c((P_1, \ldots, P_r))$	Content of the ordered set partition $(P_1, \ldots, P_r)$, page 211
$c(w)$	Content of the word w, page 209
C_0	Object set of the category (or quiver) C, page 125
C_1	Arrow set of the category (or quiver) C, page 125
C_n	The Catalan monoid of degree n, page 256
D	Standard duality functor, page 277
d_n	Number of derangements of an n-element set, page 218
$E(X)$	Set of idempotent elements in X, page 4
$F(B)$	Free left regular band on the set B, page 209
$G(M)$	Groupoid of the inverse monoid M, page 137
G_e	Maximal subgroup at the idempotent e, page 4
$GL(V)$	General linear group of the vector space V, page 79
$I(M)$	Minimal ideal of the monoid M, page 191
$I(m)$	Ideal of elements $s \in M$ with $m \notin MsM$, page 7
$I(P, \Bbbk)$	Incidence algebra of the poset P over $\Bbbk$, page 291
I_e	Ideal of non-units of eSe for a semigroup S and idempotent e of S, page 5
I_n	Symmetric inverse monoid of degree n, page 26
I_X	Symmetric inverse monoid on the set X, page 26
$I_{n,r}$	Set of injective mappings from $[r]$ to $[n]$, page 62
J_m	The $\mathscr{J}$-class of the element m, page 7
$K_0(\Bbbk M)$	Grothendieck group of M over $\Bbbk$, page 100
L_m	The $\mathscr{L}$-class of the element m, page 7
$M(C)$	Monoid associated with the category C, page 128
M/I	Rees quotient of the monoid M by the ideal I, page 14
$M/\equiv$	Quotient of the monoid M by the congruence $\equiv$, page 4
m^*	Inverse of m in an inverse monoid, page 25
m^+	The unique idempotent in the $\widetilde{\mathscr{R}}$-class of m in a $\mathscr{J}$-trivial monoid, page 250
m^-	The unique idempotent in the $\widetilde{\mathscr{L}}$-class of m in a $\mathscr{J}$-trivial monoid, page 250
$M_n(R)$	Ring of $n \times n$-matrices over a ring R, page 41
$M_{\geq X}$	Contraction of the left regular band M to $X \in \Lambda(M)$, page 265
PT_n	Monoid of partial transformations of degree n, page 26
PT_X	Partial transformation monoid on the set X, page 26
$Q(A)$	Quiver of the algebra A, page 245
Q^*	Free category on the quiver Q, page 126
R_m	The $\mathscr{R}$-class of the element m, page 7
$s^{\omega+1}$	The product $s^\omega s$ for an element s of a finite semigroup, page 5
s^ω	Unique idempotent positive power of an element s of a finite semigroup, page 5

S^{op}	Opposite semigroup of the semigroup S, page 4
S_n	Symmetric group of degree n, page 285
S_λ	Specht module associated with the partition λ, page 286
T_n	Full transformation monoid of degree n, page 7
T_Ω	Full transformation monoid on the set Ω, page 192
$T_{n,r}$	Set of mappings (transformations) from $[r]$ to $[n]$, page 62
$V^{\otimes r}$	Tensor power (of degree r) of the module V, page 55
$X(M)$	Character table of the monoid M, page 108
$Z(A)$	Center of the algebra A, page 276

Subject Index

© Springer International Publishing Switzerland 2016
B. Steinberg, *Representation Theory of Finite Monoids*,
Universitext, DOI 10.1007/978-3-319-43932-7

Specht module, 286
split semisimple, 216
stability, 10
standard duality, 277
standard tableau, 66, 286
state space, 206
stationary distribution, 206
Stirling number
  second kind, 144
subgroup, 3
submonoid, 3
  generated, 4
subsemigroup, 3
support, 206
symbolic dynamical system, 178
symmetric inverse monoid, 26
synchronizing, 196
synchronizing group, 195
syntactic congruence, 188
syntactic monoid, 178, 188

T
tableau, 66, 286
  standard, 66, 286
tabloid, 67, 287
  of shape λ, 287
tensor product
  of algebras, 119, 155
  of modules, 55, 119, 155
top-to-random shuffle, 208
trace
  of a projective module, 42
transformation module, 192
transformation monoid, 192
transient, 32
  orbit, 32
transition matrix, 206

transitive, 191, 283
triangular Grothendieck ring, 100
triangularizable
  module, 100
  monoid, 161
trivial module, 55
trivial representation, 55
Tsetlin library, 208
two-sided ideal, 7

U
unit, 4
unitary representation, 289

V
virtual character, 107, 180

W
weakly irreducible, 253
wreath product, 35

Y
Young diagram, 285
Young subgroup, 287
Young symmetrizer, 286
Young tableau, 286
  standard, 286

Z
zero, 4
zero element, 4
zeta function
  Ihara, 179, 186
  language, 177
  mapping, 180
  poset, 292
  shift, 178

Printed in the United States
By Bookmasters